Pharmacological Properties of Native Plants from Argentina

María Alejandra Alvarez

Pharmacological Properties of Native Plants from Argentina

Springer

María Alejandra Alvarez
Department of Biotecnología Vegetal
CONICET/ CEBBAD
Universidad Maimónides
Ciudad de Buenos Aires, Argentina

ISBN 978-3-030-20197-5 ISBN 978-3-030-20198-2 (eBook)
https://doi.org/10.1007/978-3-030-20198-2

This Springer imprint is published by the registered company Springer Nature Switzerland AG
The registered company address is: Gewerbestrasse 11, 6330 Cham, Switzerland

Preface

There are a large number of native plants used by traditional medicine in Argentina. The intention of this book is not to describe all the medicinal native species from Argentina but those with pharmacological studies that validate their pharmacological properties. To carry out this selection, the work by Barboza et al. (2009) was of fundamental importance. The websites of the Darwinion Institute (www.darwin.edu.ar) and Flora Argentina (www.floraargentina.edu.ar) have been consulted to update the scientific names and corroborate the distribution data and botanical characteristics found in other sources. The botanical terms used correspond to the glossary compiled by P.P.J. Herman. Each chapter begins with a brief description of the geographical, ecogeographical, or phytogeographical region in which the selected species grow. It has not been the intention of this book to describe the medicinal flora of each of these regions but to locate the selected species within the context of one of the several regions where they could be found. For this reason, for each species, all the provinces in which it is possible to find it are specified.

This work would not have been possible without the invaluable collaboration of Dr. Chana Pilberg who gave me as a gift a large part of her botanical library; her friendship is one of my greatest privileges. I must also thank my sister, Patricia G. Alvarez, for generously giving me some of the photographs that illustrate this book. I also thank all the colleagues who shared with me their publications and my colleagues from the Pharmacy and Biochemistry School and CEBBAD at Maimónides University, and Universidad de Buenos Aires for their constant support. Finally, I thank my family and in particular my daughters for their patience and affectionate collaboration.

Buenos Aires, Argentina — María Alejandra Alvarez
May 2019

References

Barboza GE, Cantero JJ, Núñez C, Pacciaroni A, Ariza Espinar L (2009) Medicinal plants: a general review and a phytochemical and ethnopharmacological screening of the native Argentine Flora. Kurtziana 34(12):7365

Preface

References

Book Introduction

The aim of this book is to offer information about native plants with pharmacological properties from Argentina to students, researchers, and graduates interested in the fields of Ethnobotany, Pharmacognosy, Phytochemistry, Pharmacy, and Medicine. It describes some of the native species used in folk medicine whose pharmacological activities have been experimentally tested. Chapter 1 summarizes the characteristics of the geographical, ecological, and phytogeographical regions from the country. Also, a brief reference about the legal regulatory framework of medicinal plants in Argentina is included. Chapter 2 reviews some aspects of ethnobotany in Argentina. The following Chaps. 3, 4, 5, 6, 7, 8, and 9 describe native medicinal plants in their botanical aspects, ethnomedicinal uses, chemical activity, and toxicity and, when appropriate, in the establishment of *in vitro* cultures and their legal status. Each of those chapters begins with a brief description of the region in which the species grow. The following species are described: from Cuyo (Chap. 3), *Aloysia gratissima* (Gillies & Hook. ex Hook.) Tronc., *Lippia integrifolia* (Griseb.) Hieron., *Minthostachys mollis*, *Acantholippia seriphioides* (A. Gray) Moldenke, and *Achyrocline satureioides* (Lam.); from the Pampa (Chap. 4), *Erythrina crista-galli* L. var. crista-galli, *Phytolacca dioica* L., *Salix humboldtiana* Wild, *Grindelia pulchella* Dunal, *Larrea cuneifolia* Cav., *Larrea divaricata* Cav., and *Larrea nitida* Cav; from Mesopotamia (Chap. 5), *Cecropia pachystachya* Trécul (Cecropiaceae), *Anadenanthera colubrina* (Vell.) Brenan var. cebil (Griseb.) Altschul, *Ilex paraguariensis* A. St.-Hil var. paraguariensis, *Tabebuia impetiginosa* (Mart. ex DC.) Standl., *Allophylus edulis* (A. St.-Hil., A. Juss., and Cambess.) Hieron. ex Niederl., *Passiflora caerulea* L., *Blechnum occidentale* L., and *Maytenus ilicifolia* Mart. ex Reissek; from the Puna (Chap. 6), *Acacia caven* (Molina) Molina var. caven, *Chenopodium ambrosioides* L., *Anemia tomentosa* (Savigny) Sw. var. anthriscifolia (Schrad.) Mickel, *Plantago australis* Lam. subsp. Australis, and *Alternanthera pungens* Kunth; from the Yungas (Chap. 7), *Lepidium didymum* L., *Clinopodium gilliesii* (Benth.) Kunze., *Smilax campestris* Griseb., and *Smallanthus macroscyphus* (Baker ex Martius) A. Grau; and from Chaco and Espinal (Chap. 8), the medicinal species *Adiantum raddianum* C. Presl., *Aloysia citriodora* Palau, *Bauhinia forficata* subsp. pruinosa (Vogel) Fortunato & Wunderlin, *Caesalpinia gilliesii* var. gilliesii

(Wall. ex Hook.) D. Dietr., *Pilocarpus pennatifolius* Lemaire, and *Jodina rhombifolia* (Hook. & Arn.) Reissek. Finally, from Patagonia Argentina, the species *Adesmia boronioides* Hook f., *Berberis microphylla* G. Forest, *Buddleja globosa* Hope, and *Verbena litoralis* Kunth var. litoralis are portrayed (Chap. 9).

Contents

About the Author

María Alejandra Alvarez completed her studies at the Universidad de Buenos Aires, Argentina. She graduated with degrees in Biochemistry and Pharmacy in 1981 and 1994, respectively, and obtained her MSc in Industrial Microbiology and Biotechnology in 1986 and her PhD in Plant Biotechnology in 1993. She was lecturer in graduate and postgraduate courses of Botany, Pharmacobotany, Biotechnology, Food Biotechnology, and Galenic Pharmacy at the Universidad de Buenos Aires, Universidad Nacional de La Plata, and Universidad Maimónides.

She started working on the production of compounds of pharmaceutical interest in *in vitro* plant cultures in 1986, being the author of numerous scientific articles. Her research interests include chemotaxonomy, medicinal plants, production of secondary metabolites, and molecular farming. She was a researcher at the Universidad de Buenos Aires and Centro de Ciencia y Tecnología Dr. César Milstein (CONICET/Fundación Pablo Cassará). Currently, she is a member of the National Council of Research and Technology (CONICET) in Argentina, professor of Pharmacobotany and Pharmacognosy and director of the Plant Biotechnology Group at the Universidad Maimónides, and professor of Biology at Colegio Divino Corazón. She was the editor of the book *Genetic Transformation* (2011), Intech Open Access ISBN: 978-953-307-364-4, and the author of the book *Plant Biotechnology for Health: From Secondary Metabolites to Molecular Farming* (2014), Springer International Publishing AG, ISBN 978-3-319-05770-5.

About the Author

Abbreviations

2,4-D	2,4-dichlorophenoxyacetic acid
2iP	2 inositol phosphate N6-(2-isopentenyl) adenine
ALT	Alanine aminotransferase
ANMAT	National Administration of Drugs, Foods and Medical Technology
AST	Aspartate aminotransferase
BAP	6-benzilaminopurine
DPPH	(2,2-difenil-1,2-picrilhidrazil)
DW	Dry weight
EOs	Essential oils
FNA	Argentina National Pharmacopeia
FW	Fresh weight
IAA	Indole-3-acetic acid
IBA	Indole-3-butyric acid
Kin	Kinetin
m.a.s.l.	Meters above sea level
m.b.s.l.	Meters below sea level
MBC	Minimum bactericidal concentration
MIC	Minimum inhibitory concentration
MPs	Medicinal plants
MS	Murashige and Skoog medium
MSG	MS medium with Gamborg's vitamins
MSRT	MS medium with Khanna and Staba vitamins
NAA	1-Naphtalene acetic acid
PIC	Picloram
ROS	Reactive oxygen species
TDZ	Thidiazuron

Argentina Provinces

BAI	Buenos Aires
CAT	Catamarca
CHA	Chaco
CHU	Chubut
COR	Córdoba
COS	Corrientes
DFE	Distrito Federal
ERI	Entre Ríos
FOR	Formosa
JUJ	Jujuy
LPA	La Pampa
LRI	La Rioja
MEN	Mendoza
MIS	Misiones
NEU	Neuquén
RNE	Río Negro
SAL	Salta
SCR	Santa Cruz
SDE	Santiago del Estero
SFE	Santa Fe
SJU	San Juan
SLU	San Luis
TDF	Tierra del Fuego
TUC	Tucumán

List of Figures

List of Tables

Chapter 1
Introduction: Native Plants of Argentina – A General Overview

1.1 Introduction

Plants are the perfect machinery that transforms light into chemical energy and produce also oxygen, a vital element for most of the organisms on Earth. They are a source of food, fodder fuel, cellulose, paper, timber, textile fibers, essences and flavors, and medicines. They are also natural pesticides, are helpful to mitigate environmental pollution, and have a vital role to fight climate change. Plants are also central to scientific research on basic aspects of plant life and on challenges related to agriculture, health, and environment. Besides, plants have been part of ritual, religious, and festive practices.

1.2 República Argentina

Argentina is located in the Southern tip of South America. It shares borders with Bolivia and Paraguay to the North; Chile to the West; Brazil, Uruguay, and the South Atlantic Ocean to the East; and Chile and the South Atlantic Ocean to the South. Their endpoints are North, 21° 46' S, 66° 13' W.; South, 55° 03' S, 66° 31' W; East, 26° 15' S, 53° 38' W; and West, 50° 01'S, 73° 34' W. The Antarctica Argentina is located between the meridians 25° W and 74° W and the parallels 60° S and 90° S (South Pole). The country has a total surface area of 3.761.274 km^2, 2.791.810 km^2 in the American continent and 969.464 km^2 in the Antarctica. The population is around 40.117.096 inhabitants with a population density of 10.7 inhabitants/km^2 (*Instituto Geográfico Nacional República Argentina*). The country is composed of 23 provinces and the Autonomous City of Buenos Aires which is the capital city of the country. It is also divided in seven geographical regions: Northwest, Mesopotamia, the Gran Chaco, the Sierras Pampeanas, Cuyo, Región Pampeana, and Patagonia. The country has various ecosystems with a substantial biodiversity due to its large size, geographic variety, and diverse climate types.

M. A. Alvarez, *Pharmacological Properties of Native Plants from Argentina*,
https://doi.org/10.1007/978-3-030-20198-2_1

1.2.1 Geographical Regions

The geographical regions from Argentina are the Northwest region, Mesopotamia, the Gran Chaco, Cuyo, the Pampas, Patagonia, and South Atlantic Islands and Antarctica.

The *Northwest region* includes the provinces of Catamarca (CAT), Jujuy (JUJ), La Rioja (LRI), Salta (SAL), Santiago del Estero (SDE), and Tucumán (TUC). The climate is varied, with rainfall diminishing from North to South and from East to West. It comprises two subregions, the Puna and the Yungas. The Puna, to the high Andean west, is dry and with a great temperature oscillation, mostly cold, frequently decreasing below freezing point at night. The Yungas, to the East, is tropical, very hot, and humid.

Mesopotamia includes the provinces of Misiones (MIS), Entre Ríos (ERI), and Corrientes (COS). The climate is mostly subtropical, with a hot and very humid tropical climate to the North, gradually becoming temperate and semi-humid to the South.

The *Gran Chaco* includes the provinces of Chaco (CHA), Formosa (FOR), and Santiago del Estero (SDE). The region has a very hot subtropical to tropical climate, characterized by humid summers and mild dried winters. It has heavy seasonal rainfalls and periodic droughts.

Cuyo includes areas from the provinces of San Juan (SJU), San Luis (SLU), and Mendoza (MEN). It has an arid or semiarid climate with an average annual precipitation of about 100 to 500 millimeters (4 to 20 in). The diurnal temperature range is very large with extremely hot temperatures during the day followed by cold nights.

The *Región Pampeana* could be divided into *Llanura pampeana* (Pampa's plains) and the *Sierras Pampeanas* (Pampa's hills). The *Llanura pampeana* covers the provinces of Buenos Aires (BAI), La Pampa (LPA), Córdoba (COR), Santa Fe (SFE), and Entre Ríos (ERI). The climate is temperate, with hot, stormy summers, and cool winters. The *Sierras Pampeanas* comprises sectors from the provinces of San Luis (SLU), San Juan (SJU), Córdoba (COR), La Rioja (LRI), Catamarca (CAT), Santiago del Estero (SDE), and Tucumán (TUC). They have a temperate and semiarid climate, with warm summers and cool winters. The Northern-East slope is covered by rainforest, generating high humidity. The area inside COR and SLU has a Mediterranean-type climate, with intense summer rainstorms and snowy winters. The eastern slopes that receive the moist winds from the Atlantic Ocean have more rainfalls.

Patagonia includes the provinces of Neuquén (NEU), Río Negro (RNE), Chubut (CHU), Santa Cruz (SCR), and Tierra del Fuego (TDF). The region is very windy, with mild summers, cold to very cold winters, and heavy snowfall and frost, especially in mountainous zones. Precipitation steeply diminishes from west to east.

South Atlantic Islands and Antártida Argentina. Argentina is a founding signatory member of the Antarctic Treaty and is also a permanent consulting member. There are 13 Argentinean bases (6 permanent and 7 seasonal) in Antarctica with activities coordinated by the *Instituto Antártico Argentino* and the *Dirección Nacional del Antártico*. There the temperatures range from 0 °C in the summer to −60 °C to −80 °C in winter.

1.2.2 Eco-regions

An eco-region is defined as a relatively large geographical area characterized by a unique geology, weather, soil, hydrology, and plant and animal species. Argentina is divided into eighteen eco-regions, fifteen of them are continental, two marine, and one Antarctic. The eco-regions are *Altos Andes* (High Andean Complex), *Puna* (Puna grassland), *Monte de sierras y bolsones* (Argentine Northwest Monte and Thistle of the Prepuna), *Selva de Yungas* (Yungas), *Chaco seco* (Arid Chaco), *Chaco húmedo* (Humid Chaco), *Bosques patagónicos* (Patagonia forests), *pastizales y matorrales* (grasslands and shrublands), *Delta e Islas del Paraná* (Paraná flooded savanna), *Espinal* (Espinal), *Estepa patagónica* (Patagonian steppe), *Esteros del Iberá* (Iberá Wetlands), *Monte de llanuras y mesetas* (Argentina Low Monte), *Pampas* (Pampa), *Selva Paranaense* (Alto Paraná Atlantic forests), *Islas del Atlántico Sur* (South Atlantic islands), *Mar Argentino* (Argentina sea), and *Antártida* (Antarctic Peninsula) (Burkart et al. 1999; Diminich et al. 2018).

The *High Andean Complex* or *Altos Andes* is the high mountain sector in the West of Argentina, from the border with Bolivia to the North of the province of NEU, across the provinces of JUJ, SAL, TUC, CAT, LRI, SJU, and MEN. To the North it includes the summits and slopes of the Eastern Andes mountain range. Towards the South it includes some sections of the Andes foothills (Matteucci 2012). The climate is cold with everlasting snow. The rains are around 100 to 200 mm. Vegetation is grassy or shrubby, highly adapted to the extreme weather and altitude.

The *Puna grassland* or *Puna* extends from the Northwestern border of the country (province of Jujuy) to the North of the province of SJU. It has a cold desert climate characterized by cushion bog vegetation. Its plateau is crossed by mountains with numerous volcanoes.

Argentine Northwest Monte and Thistle of the Prepuna extends from JUJ to the North of MEN. It has extensive and pronounced slopes that connect with the Puna to the North and with the High Andes to the South. At the foothills are the *Quebrada de Humahuaca*, the *Quebrada del Toro*, the *Calchaquí Valleys*, and the rivers Jáchal and Bermejo. The climate is subtropical dry to the North; in their valleys and canyons, the rains concentrated in the summer season, while in the South, rains are distributed throughout the year. The radiation is intense, and the cloudiness is low. Temperature changes are very marked, both during the day and between seasons. Soil is predominantly sandy and poor in organic matter.

The *Yungas* are located to the north of the country with a discontinuous distribution in the provinces of SAL, JUJ, TUC, and CAT. The climate is warm and humid to sub-humid. The conditions of temperature and humidity are variable according to altitude and latitude. The region has a high biodiversity with evergreen forests.

Arid Chaco extends over the western half of FOR and CHA; the eastern half of SAL; almost all of SDE, North of SFE, and COR; and sectors of CAT, LRI, and SLU. The climate is continental and subtropical, with areas that attain the maximal absolute temperature of the country. The type of vegetation is the xerophilous forest, whose trees become lower and narrower towards the southwest. There are abundant mountain forests, savannas, and grasslands.

Humid Chaco occupies the eastern half of FOR and CHA and the extreme northwest of COS and North of SFE. The climate is subtropical warm. The average annual temperature decreases from North to South and from 23 °C in the border with Paraguay to about 18 °C in the center of the Province of SFE. The vegetation has greater diversity than in the Arid Chaco.

Patagonia forest is the driest part of the Andean-Patagonic forest; it extends from NEU to TDF. The landscape is mountainous, with abrupt and steep reliefs, snowy peaks, glacial valleys, volcanic geo-forms, numerous water courses, and lakes that drain through different rivers to the slopes of the Atlantic or the Pacific oceans. The climate is temperate to cold and humid, with heavy snowfall, rains, strong winds, and frost most of the year. The dominant plant formation is the humid, semidry temperate forest.

Grassland and shrublands are located in South MIS and Northeast COS being mostly pastures and grasslands corresponding to the phytogeographical Paraná flooded savanna. The dominating flora is herbaceous.

Paraná flooded savanna borders the southern Paraná River, from CHA to BAI. It runs through the Humid Chaco and the Humid Pampas eco-regions, ending up in the Río de la Plata estuary. It has a landscape of low and floodable islands, delimited by the lateral arms and main channels of rivers (e.g., Paraná Guazú, Paraná de las Palmas, Paraná Miní, Barca Grande, Luján) and extensive coastal waters. From the original habitat, only patches including strips of forest and shrubs, scrub and pasture, and hydrophilic and aquatic flora still remain.

Espinal is located in central Argentina involving the south of the Province of COS, northern half of ERI, a central belt of SFE and COR, central and southern SLU, and half East of LPA and south of BAI. The predominant landscape is flat plain to gently undulating, occupied by low forests, savannahs, and grasslands, nowadays largely adapted to agriculture.

The *Patagonian steppe* is around 730.000 Km^2 extended from southwest MEN, center of NEU, southwest of RNE, and most of CHU and SCR. The soils are rocky and sandy, lacking rich organic materials. The weather is temperate-cold, with snow in winter and frosts almost at any time of the year. Precipitations are scarce, ranging from 100 to 300 mm per year, and are concentrated on the coldest months, from April to September. In spring and summer, strong West winds blow almost constantly. There are low shrubs with small leaves and thorns and grasslands.

Iberá Wetlands are a deposit of stagnant water that occupies the northern center of the province of COS. It is one of the most important fresh water reservoirs of the continent. The wetlands, with a deep no higher than 3 m, are covered by aquatic plants (aguapés, poppies, and sequins of water) forming floating islands with up to 2 m of thickness. It is protected in the *Iberá* Provincial Reserve.

The *Argentina Low Monte* is the driest region of the country located in North-central Argentina covering the eastern foothills of the Andes. It extends from the Andean mountain range in MEN, along NEU and LPA, to the coast of the Atlantic Ocean of RNE and northeast of CHU. It shares with the eco-region of the *Monte de Sierras and Bolsones* the most arid characteristics of Argentina but differs on its prevailing landscape of plains and extensive plateaus. The climate is temperate-arid with scarce precipitations. Average annual temperatures are on the order of 10–14 °C. The vegetation is poorer than in the *Mount of Sierras and Bolsones*.

The *Pampa* represents around 60% of the grasslands of the country; it is extended from BAI (except the southern section), northeast of LPA, and south of COR, SFE, and ERI. The rains are distributed during the year. Average annual temperatures range from 15 °C in the South to about 18 ° in the North.

The *Alto Paraná Atlantic forests* extend from the center and North of the province of MIS through ERI as a streamside zone of narrow hills by the rivers from the basin of the Río de la Plata. The weather is warm and humid, with more intense rains in the summer. The average annual temperature is on the order of 20 °C. The red soils, characteristic of the eco-region, are a consequence of the transformation of the basaltic material under warm and humid conditions. It has 2000 vascular plants, and the Argentinean sector exhibits the highest biological diversity of all the country eco-regions. The dominant vegetation is the subtropical forest.

In the *South Atlantic islands*, due to the extreme climatic conditions and the rocky nature of the terrain, the soils are poorly developed. There are no natural forests, and the dominant vegetation is steppe grasses with tundra bushes. This eco-region is remarkably poor in phanerogams, several of which are endemic. Among the most prominent plant communities are the Tussock grass pastures, which occupy the coastal areas of Malvinas and South Georgia Islands, with different species of grasses and mosses, among other plants.

Mar Argentino includes the Argentina Continental Platform which can be subdivided into a coastal subregion, represented by the strip of coast up to 40 m depth and the subregion of the external platform, which extends from the depths of 40 m to 200 m. On the coasts of CHU, SCR, and TDF, marine vegetation develops with a huge diversity of green, red, and brown algae on the seabed.

The *Antarctic Peninsula* has an extremely cold climate, with average temperatures below zero and with snow and ice all year round. Two types of climate are distinguished: the glacial one, dominant in the sectors covered by ice, and the insular one that includes the Antarctic Peninsula and adjacent islands. Winds are very strong; temperature is always very low, reaching – 42 °C on polar night and 0 °C in the hottest months. The flora is very scarce and limited to sectors near the sea, which lose the snow cover during the brief Antarctica summer. There are lichens or moss cushions. The only native vascular plant species are *Colobanthus quitensis* (with tiny white flowers) and *Deschampsia antarctica* (Antarctic grass) usually among moss communities. A nitrophilous alga (*Prasiola crispa*) grows associated with penguin breeding ground. There are two introduced species (*Poa* spp.), which grow near to the permanent bases.

1.2.3 Phytogeographical Regions

Phytogeography can be defined as the geographical distribution of plants according to the temperature, humidity, light intensity, day length, soil composition, and biological communities, among other characteristics, of each region. With the exception of some species that grow in very hostile environments, plants usually grow with other individuals, from the same or other species, establishing plant communities or associations with all their members having similar ecological needs. There have been several attempts to organize the Argentine vegetation in phytogeographical regions, as is shown in the works of Lorentz (1876), Holmberg (1898), Hauman (1920, 1931), Frenguelli (1941), Castellanos and Pérez Moreau (1941, 1945), Cabrera (1953, 1958, 1971), and Cabrera and Willink (1980).

Cabrera (1971) divided Argentina into three phytogeographical regions: Neotropical, Antarctic, and Oceanic. Each region is categorized in domains, which are divided in provinces that in turn are subdivided into districts. For each domain Cabrera specifies its geographical location and for each province the location and characteristic vegetation. The Neotropical region was divided in the Amazonian (subdivided in the Yungas and Paranaense provinces), Chaqueño (subdivided in the Chaqueña, Espinal, Prepuna, Monte, and Pampeana provinces), and Andino Patagónico (subdivided in Altoandina, Puneña, and Patagónica provinces) domain. The Antarctic region was subdivided into the sub-Antarctic (subdivided in the sub-Antarctic and insular provinces) and the Antarctic (Antarctic province) domain and the Oceanic region without subdivisions (Cabrera 1971) (Fig. 1.1).

1.2.3.1 Neotropical Region

Amazonian domain: Despite its small area, the Amazonian domain is the largest floristic wealth in Argentina.

Yungas province: it has many genus and species in common with the Paranaense Province, such as *Nectandra*, *Cedrela*, *Blepharocalyx*, *Jacaranda* sp., etc. However, unlike the Paranaense province, it has few bamboos and no palm trees. It presents exclusive genera such as *Tipuana*, *Juglans*, *Phoebe*, and *Myroxylon*.

Paranaense province: more than 200 arboreal species have been counted in the Misiones Forest. In the district of the Mixed Jungle are arboreal communities of laurel forest (*Nectandra saligna*), Guatambú (*Balfourodendron riedelianum*), palo rosa (*Aspidosperma polyneuron*), and palmito (*Euterpe edulis*). Also, there are arboreal species such as anchico colorado (*Parapiptadenia rigida*), canchararana (*Cabialea oblongifolia*), and cocó (*Allophylus edulis*) among others. The pindó (*Syagrus romanzoffiana*) palm are scattered and distributed into the forest. In the northeastern sector of the MIS jungle, there are Paraná pine (*Araucaria angustifolia*) communities.

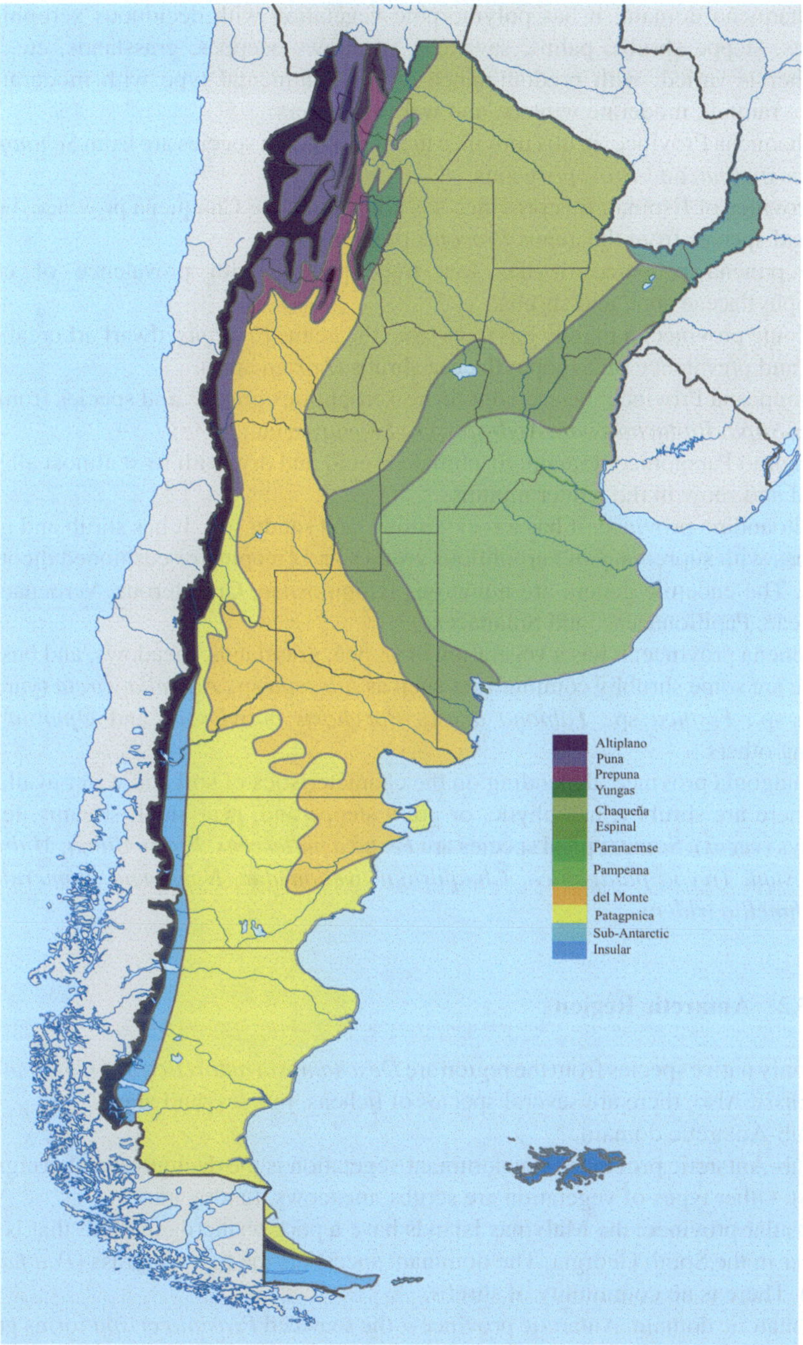

Fig. 1.1 Phytogeographical regions from Argentina. (Adapted from Cabrera (1971))

Chaqueño domain: it has polymorphic vegetation with deciduous xerophilous forests, steppe shrubs, palms, savannas, meadows, steppes, grasslands, etc. The weather is varied, with predominance of the continental type with moderate to scarce rainfall, moderate winters, and warm summers.

Chaqueña Province: in this province the predominant species are from *Schinopsis*, *Aspidosperma*, and *Prosopis* genus.

Province of Espinal: this province has connections to Chaqueña province, but in Espinal species from the genus *Prosopis* predominate.

Prepuneña province: it has few tree species, with prevalence of cacti, Zygophyllaceae spp., and shrubs.

Monte province: it mainly has trees from the genus *Prosopis*, dwarf arboreal species, and prevalence of Zygophyllaceae shrubs (*Larrea* spp.).

Pampeana Province: here predominate xerophilous grasses and species from the genus *Stipa*, *Piptochaetium*, *Andropogon*, *Leonurus*, etc.

Andino Patagónico domain: its climate is cold and dry, with frost almost all year round and snow in the winter months.

Altoandino province: it has a rocky, stony, and sandy soil. It has shrub and grass steppes with supremacy of xerophilous grasses and creeping or cushioned dicotyledons. The endemic genera are numerous (Compositae, Cruciferous, Verbenaceae, Poaceae, Papilionaceae, and Solanaceae).

Puneña province: it has a vegetation of steppe, grasslands, meadows, and bushes. There are some shrubby communities such as *Adesmia* sp., *Azorella yareta* (yareta), *Stipa* sp., *Festuca* sp., *Fabiana densa*, *Baccharis boliviensis*, and *Opuntia* sp., among others.

Patagonia province: depending on the characteristics of soil and water availability, there are shrubby, halophytic, or grass steppe and, near small streams, fertile valleys (*vegas*). Some typical species are *Festuca pallescens*, *F. gracillima*, *Mulinum spinosum*, *Trevoa patagonica*, *Chuquiraga avellanedae*, *Nassauvia glomerulosa*, and *Junellia tridens*.

1.2.3.2 Antarctic Region

The only native species from the region are *Deschampsia antarctica* and *Colobanthus quitensis*. Also, there are several species of lichens, mosses, and algae.

Sub-Antarctic domain.

Sub-Antarctic province: The dominant vegetation is the deciduous and evergreen forest. Other types of vegetation are scrubs, meadows, tundra, etc.

Insular province: the Malvinas Islands have a poor phanerogam flora that is still poorer in the South Georgia. The dominant species is the tussock grass (*Poa flabellata*). There is no community of shrubs.

Antarctic domain: Antarctic province—the seaweed *Prasiola crispa* forms green wavy, snow-free emerald mats. There are also mosses and lichens grass.

1.2.3.3 Oceanic Region

It is the most extensive region that covers all the oceans. In Argentina, it extends through the epicontinental sea, from the mouth of the Río de la Plata to the Antarctica.

The Cabrera classification of the phytogeographical regions is still the most popular. However, there are other classifications that should be considered.

For Takhtajan (1986), the terrestrial flora of Argentina could be distributed in three phytogeographical regions, two within the Neotropical (Brazilian and Andean) and one in the Holantarctic (Patagonian-Chilean) kingdoms. According to Takhtajan, the region and kingdom categories are equivalent to the domain and region categories defined by Cabrera (Matteucci et al. 2016). In a detailed work, for each of the phytogeographical regions defined by Cabrera, Apodaca et al. (2015) established the eco-regions involved, its dominant vegetation, area (in km^2), general weather, annual rain (expressed in mm), average annual temperature (expressed in °C), and species, genus, and/or families from its endemic and characteristic plant species. They also included a concise description of international, national, and provincial protected areas. They recognized three main biomes in Argentina: steppe (that include seven of the phytogeographic provinces), woodland (Chaco, Espinal, sub-Antarctic), and forest. In the steppe predominate the Gramineae; according to the accompanying species, the steppe could be shrubby, herbaceous, etc. In the woodlands prevail trees, and in the forest prevail tall trees along with copious lianas and epiphytes and stratified vegetation (Paranaense, Yungas) (Apodaca et al. 2015).

Ribichich (2002) disclosed weaknesses and discrepancies in the work of Cabrera (1971) and Cabrera and Willink (1980), particularly related to the imprecise use of the terms endemic and endemism, the irregular use of the taxonomic categories to classify phytogeographical areas in different hierarchical levels, the borderlines of the Argentinean provinces and geographical regions, and the ambiguities or incoherencies related to the characterization of the phytogeographical territories. Besides, Ribichich compares the phytogeographical systems from Cabrera and Takhtajan.

Recently, Oyarzábal et al. (2018) published a physiognomic-floristic map of spontaneous vegetation of Argentina based in vegetation surveys published in the last decades. For each phytogeographic province identified by Cabrera (1976), they reviewed the charts and available descriptions and mapped their internal heterogeneity. However, they made two modifications on Cabrera classification; on the one hand, they incorporated the documented ecotone Monte-Patagonia (León et al. 1998; Cingolani et al. 2000). On the other hand, they established slightly different phytogeographic limits (Cabrera 1976; León et al. 1998). The resulting map had 50 vegetation units as subdivisions of the phytogeographic ecotone and provinces described by Cabrera and a brief physiognomic-floristic description of each ecotone, province, and vegetation unit. A nomenclature of each vegetation unit based on dominant type of spontaneous vegetation and characteristic species was also proposed.

1.3 Medicinal Plants

The use of medicinal plants by early civilizations has been recorded for as old as 5000 years ago as is shown in Sumerian clay slabs containing the recipes for drug preparations (Sumner 2000), the clay tablets from ancient Mesopotamia dated back to around 2600 BC (Scurlock 2014), the registers from the Chinese emperor Shen Nung ca. 2500 BC (Zhu 1998), the Code of Hammurabi from 1800 BC (Sigerist et al. 1955), the Egyptian Ebers Papyrus dated 1500 B.C. (Ebers 1889), and the Sanskrit text *Charaka Samhita* on Ayurveda medicine dated 700 B.C. (Van Loon 2003). In occident, Hippocrates (460–377 BC) recorded and categorized by their physiological action over 200 medicinal plants in the *Book of* [Medical] *Aphorisms* (Castiglioni 1985), Dioscorides wrote the *De Materia Medica* (77 AD) with over 900 drugs mostly of plant origin (Riddle 1985), and Galen in the *Treatise of Simple Medicines* emphasized on the properties of plants, animals, and minerals as simple medicines (130–200 AD) (Petit 2017; Malapani 2016). In Egypt, the *Maimonides Medical Writings* (1100 AD) compiled medical theory and herbal information, especially in the 7th volume *Treatise on Poisons and Their Antidotes* (Rosner 2002). In America, the Aztec and Maya Indian cultures of Mexico and Central America also used natural substances as medicines. As is referred by Evans, postclassic Mesoamericans *developed a large and effective pharmacopoeia, formulae for medicines concocted from animals, minerals, and especially plants* (Evans 2004). The ancient Aztec healers exploited at least 132 medicinal herbs to treat ailments ranging from pimples and nosebleeds to gout and epilepsy and respiratory and gastrointestinal infections (Berdan 2005). After the arrival of Columbus, the Spanish Crown dominated and exploited the medicinal plants market from the Americas up to the eighteenth century (Miller 2017). Nowadays, each country has a particular approach to medicinal plants; some countries have specific regulations and national pharmacopeia that included plants or their derivatives, while others do not have regulations about the use of traditional medicines of plant origin.

The World Health Organization (WHO) confirmed that traditional medicine (TM) is still used by a high proportion of the world's population. It has been estimated that more than 80% of the people in developing countries depend on herbal medicine for basic healthcare needs. In Western countries, where chemistry is the backbone of the pharmaceutical industry, about 25% of the bioactive molecules are of plant origin. The WHO has released the *WHO guidelines on good agricultural and collection practices (GACP) for medicinal plants* as a means to recommend practices to ensure good quality, safe herbal medicines, and ecologically sound cultivation practices for future generations (WHO 2003).

1.4 Native Medicinal Flora from Argentina

Traditional medicine is a set of empirical practices from a group, which are transmitted from generation to generation with the intention of solving health problems from a specific population. The recognized native medicinal flora of Argentina comprises around 1529 taxa grouped into 175 families and 688 genera of vascular plants (pteridophyte, gymnosperm, and angiosperm). Barboza et al. (2009) have made an exhaustive analysis of medicinal plants in Argentina. They have found that the aerial parts (leaf, stem bark, etc.) and leaves are the most widely parts used in medicinal preparations. Stems, underground organs, and flowers are used in a lesser extent. Also, they found a high amount of references without specifications on the parts used in folk medicine. More information related to medicinal Dicotyledoneae than about Monocotyledoneae and Pteridophyta (58.92%) were found, being Asteraceae, Fabaceae, and Solanaceae the three best representative families, followed by Verbenaceae, Lamiaceae, Myrtaceae, and Apiaceae. From the available data, the anti-inflammatory, antioxidant, and antimicrobial activities were the most frequently cited; however, around 58% of the medicinal plants were not yet studied on their pharmacological properties. Finally, they found that most of the available information is referred to species from the provinces of SAL, JUJ, TUC, COS, MIS, ERI, and CHA. Less cited are species from COR, BAI, and CAT. SDE and CHA are the provinces with more medicinal plants and TDF the one with the lesser (Barboza et al. 2009). In the following chapters, some of those species having pharmacological studies that support their popular use will be described.

1.5 Regulatory Framework on Medicinal Plants in Argentina

Since 1999 Argentina has a legislation that normalizes and regulates the use of medicinal plants. The National Administration of Drug, Food and Medical Technology (ANMAT) is the body that aims to strengthen the regulatory, inspection, and surveillance capacities about health products, including those referring to medicinal plants and medicines of plant origin, at national and provincial levels. The law N° 16,463, Resolution 1817/2013, from the Ministry of Health, approved a glossary of terms related to phytomedicines, established the regime for registering herbal medicines, defined a simplified regime for registering traditional herbal medicines, and established the requirements that these medicines must fulfill to be registered, their methods of elaboration, and the specifications for labeling and advertising. The Argentine National Pharmacopoeia (FNA) is a national law that contains the information about plants and their products that are authorized as medicinal in the country. The last printed versions are the FNA VI edition (FNA VI) and the last revision, Volume I of the FNA VII. Volumes II, III, and IV of FNA VII were approved in November 2013 and published in the Official Gazette. The FNA VIII edition has been sent to the National Congress for being sanctioned. However, the FNA VI edition is still consulted. The FNA VII established that a Vegetable Drug is *a plant or their parts, entire, ground or powdered (flowers, fruits, seeds, tubers, bark,* etc.*) fresh or dried, as well as juices, gums, latex, essential or fixed oils and others similar components, which are used pure or mixed in the elaboration of medicines* (Argentina National Pharmacopeia 2013).

References

Apodaca MJ, Crisci JV, Katinas L (2015) Las provincias fitogeográficas de la República Argentina: definición y sus principales áreas protegidas. In: El deterioro del suelo y del ambiente en la Argentina, vol 1. Editorial Dunken, Argentina, pp 79–101

Argentina National Pharmacopeia 7th edition (2013). http://www.anmat.gov.ar/webanmat/fna/pfds/Farmacopea_Argentina_2013_Ed.7.pdf

Barboza GE, Cantero JJ, Núñez C, Pacciaroni A, Ariza Espinar L (2009) Medicinal plants: a general review and a phytochemical and ethnopharmacological screening of the native Argentine Flora. Kurtziana 34(12):7365

Berdan FF (2005) The Aztecs of Central Mexico: an Imperial Society. Thomson Wadsworth, Belmont

Burkart R, Bárbaro NO, Ro S, Gómez DA (1999) Eco-regiones de la Argentina. Administración de Parques Nacionales. Ed. Programa Desarrollo Institucional Ambiental. Presidencia de la Nación. Bs. As, Argentina. 42 pp

Cabrera AL (1953) Esquema Fitogeográfico de la República Argentina. Rev Mus La Plata (N.S.) Bot. 8:87–168

Cabrera AL (1958) Fitogeografía. In: de Aparicio F, Difrieri HA (eds) La Argentina: Suma de Geografía, vol 3, pp 101–207

Cabrera AL (1971) Fitogeografía de la República Argentina. Boletín de la Sociedad Argentina de Botánica 14:1–42

Cabrera AL (1976) Regiones fitogeográficas argentinas. In: Kugler WF (ed) Enciclopedia Argentina de Agricultura y Jardinería. 2nd edn. Acme, Buenos Aires, Argentina. 2(1):1–85

Cabrera AL, Willink A (1980) Biogeografía de América Latina. 2nd edition Serie de Biología. Secretaría General de la Organización de los Estados Americanos. EEUU, Washington, D.C., p 120

Castellanos A, Pérez Moreau RA (1941) Contribución a la Bibliografía Botánica Argentina. Lilloa 7

Castellanos A, Pérez Moreau RA (1945) Los tipos de vegetación de la República Argentina. Monogr Inst Estud Geogr Univ Nac Tucumán 4:1–154

Castiglioni A (1985) A history of medicine. Jason Aronson Inc. Publishers, New York

Cingolani AM, Bran D, López C, Ayesa J (2000) Comunidades vegetales y ambiente en el ecotono boreal entre los distritos patagónicos Central y Occidental (Río Negro, Argentina). Ecol Austral 10:47–61

Diminich MC, Franco RD, Fernández EM, Zorzi F (2018) Capítulo 5 Ecosistemas. In: Biología 3. Flujo de la información genética, los procesos evolutivos, de los individuos a los ecositemas, sistemas de relación y autorregulacion. Doce Orcas Ediciones, Ciudad Autónoma de Buenos Aires, pp 94–109

Ebers G (1889) The Papyrus Ebers. A translation. https://archive.org/details/papyrusebersdie00ebergoog/page/n17

Evans ST (2004) Ancient Mexico and Central America. Thames and Hudson, New York

Frenguelli J (1941) Rasgos principales de Fitogeografía Argentina. Rev Mus La Plata (N S) Bot 3

Hauman L (1920) Ganadería y Geobotánica en la Argentina. Rev Centr Estud Agron y Vet Bs As 102:45–65

Hauman L (1931) Esquisse phytogeographique de l'Argentine subtropicale. Bull Soc Roy Bot Belgique 64:20–79

Holmberg E L (1898) La Flora de la República Argentina, en Segundo Censo de la República Argentina, 1 Buenos Aires, Taller tip. de la Penitenciaria nacional

León RC, Bran D, Collantes M, Paruelo JM, Soriano A (1998) Grandes unidades de vegetación de la Patagonia extra andina. Ecol Austral 8:123–141

Lorentz PG (1876) Die Vegetationsverhältnisse der Argentinischen Republik. In: Richard Napp: Die Republik Argentinien. Buenos Aires: Sociedad Anónima, calle Belgrano 180. pp. 86–149

Malapani A (2016) Galen on pharmacology: his scientifi c way of thinking and contribution to the pharmacology of Cyprus. Hist Med 3(3):230–242

Matteucci, SD (2012) Chapter 1: Ecorregión Altos Andes. In: Ecorregiones y complejos ecosistémicos argentinos (Jorge Morello, Silvia D. Matteucci, Andrea F. Rodríguez, Mariana Silva). pp. 1–88. Orientación Gráfica Editora S.R.L, Buenos Aires, Argentina

Matteucci SD, Silva ME, Rodríguez AF (2016) Clasificaciones de la tierra: ¿províncias fitogeográficas, ecorregiones o paisajes? Fronteras 14:1–16

Miller C (2017) The apothecary in colonial America. Herbal Med Open Access 3(2):9. https://doi.org/10.21767/2472-0151.100029

Oyarzábal M, Clavijo J, Oakley L, Biganzoli F, Tognetti P, Barberis I, Maturo HM, Aragón R, Campanello PI, Prado D, Oesterheld M, León RJC (2018) Unidades de vegetación de la Argentina. Ecol Austral 28:040–063

Petit C (2017) Galen, pharmacology and the boundaries of medicine: a reassessment. In: Lehmhaus L, Martelli M (eds) Collecting recipes. Byzantine and Jewish pharmacology in dialog. Walter De Gruyter Inc, Boston/Berlin

Ribichich AM (2002) El modelo clásico de la fitogeografía de Argentina: un análisis crítico. Interciencia 27(12):669–675

Riddle JM (1985) Dioscorides on pharmacy and medicine. University of Texas Press, Austin

Rosner F (2002) The life of Moses Maimonides, a prominent medieval physician. Einstein Quart J Biol Med 19:125–128

Scurlock (2014) Sourcebook for ancient Mesopotamian medicine. SBL Press, Atlanta

Sigerist HE (1955) A history of medicine, vol 1. Oxford University Press, New York

Sumner J (2000) The natural history of medicinal plants. Timber Press, London

Takhtajan A (1986) Floristic regions of the world. University of California, Berkeley, p 522

Van Loon G (ed.) (2003) Charak Samhita. Handbook on Ayurveda. Vol 1 pp. 621. P.V. Lulu Inc. Morrisville, North Carolina, USA

WHO (2003) WHO guidelines on good agricultural and collection practices (GACP) for medicinal plants. WHO, Geneva, p 80

Zhu YP (1998) Chinese Materia Medica: chemistry, pharmacology and applications. Harwood Academic Publishers, Amsterdam

Chapter 2
Ethnobotany in Argentina

2.1 Introduction

Ethnobotany is the scientific approach to the relationships between people and plants into the frame of their sociocultural context. In this chapter a review of some aspects from ethnobotany in Argentina from the first references made by the members of the Society of Jesus (*Compañía de Jesús*) up to this day is introduced. Most of the studies already made were focused from a regional or ethnic point of view, although some combine both aspects.

2.2 Ethnobotany

In ancient times, naturalists, travelers, and missionaries collected information about habits and practices of societies in distant and new discovered communities or tribes. That information was often preserved in monasteries, abbeys or, in the New World, in catholic reductions called missions. The written documentations on the therapeutic uses of plants in the ancient Mediterranean world currently known are the *Hippocratic Collection*, *Historia Plantarum* by Theophrastus, and *De Materia Medica* by Dioscorides. The last one, often dated to ca. 70 AD, has been the basis of most of the literature on *materia medica* in the Mediterranean and Western tradition up to the surge of modern pharmacy. The study of useful plants is probably one of the oldest sciences in the world. Moreover, people that gathered the information about plant properties were always respected in their communities for having a knowledge that was hidden to the majority. Even before the writing was developed, the man left records of the many uses, including medicinal and religious, of plants. Traditionally, ethnobotanists were looking into documents and archeological sites for connections between ancient civilizations and settlements with plants. Those data revealed the migration habits, commercial routes,

M. A. Alvarez, *Pharmacological Properties of Native Plants from Argentina*,
https://doi.org/10.1007/978-3-030-20198-2_2

agricultural practices, and the origin and dispersion of those plant species. Similarly, ethnobotanists study the relationship between people and plant species of today compiling the knowledge safeguarded by indigenous and rural people (Álvarez 2014). Nonetheless, ethnobotany is not only restricted to the interactions between traditional cultures and plants. In a more broad and modern approach, ethnobotany comprises the study of both traditional and nontraditional populations and their connections with their botanical environment. Then, ethnobotany is not confined to the use of plants by ancient communities but also to the relationship that modern societies have with them. Recently, some authors are referring to the *ecological dimension* of ethnobotany as is the case of the relation with plants among immigrants (Kujawska and Pieroni 2015; Scarpa et al. 2016), in urban environments (Hurrell 2014), gardens (Kimber 2004; Calvet-Mir et al. 2014), and in fairs and markets (Ladio et al. 2013; Albuquerque et al. 2017). Some new questions are: How does the botanical background of a group change with migration or the arrival of immigrants? When the communities age and young people move, are the knowledge moving with them or it remains nested in the old community? How do traditional communities manage horticultural and agricultural techniques without harming the ecosystem? How was the environment of the communities modified by human interaction along time? What represent plants to people? (Korstanje 2016; Albuquerque et al. 2017). Regularly, in traditional communities plants are more than food and health; they are also part of their religion, magic, and rites. In brief, they are part of their culture and everyday life. However, it is usual to find reports describing the medical use of plants avoiding the underlying significance of that use. In that case, the current consensus is to call that type of studies as *medical ethnobotany* (Pake 1987). Consequently, with a broad-minded analysis, ethnobotany focus is turning to the study of the dynamic interrelationship between human societies and plant communities within particular sociocultural and environmental contexts (Rodríguez et al. 2006; Bonomo et al. 2011).

From the scientific point of view, ethnobotanical research evolved from inventories of useful plants to detailed quantitative studies supported by solid methods and analysis. Those methods should aim at collecting high-quality, representative data that have statistical strength. First of all, ethnobotany should accurately identify plant species. In doing that, the researcher will allow for the recovery of all the information linked to that species over the years. Not only the scientific name of the species must be properly recorded, but also herbarium specimens must be collected and appropriately deposited. Misidentifications will produce strong inaccuracies on current and future research (Blanché et al. 1996; Arenas and Kamienkowski 2014). Second, information must be analyzed; cultural domain analysis, ethnographic methods, participatory mapping and ground-truthing, plot, transects and tree trials, etc. are approaches that can be used to manage and analyze the collected information (Caruso et al. 2015).

The great importance of traditional knowledge is recognized in international initiatives as the Ecosystem Assessment (Reid et al. 2006), the Economics of Ecosystems and Biodiversity (Brondizio et al. 2010), or the Intergovernmental Platform Scientific and Technological Advice on Biodiversity and Ecosystem Services (IPBES) (de Santanaya 2014). Certainly, its inclusion in the United Nations

Convention on Biological Diversity (1992) was a landmark in its recognition and appreciation. Therefore, it has been recommended to all states to deepen scientific research, especially with regard to biological inventories and methods of extraction, identification, and standardization of substances from plants (OMS 1993). On the other hand, the ethical dimension of research must be considered; the Code of Ethics from the International Society of Ethnobiology proposes that "all ethnobiological research should be collaborative; it establishes a community-centered approach, and calls for all research to strengthen community goals" (http://ethnobiology.net/code-of-ethics/).

2.3 Ethnobotany in Argentina

The oldest reference about medicinal plants from Latin America was a Royal decree dated on the year 1570 signed by King Felipe II from Spain. There, he established that "... all herbs, trees, plants, or seeds of any medicinal value that can be found in those places will be sent to this Kingdom." Hence, the first collection on medicinal plants from the New World was published in 1597 under the title *Introduction and memories of the description of the Indians that His Majesty ordered to do for the good government and ennoblement of them* (Trillo et al. 2011).

At least empirically, ethnobotany in Argentina dates from the seventeenth century, when the Jesuits settled the reductions or missions at the northeast of the country in the Mesopotamia and also in the neighboring regions from Brazil, Bolivia, and Paraguay. Besides their religious, social, and economical purposes, the Jesuits also gathered information about native traditions, language, and customs which also included the characteristics of the native plants and their medicinal uses. However, there is scarce documentation about their findings. Some of them are the publications of three missionaries from the Society of Jesus, Pedro Lozano, Florian Paucke, and Pedro Montenegro. Lozano was designated historian from the Society (*historiograpus provinciae*) from 1730; hence, as an endorsed historian, he offered an official chronicle following the constraints of the order. He wrote the *Descripción Chorográfica del Gran Chaco Gualamba* and other works of historical value, *La historia de la conquista*, *La historia de la Compañía*, and *La historia de las revoluciones del Paraguay*. The book *Descripción Chorográfica del Gran Chaco Gualamba* was published in 1733 in Córdoba, Spain, and reedited in 1941 in Tucumán, Argentina (Lozano 1941). It was accompanied by a map of Gran Chaco by father Antonio Machoni that could also be found in the reedition. Florián Paucke was a missionary in the *mocovíes* reductions of *San Francisco Javier* and of *San Pedro*, from 1749 up to the eviction of the Society in 1767. His work *Hacia allá y para acá: Una estada entre los indios Mocobíes 1749–1767* was written between 1778 and 1779, but first published in 1944. His style differs from Lozano in the close approach to the indigenous and locations he made; consequently, his work is considered by some authors comparable to that of an ethnographer. The final version of his work, which included watercolor illustrations, was written from exile with the subtitle *There we went pleasant and happy, here we returned bitter and saddened* (Paucke s/f 2010; Rosso and Cargnel 2012; Rosso and Medrano 2013). Brother Pedro de Montenegro, a physician that entered the Society and had established in the Paraguay province, wrote the *Materia Médica Misionera* (1710). It was described by Lozano as "an incredible surgeon (*chirurgus*), nurse (*infirmarius*), and herbalist." His *Materia Medica* contains a rich description of native plant species, their therapeutic properties, and mode of administration (Montenegro 1945). *Materia Médica Misionera* is a codex of 458 illustrated pages with 136 figures, drawn by pen and ink. On the cover there is a picture of Our Lady of Sorrows, patron saint of the sick, and at the bottom the date "Year of 1710." The first three sections refer to the botanical nomenclature, to the properties of the plants, the time of collect, and how to preserve them, their healing virtues, and how to take advantage of them. The fourth part is a medical study of diseases that are cured by herbs, roots, and barks (Perkins de Piacentino 2007). Hence, the *Plantae Diaphoricae Florae Argentinae* (Hieronymus 1882) is considered as the first compilation of medicinal, poisonous, or useful plants that grew in Argentina (native or exotic). A remarkable characteristic of that book is that the author only included species whose scientific names were known in those days, producing a valuable scientific enumeration of species, their characteristics, and uses. Other scientists also made a great contribution to plant knowledge related to the primitive inhabitants of the country. Aimé Bonpland, a French explorer, physician, and botanist, traveled with Alexander von Humboldt to Spain, Venezuela, Colombia, Ecuador, Cuba, México, and the United States (1799–1804). They produced an herbarium with approximately 6000 samples, 4500 of them new (described in *Nova genera et species plantarum*, from Humboldt, Bonpland, and Kunth) that are deposited in Paris. Later on, he traveled to South America, arriving to Argentina in 1817. His archive, with more than 2050 documents (scientific, epistolary, historical) is in the *Museo de Farmacobotánica, Juan A. Domínguez*, in the Faculty of Pharmacy and Biochemistry, University of Buenos Aires. Among his manuscripts are the Botanical Journals where he described 2881 plants that he collected in the Río de la Plata region.

Juan A. Domínguez was a pharmacist committed to the study of phytochemistry and botany. He was the author of *Contribuciones a la Materia Médica Argentina* (1928) and a series of publications on medicinal plants, particularly those used by indigenous from Argentina. His herbarium is part of the *Museo of Farmacobotánica Juan A. Domínguez* collection (Amorín 1990; Giberti 2008; Krapovickas 2008).

Raúl N. Martínez Crovetto was the first Argentinean researcher that introduced the expression ethnobotany in his articles. His work was focused on both indigenous and rural people from different locations in the country, mainly from the northeast region. His approach was qualitative, meticulously descriptive and interpretive (Pirondo and Keller 2012). Among its prolific series of publications are the *Estudios etnobotánicos I-V*, the last one published post-mortem by his collaborators in Bonplandia, the journal he founded. The library of the *Instituto de Botánica del Nordeste* preserves numerous documents belonging to him (Pirondo and Keller 2012).

Domingo Parodi, Eduardo Matoso, Nicolás Rojas Acosta, Lorenzo R. Parodi, and Arturo Ragonese were also remarkable contributors to the ethnobotanical studies in Argentina.

Currently, there are profuse investigation covering aspects such as the relationship between people and plants in connection with their ethnical origin, history, and cultural landscape (Arenas 2009; Arteaga 2010; Capparelli et al. 2011; Rosso 2013; Kujawska 2016; Scarpa and Rosso 2018). I will make here a brief review of some of the available literature in order to bring a general glance about the work already made. For more information there is a great amount of material that could be found in the articles cited in this chapter.

I consider the work of Barboza et al. (2009) of a paramount relevance, since they have performed an exhaustive and detailed review of native medicinal plants including bibliometric, phytochemical, and etnopharmacological screenings. In that publication are described around 1520 taxa. Among the species listed, only about half of them have chemical and pharmacological studies carried out. Consequently, there is still a lot of work to be done in order to substantiate the accuracy of their traditional use. From the article, it is clear that the uppermost diversity of plants with medicinal properties is located at the Northwest, Mesopotamia, and Gran Chaco regions, being SAL and JUJ the provinces with the highest number of medicinal taxa. In consequence, reports are more numerous from the central and northern part of the country than from the south.

There are records compiling plant catalogues, references about plant handling, methods of collection, uses, part used, preparations, route of administration, conservation, and the social-cultural context of the consumers.

2.3.1 *Ethnobotany in Indigenous and* criollos *Communities*

There are a large number of publications specifically related to the ethnobotany of indigenous communities (Arenas 1995). I will refer here only to some of them; a list of ethnobotanical research related to communities from Argentina (indigenous and criollos) is displayed in Table 2.1. In 1967 Martínez Crovetto gathered information about the ethnic group *mocovi*, or *moqoit*, from the southwest of the province of CHA. Those records were included in several articles describing the medicinal plants of *mocovis*, their scientific name, specific use, part of the plant used, elaboration, and administration (Martínez Crovetto 2014; Scarpa and Rosso 2014). A comparison was later made between the contemporary alimentary ethnobotany and the one of *mocovis* during the eighteenth and twentieth centuries (Rosso and Scarpa 2017) founding the highest similarities with the information recorded in the twentieth century. *Bromelia hieronymi*, *Aechmea distichantha*, *Araujia odorata*, and *Sarcomphalus mistol* were the more consumed species, preferably as raw fruit. In those communities, the edible use of the fresh fruits of *Scutia buxifolia*, the flowers of *Glandularia peruviana*, and the use as salt of cfr. *Sporobolus spartinus* ashes were there for the first time recorded. In the case of *chorote*, hunter-gatherer and fishermen who live in the Northwest of Argentina, a description was made about the edible wild plants they collected as well as a detailed description of the used parts, their elaboration, mode of conservation, and storage. Some of the species were also consumed by other groups of the same linguistic family (*Mataco-Macá*), and only a few ones were

Table 2.1 Ethnobotanical studies related to ethnic groups from Argentina (indigenous and criollos)

Communities	Research subject	References
Aónikenk, *Selk'nam*, *Kawésqar*, *Yagan*, and *Haush*	Ethnobotany mushroom	Díaz (2010)
Chorote	Alimentary ethnobotany	Arenas and Scarpa (2007), Scarpa and Pacor (2017)
	Medical ethnobotany	Scarpa (2009)
	Ethnobotany and religion	Scarpa and Pacor (2015)
Criollos	Ethnobotany	Scarpa et al. (2016)
	Medical ethnobotany	Scarpa and Rosso (2019)
Huarpes	Medical ethnobotany	Montani (2013)
Mapuches	Alimentary and medical ethnobotany	Molares and Ladio (2015)
Mataco-ihoko'tax and toba-pilagá	Alimentary ethnobotany	Arenas (2000)
Mbya	Medical ethnobotany	Remorini and Sy (2002)
Mocovi	Ethnobotany	Martínez Crovetto (2014), Scarpa and Rosso (2014)
	Alimentary ethnobotany	Rosso and Scarpa (2017)
	Historical ethnobotany	Rosso (2013)
Pilagá	Medical ethnobotany	Filipov (1994)
Quilmes	Medical ethnobotany	Ceballos and Perea (2014)
Ranqueles	Ethnobotany	Steibel (1997)
Selknam	Archeobotany	Caruso et al. (2008)
Toba-pilagá	Ethnophytogeography	Scarpa and Arenas (2004)
Toba	Sorcery and funeral practices	Arenas (2013)
Wichi	Ethnobotany, phytonomy of herbs and climbing plants of the semiarid Chaco, Salta Province	Suárez (2011), Suárez (2014)

exclusive to them (Arenas and Scarpa 2007). As the knowledge about the *chorote* ethnobotany is scarce, that work is a contribution of great relevance; however, it was only referred to *chorotes* living in Argentina, not in Paraguay. Scarpa and Pacor (2015) have made an analysis about the religious symbolism of *Anisocapparis speciosa* (Capparaceae) (*axis of the world* and *inexhaustible regeneration*) for *chorotes* and its medicinal use. It was interesting to found that *chorotes* and *criollos* from the same region shared approximately half of their ethnomedicine. Maybe the incorporation by the *chorotes* of some species used by *criollos* is a result of the restrictiveness of their ethnomedicine (Scarpa 2009). On the other hand, it was reported a cultural change in *chorote* society as a consequence of the growing stigma and prejudices involved in the collection of edible plants (Scarpa and Pacor 2017).

As for the *wichis* from semiarid CHA, the phytonymy of herbs and climbing plants was made (Suárez 2011). The vernacular name of 90 species was recorded along with their translation to Spanish and also a semantic and linguistic analysis in order to establish nomenclature patterns.

A portrayal of the *Mapuche* cosmovision, the participation of shamans and other mediators, and the description of specific healing rites are portrayed in the article of Albornoz et al. (2004). A record of their medicinal species, with their scientific and vernacular name and their uses, was included. A similar analysis was made for the *mbya-guaraní* communities from Misiones (Remorini and Sy 2002).

An inventory of the species of ethnobotanical interest for the *Aónikenk*, *Selk'nam*, *Kawésqar*, *Yagan*, and *Haush* communities from West TDF-Patagonia with their uses and medicinal properties was made based on bibliographic information (Díaz 2010). A number of 68 taxons were registered, mainly used as food, medicine (analgesic, digestive, disinfectant, diuretic, etc.), and to a lesser extent, as fuel or as part of rituals. The different ways of plant use among the *Selknam* from Tierra del Fuego were discovered from carbonized and non-carbonized wood remains recovered at Ewan site. Woody species in that hunter-gatherer group were particularly used for construction and as fuel. Seeds of *Empetrum rubrum* were found in one of the archeological sites (Caruso et al. 2008). A record of the species of ethnomedical interest was also made for the *pilagá* from Central CHA (Filipov 1994), *ranqueles* from LPA (Steibel 1997), and *quilmes* from TUC (Ceballos and Perea 2014).

Also, a historical analysis from non-medicinal plant species used by *criollo* groups was made from the information presented by the Argentinian Government at the Universal Exhibition of Paris of 1889. The author identified the most employed plant families of that time and their uses (Scarpa 2017). The more reported uses were, from most cited to less cited, as dyes, utensils, furniture, food, transport, tanning, home building, other constructions, and fuel.

The medical ethnobotany from Criollos from the Humid Chaco region was analyzed considering the data from the 1921 National Folklore Survey. The remedies for the digestive system, respiratory system, and vulneraries were the most represented (Scarpa and Rosso 2019).

2.3.2 *Ethnobotany from Different Regions from Argentina*

The ethnobotany from different regions from Argentina was also profusely analyzed (Table 2.2). There are numerous catalogues of medicinal species from the province of BAI with its corresponding taxonomic classification, botanical characteristics, uses, and medicinal preparations (D'Alfonso et al. 2011; Hurrell et al. 2015). Besides, agricultural practices in the conservation and selection of vegetables

Table 2.2 Ethnobotanical research in different regions of Argentina

Region	Research subject	References
Buenos Aires		
Azul	Medicinal ethnobotany	Vercelli et al. (2013)
Tandilia system	Botanical knowledge	D'Alfonso et al. (2011), Hilgert et al. (2010)
Metropolitan area	Urban ethnobotany	Hurrell et al. (2015), Hurrell et al. (2013), Puentes and Hurrel (2015)
Buenos Aires-La Plata conurbation	Urban ethnobotany	Hurrell et al. (2016), Puentes and Hurrel (2015)
La Plata	Ethnobotany	Doumecq and Arenas (2018), Puentes and Hurrel (2015)
Isla Santiago	Medical ethnobotany	Hernández et al. (2013)
Parque Costero del Sur	Alimentary and medical ethnobotany	Echenique et al. (2018)
Laguna de Mar Chiquita	Medical ethnobotany	Arias Toledo and Trillo (2014)
Hurlingham	Urban ethnobotany, urban phytotherapy	Bach et al. (2014)
Paraná Delta	Archeobotany	Bonomo et al. (2011)
Misiones	Medical ethnobotany	Stampella et al. (2018), Zamudio et al. (2010), Kujawska and Pieroni (2015), Kujawska (2016)
	Ethnopharmacology	Amat and Vajía (1991)
Formosa	Historical ethnobotany	Anconatani and Scarpa (2015), Scarpa and Anconatani (2017)
Chaco	Medical ethnobotany	Arenas (1997), Scarpa (2004), Marinoff et al. (2006)
Corrientes	Urban ethnobotany	Pirondo et al. (2011)
La Pampa	Alimentary ethnobotany from peasants	Muiño (2012)
Sierras Pampeanas	Archeobotany	Salazar (2015)
	Medical ethnobotany from monocotyledonous	Novoa et al. (1998)
Patagonia	Urban ethnobotany	Ladio et al. (2013), Molares and Rovero (2016)
Yungas	Alimentary and medical ethnobotany	Hilgert (1999), Hilgert and Gil (2006)
Salta	Medical ethnobotany	Martínez et al. (2004), Hilgert (2001)
Upper Bermejo Basin	Plants involved in the manufacture of *yista*	Hilgert (2000)
North West Region (NOA)	Archeobotany research (0–600 a.C.)	Oliszewski (2005)
Jujuy	Medical ethnobotany	Acosta et al. (2015), Romeo (2015)
	Urban ethnobotany	Acosta et al. (2017)
Tilcara	Ethnobotany	Barbarich and Suárez (2018)

(continued)

Table 2.2 (continued)

Region	Research subject	References
Los Juríes, Santiago del Estero	Medical ethnobotany	Riat and Pochettino (2015)
Puna	Archeobotany	Rodríguez et al. (2006)
Córdoba	Medical ethnobotany	Martínez (2005), Arias Toledo et al. (2007), Martínez (2008a, 2008b), Arias Toledo (2009)
	Veterinary ethnobotany	Martínez and Luján (2011)
	Cultural landscape	Trillo (2016)
	Diachronic ethnobotany	Fernández and Martínez (2019)
	Ethnobotany	Arias Toledo et al. (2009)
Sierra de Comechingones	Medical ethnobotany	Goleniowski et al. (2006)
Río Cuarto	Medical and urban ethnobotany	Madaleno and Montero (2012)
Chaco forest	Alimentary and ethnomedicine	Arias Toledo et al. (2007)
Pampa-Cuyo	Medical ethnobotany	Muiño (2011)
Andean Argentina	Archeobotany and ethnohistory	Caparelli et al. (2005)
Rio Negro, Patagonia	Medical ethnobotany	Ochoa et al. (2010)
	Ethnobotany	Lozada et al. (2006)
Tierra del Fuego	Archeobotany	Caruso et al. (2008)

(Ahumada et al. 2010), traditional botanical knowledge in urban zones (Pochettino et al. 2008), marketing of medicinal plants (Arenas et al. 2011; Hurrell et al. 2013), and the use of forest species as food and medicine, (Sharry et al. 2011) among other subjects. Were also studied. The current sociocultural changes in the metropolis of Buenos Aires and the province are reflected in research related to more recent immigration. Such is the case of two articles on urban ethnobotany. One of them revealed an increase in local biocultural diversity by analyzing the presence of medicinal plant products (e.g., *Stevia rebaudiana*, *Achyrocline alata*, *Cuphea lysimachioides*, *Euphorbia serpens*, etc.) brought by immigrants from Paraguay (which represents 30.5% of the immigrants in Argentina) to the metropolitan area of BAI (Hurrell et al. 2016). The others refer to the traditional Chinese phytotherapy insertion in ciudad de Buenos Aires. For each species their scientific and vernacular names in Spanish, Chinese, and Latin are included along with their distribution, use linked with tradition, biological activity, and effects (Hurrel and Puentes 2017).

The province of COR is proliferous in investigations related to its flora. The changes in the knowledge and management of medicinal plants were evidenced by various articles. A study made in a rural community revealed that people used and cared more medicinal plants than wild edible plants, with a marked influence by the sociocultural context (Arias Toledo et al. 2007). From the individuals studied, those with better awareness were people that had acquired that knowledge through formal education and people, old or native residents, that had an empirical knowledge about nature. A third and extensive group was composed by people without any familiarity with medicinal plants. The same indifference was found related to the use of native plants as food, which was attributed to the prevalence of meat as the main ingredient of the diet for Argentinean people, neglecting the ingestion of vegetables, and to the extended association of the use of wild plants as food with poverty (Pelto et al. 1989; Aguirre 2005). The knowledge and use of medical plants by residents living near natural protected areas in *Sierras de* COR, which has a higher variety of species, did not show differences with people living in remote places. Both groups used the same species in cases of digestive, respiratory, or circulatory problems (Furlan et al. 2011). In the *chaqueña* phytogeographical region from COR, the familiarity about medicinal plants was higher in older residents than in young people, which is probably related to the modifications in lifestyle, the access to medical assistance, and a growing detaching of rural practices. On the contrary, no differences were found related to gender (Arias Toledo et al. 2009).

In MIS, which also has a great amount of literature referred to its medicinal plants, studies related to the behavior of natives, criollos, and descendant of immigrants coming in the first moiety of the twentieth century were made. There were found patterns of medicinal pluralism among peasant descendants from Polish immigrants in MIS (in the settlements Wanda and Gobernador Lanusse). Their determination to retain their traditional healing practices was mixed with the apprehension of knowledge from natives and criollos through social cohabitation and mixed marriages. The study revealed that they used medicinal plants (22 species known from the home country and other native species), restored their health at home, and required the assistance of healers as *curanderos*, *hueseros*, and *naturistas*. Medicalization was restrained to childbirth and fractures or when life was threatened (Kujawska 2016). Polish and criollo community from MIS diverged on the use of wild honeys produced by insects of the order Hymenoptera (bees, wasps, and bumblebees) as food and medicine. All the investigated honeys were used by both communities as food, but only a few of them are considered also as a medicine (honey from bee, *yateí*, *mirí*, *carabozá*). The ethnospecies *borá*, *mandasaia*, and *iratin* were exclusively mentioned as medicinal by the *criollo* population. Home remedies based on honey are most frequently used to treat illnesses related to the respiratory system in both populations. Polish and *criollos* prepared home remedies in a similar way but showed differences in the frequency and mode of application (Zamudio et al. 2010). As for the relevance regarding the cultural context, it was seen that the main *Citrus* spp. from the region varies according to the cultural context; for criollos the more relevant species are *limón mandarina* and *naranja dulce*, whereas for residents of the *Yabotí* biosphere is *lima ácida* (Stampella et al. 2018). The medicinal use of *Citrus* among the criollos revealed that the juice, tea, or decoctions made with non-processed endocarp and leaves (7 ethnospecies, 5 taxa) alone or combined with other plants were used to treat flu, fever, hypertension, cough, and digestive or nerve disorders.

An alimentary ethnobotanical analysis from peasant descendants from immigrant settlers in LPA registered 62 taxons (4.7% of the wild species, 31% native taxons) that were used to prepare meals, sweets, and beverages. The knowledge about those plants was higher in older people (Arias Toledo et al. 2009; Muiño 2012).

In the Yungas the edible plants from the *Quebrada de Humahuaca* were recorded including their scientific and vernacular name, parts of the plants employed, methods of collection and application, cooking, commercial relationships with neighboring settlements, and their geographical localization (Hilgert 1999). In the Northwest (Korstanje 2016), in peri-urban neighborhoods from the city of SAL, the habit of chewing leaves of *Erythroxylon coca* (*coqueo*) was extended to a 54% of the population which is related to ancestral cultural customs not only restricted to the descendants of the original natives. Other medicinal plants that were reported to be commonly used were *quimpe* (*Lepidium didymum*), *tusca* (*Acacia aroma*), *paico* (*Chenopodium ambrosioides*), *molle* (*Schinus areira*), *ruda* (*Ruta chalepensis*), and *hediondilla* (*Cestrum parqui*), being taken in cases of gastrointestinal, respiratory, and kidney alterations in different preparations. In opposition with the finding of Arias Toledo et al. (2009), no differences in knowledge about medicinal plants were found related to age (Oliva 2016). In the *Calchaquí* Valley (Salta), traditional healers managed a phytotherapy of 42 species specific for regional diseases related to the environment and the rural activities of the population, such as osteoarticular and cardiocirculatory problems. The article of Martínez et al. (2004) brings a deep description of species, parts of the plants used, method of preparation, medical uses, and route of administration. In Patagonia, the commercialization of weeds as food and medicinal plants in urban fairs from Bariloche was analyzed, paying particular attention to the cosmovision, attitudes, and behavior of the horticulturalists. There, as was referred for the province of COR (Ladio et al. 2013), the use of edible wild plants was initially associated to poverty, a perception that slightly changed after an instructive workshop. Another study made in CHU, in a region inhabited predominantly by descendants of the *Mapuche-Tehuelche* indigenous people, revealed that the most versatile species (*Acantholippia seriphioides*, *Valeriana* spp., *Mentha*, and *Artemisia absinthium*) were the most relevant in folk medicine since they can treat a variety of medical conditions (Richeri et al. 2013).

There are also ethnomedical studies related to veterinary medicinal plants. In CAT, it was found 43 species from 30 botanical families with 62 medicinal uses such as healing, digestive, antiparasitic and oxytocic. Also, plants were involved in cultural practices of religious-ritualistic nature to heal animals (Martínez and Jiménez-Escobar 2017). Similar studies were made in *Sierras de* COR with the additional comparative analysis with the use of those plants in human medicine. A total of 127 medicinal uses were registered, corresponding to 70 species of plants belonging to 39 botanic families (Martínez and Luján 2011).

Ethnobotanical studies focused on a particular species or biological activity were also profused. Some examples are the ethnobotanical studies about cardón (*Trichocereus atacamensis*) from Tilcara in JUJ (Barbarich and Suárez 2018),

Hypericum robsonii spec. nova sect. *Trigynobrathys* (Hypericaceae) from MIS (Keller and Crockett 2015), *ligas* (Loranthaceae), and *Synandrospadix vermitoxicus* (Scarpa and Montani 2011; Arenas 2016). As for the ethnobotanical studies focused on the biological activities, there are reports about antimicrobial and antifungal activities, among others (Demo et al. 2008; Rondina et al. 2010).

References

Acosta ME, Vignale ND, Ladio AH (2015) ¿Qué saben sobre plantas empledas en medicina tradicional los niños de una escuela primearia de S. S. De Jujuy, Argentina? Gaia Scientia 2015. Ed Esp Etnobiol Arg 9(3):90–104

Acosta ME, Ladio AH, Vignale ND (2017) Medicinal plants traded in San Salvador de Jujuy city (Argentina) and its botanical quality. Bol Latinoam Caribe Plant Med Aromat 16(1):34–52

Aguirre P (2005) Estrategias de consumo: qué comen los argentinos que comen. Centro Interdisciplinario para el Estudio de Políticas Públicas (CIEPP) - Miño y Dávila (eds.), Argentina

Ahumada A, Ciampagna ML, Vera Bahima J, Garat JJ, Otero J (2009) Prácticas Culturales em la selección y conservación de hortalizas locales en el cinturón hortícola platense. Trad and Transformations Ethnobot 472–478

Albornoz A, Farías V, Montero G, Negri A (2004) Introducción a la complejidad herbolaria de la medicina tradicional mapuche. www.unisi.it/ricerca/centri/cisai/Albetal.doc

Albuquerque UP, Ramos MA, Soares Ferreira W, Muniz de Medeiros P (2017) Ethnobotany for beginners, Springer briefs in plant science. Springer International Publishing AG 2017

Álvarez MA (2014) Chapter 2: Plants for health. In: Plant biotechnology for health. From secondary metabolites to molecular farming. Springer, Cham/Heidelberg/New York/Dordecht/ London, p 161

Amat AG, Vajía ME (1991) Plantas Medicinales y Etnofarmacología en la Provincia de Misiones (Argentina). Acta Farm Bonaer 10(3):153–159

Amorín JL (1990) Juan Aníbal Domínguez (1876-1946). Dominguezia 8(1):35–40

Anconatani LM, Scarpa GF (2015) Etnobotánica histórica de las Misiones Franciscanas de Formosa I. Hallazgos documentales de fuentes primarias, análisis crítico y comparación con la obra Erbe medicinali del Chaco de Franzè (1925). Dominguezia 31(1):49–61

Arenas P (1995) Encuesta etnobotánica aplicada a indígenas del Gran Chaco. Hacia una nueva carta étnica del Gran Chaco VI:161–178

Arenas P (1997) Las fuentes actuales y del pasado par la etnobotánica del gran Chaco. Monograf Jard Bot Córdoba 5:17–25

Arenas P (2000) La alimentación tradicional y sus cambios entre dos etnias del Gran Chaco: los mataco-ihoko-tax y toba-pilagá. In: Estévez MG (ed) Sustentos, aflicciones y postrimerías de los Indios de América. Casa de América, Madrid, pp 31–56

Arenas P (2009) Los estudios sobre medicina y farmacopea vernácula en el gran Chaco. Rojasiana 8(2):81–100

Arenas P (2013) Hechicería, muerte y prácticas funerarias para vengar maleficios entre los toba del occidente de Formosa (Argentina). Una aproximación etnobiológica. In: Etnobotánica en zonas áridas y semiáridas del Cono Sur de Sudamérica. Ed Pastor Arenas/ CONICET pp-179–193

Arenas P (2016) Etnobotánica de *Synandrospadix vermitoxicus* (Araceae) en el Gran Chaco y en regiones aledañas. Bol Soc Argent Bot 51(2):379–399

Arenas P, Kamienkowski NM (2014) La documetnación del material vegetal incompleto o fragmentario en la investigación etnobotánica. ProBiota, Ser Tecn Didáct 26:1–17

Arenas P, Scarpa GF (2007) Edible wild plants of the Chorote Indians, Gran Chaco, Argentina. Bot J Linn Soc 153:73–85

Arias Toledo B (2009) Diversity of uses, recollection practices and gender and age differentials in the use of medicinal plants in Cordoba, Argentina. Bol Latinoam Caribe Plant Med Aromat 8(5):389–401

Arias Toledo B, Trillo C (2014) Animales y plantas que curan: avances sobre la farmacopea natural de los pobladores del área de Laguna Mar Chiquita. Revista FCEFYN 1(2):77–85

Arias Toledo B, Colantonio S, Galetto L (2007) Knowledge and use of edible and medicinal plants in two populations from the Chaco forest, Córdoba province, Argentina. J Ethnobiol 27(2):218–232

Arias Toledo B, Galetto L, Colantonio S (2009) Ethnobotanical knowledge in rural communities of Cordoba (Argentina): the importance of cultural and biogeographical factors. J Ethnobiol Ethnomed 5:40. https://doi.org/10.1186/1746-4269-5-40

Arenas PM, Cristina I, Puentes JP, Buet Costantino F, Hurrell JA, Pochettino ML (2011) Adaptogens: traditional medicinal plants commercialized as dietary supplements in the conurbation Buenos Aires-La Plata (Argentina). Bonplandia 20(2):251–264

Arteaga F (2010) AIBR. Traditional medicine in the Pampa Argentina. About the praxis and medical knowledge, folklore and catholic rituals. AIBR 5(3):397–429

Bach HG, Wagner ML, Ricco RA, Fortunato RH (2014) Sale of medicinal herbs in pharmacies and herbal stores in Hurlingham district, Buenos Aires, Argentina. Rev Bras Farm 24:258–264

Barbarich MF, Suárez ME (2018) Los guardianes silenciosos de la Quebrada de Humahuaca: Etnobotánica del “Cardón” (*Trichocereus atacamensis*, Cactaceae) entre pobladores originarios en el departamento de Tilcara, Jujuy, Argentina. Bonplandia 27(1):59–80

Barboza GE, Cantero JJ, Núñez C, Pacciaroni A, Ariza Espinar L (2009) Medicinal plants: a general review and a phytochemical and ethnopharmacological screening of the native Argentine Flora. Kurtziana 34(12):7365

Blanché C, Bonet MA, Nuntané J, Vallès J (1996) Bases de datos en etnobotánica: elaboración de los resultados. Monograf Jard Bot Córdoba 3:63–68

Bonomo M, Aceituno FJ, Politis GG, Pochettino ML (2011) Pre-Hispanic horticulture in the Paraná Delta (Argentina): archaeological and historical evidence. World Archaeol Debates World Archaeol 43(4):554–575

Brondizio ES, Gatzweiler F, Zagrafos C, Kumar MD (2010) Socio-cultural context of ecosystem and biodiversity valuation. In: Kumar P (ed) The economics of ecosystems and biodiversity (TEEB). UN Environmental Programme and the European Commission. Earthscan Press, London, pp 150–181

Calvet-Mir L, Garnatje T, Parada M, Vallès J, Reyes-García V (2014) Más allá de la producción de alimentos: los huertos familiares como reservorios de diversidad. Ambiente 107:40–53

Caparelli A, Lema V, Giovannetti M, Raffino R (2005) The introduction of Old World crops (wheat, barley and peach) in Andean Argentina during the 16th century a.d.: archaeobotanical and ethnohistorical evidence. Veg Hist Archaeobotany 14:472–484

Capparelli A, Hilgert N, Ladio A, Lema VS, Llano C, Molares S, Pochettino ML, Stampella P (2011) Paisajes culturales de Argentina: Pasado y presente desde las perspectivas etnobotánica y paleoetnobotánica. Rev. Asoc. Argent. Ecol. Paisajes 2(2):67–79

Caruso L, Mansur ME, Piqué R (2008) Voices in the forest: hunter-gatherers use of vegetable resources in the central part of Tierra del Fuego (Argentina). Darwin 46(2):202–212

Caruso E, Olwen G, Krause R, Martin G, Puri R, Rankou H, Tekguc I (2015) Conducting and communicating ethnobotanical research. A Methods Manual. Global Diversity Foundation

Ceballos SJ, Perea MC (2014) Medicinal plants used by the indigenous community of Quilmes (Tucumán, Argentina). Bol Latinoam Caribe Plant Med Aromat 13(1):47–68

D’Alfonso C, Scaramuzzino R, Farina E (2011) Actualización del catálogo de la Flora Medicinal de las sierras de Azul (Sistema de Tandilia), Provincia de Buenos Aires, Argentina. Rojasiana 10(1):29–53

Demo M, Oliva M d l M, López ML, Zunino MP, Zygadlo JA (2008) Antimicrobial activity of essential oils obtained from aromatic plants of Argentina. Pharm Biol 43(2):129–134

de Santanaya MP (2014) Etnobotánica e Inventario Español de Conocimientos Tradicionales. Conservación vegetal 18:1–4

Díaz ED (2010) Flora de interés etnobotánico usada por los pueblos originarios: Aónikenk, Selk'nam, Kawésqar, Yagan y Haush en la Patagonia Austral. Dominguezia 26(2):19–29
Doumecq MB, Arenas PM (2018) ¿qué madera es buena para leña? Conocimiento botánico local en "leñeras" del partido de La Plata (Buenos Aires, Argentina). Bol Soc Argent Bot 53(3):491–506
Echenique NG, Doumecq MB, Pochettino ML (2018) Saberes botánicos en el Talar: utilización de plantas silvestres con fines medicinales y alimenticios en el Parque Costero del Sur (partidos de Magdalena y Punta Indio, Buenos Aires, República Argentina). Gaia Sci 12(1):56–80
Fernández A, Martínez GJ (2019) Las plantas en la alimentación de pobladores rurales de los ambientes serranos de La Calera (Dpto. Colón, Córdoba, Argentina). Una perspectiva etnobotánica diacrónica. Bonplandia 28(1):43–69
Filipov A (1994) Medicinal plants of the Pilagà of Central Chaco. J Ethnopharmacol 44:181–193
Furlan V, Torres C, Galetto L (2011) Conocimiento y utilización de plantas medicinales por pobladores rurales del bosque chaqueño serrano de Córdoba (Argentina). Bonplandia 20(2):285–307
Giberti GC (2008) El archivo Bonpland en el Museo de Farmacobotánica "Juan A. Domínguez". Dominguezia 24(1):5–9
Goleniowski ME, Bongiovanni GA, Palacio L, Nuñez CO, Cantero JJ (2006) Medicinal plants from the "Sierra de Comechingones", Argentina. J Ethnopharmacol 107:324–341
Hernández MP, Novoa MC, Civitella SM, Masson D, Oviedo A (2013) Plants used in popular medicine in the Santiago Island, Buenos Aires, Argentina. Bol Latinoam Caribe Plant Med Aromat 12(4):385–399
Hieronymus J (1882) Plantae diaphoricae Florae Argentinae. Buenos Aires, 404pp
Hilgert NI (1999) Las plantas comestibles en un sector de las Yungas meridionales (Argentina). An Jard Bot Madr 57(1):117–138
Hilgert NI (2000) Especies vegetales empleadas en la insalivación de hojas de "coca" (*Erythroxylum coca* var *coca*, Erytrhoxylaceae). Darwiniana 38(3–4):241–252
Hilgert NI (2001) Plants used in home medicine in the Zenta River basin, Northwest Argentina. J Ethnopharmacol 76:11–34
Hilgert NI, Gil GE (2006) Medicinal plants of the Argentine Yungas plants of the Las Yungas biosphere reserve, Northwest of Argentina, used in health care. Biodivers Conserv 15:2565–2594
Hilgert NI, Higuera ML, Kristensen MJ (2010) La medicina herbolaria en el contexto urbano. Estudio de caso en un barrio de la ciudad de Tandil, Argentina. Bol Latinoam Caribe Plant Med Aromat 9(3):177–190
Hurrell JA (2014) Urban ethnobotany in Argentina: theoretical advances and methodological strategies. Ethnobiol Conservation 3(2):1–11
Hurrell JA, Pochettino ML, Puentes JP, Arenas PM (2013) Del marco tradicional al escenario urbano: Plantas ancestrales devenidas suplementos dietéticos en la conurbación Buenos Aires-La Plata, Argentina. Bol Latinoam Caribe Plant Med Aromat 12(5):499–515
Hurrell JA, Arenas PM, Cristina I (2015) El conocimiento botánico en zonas urbanas: potenciadores cognitivos comercializados en el área metropolitana de Buenos Aires, Argentina. Gaia Sci Ed Esp Etnobiol Arg 9(3):17–31
Hurrell JA, Puentes JP, Arenas PM (2016) Estudios etnobotánicos en la conurbación Buenos Aires-La Plata, Argentina: productos de plantas medicinales introducidos por inmigrantes paraguayos. Bonplandia 25(1):43–52
Hurrel JA, Puentes JP (2017) Plant species and products of the traditional Chinese Phytotherapy in the Ciudade Autónoma de Buenos Aires, Argentina. Ethnobiol Conserv 6:1–43
Keller HA, Crockett S (2015) *Hypericum robsonii* spec. nova sect. *Trigynobrathys* (Hypericaceae) from the Misiones Province in Argentina. Phyton 55(1):17–29
Kimber CT (2004) Gardens and dwelling: people in vernacular gardens. Geogr Rev 94(3):263–283
Korstanje MA (2016) Narrated archeobotany at Northwestern Argentina: stories domesticate plants told us in last 30 years. Cad Lepaarq XIII 25:1–29
Krapovickas A (2008) Bonpland, sesquicentenario de su muerte. Bonplandia 17(1):5–11

Kujawska M (2016) Forms of medical pluralism among the Polish Community in Misiones, Argentina. Anthropol Med 23(2):205–219

Kujawska M, Pieroni A (2015) Plants used as food and medicine by Polish migrants in Misiones, Argentina. Ecol Food Nutr 54:255–279

Ladio AH, Molares S, Ochoa J, Cardoso B (2013) Applied ethnobotany in Patagonia: the commercialization of weed for food and medicinal use in an urban fair in San Carlos de Bariloche (Río Negro, Argentina). Bol Latinoam Caribe Plant Med Aromat 12(1):24–37

Lozada M, Ladio A, Weigandt M (2006) Cultural transmission of ethnobotanical knowledge in a rural community of northwestern Patagonia, Argentina. Econ Bot 60(4):374–385

Lozano P (1941) Descripción corográfica del Gran Chaco Gualamba. Instituto de Antropología, Tucumán ed. https://pueblosoriginarios.com/textos/lozano/descripcion_corografica.html

Madaleno IM, Montero MC (2012) El cultivo urbano de plantas medicinales, su comercialización y usos fitoterapéuticos en la ciudad de Río Cuarto, Provincia de Córdoba, Argentina. Cuad Geog 50(2012–1):63–85

Marinoff MA, Chifa C, Ricciardi AIA (2006) Hydrophytas and marshy species used as medicines by the population in the North and NorthEeast of Chaco. Dominguezia 22(1):15–19

Martínez GJ (2005) Recolección y Comercialización de Plantas Medicinales en el Departamento Santa María, Provincia de Córdoba. Argentina Acta Farm Bonaerense 24(4):575–584

Martínez GJ (2008a) Farmacopea natural y tratamiento de afecciones de la piel en la medicina tradicional de los campesinos de las sierras de Córdoba (República Argentina). Dominguezia 24(1):27–46

Martínez GJ (2008b) Traditional practices, beliefs and uses of medicinal plants in relation to maternal–baby health of Criollo woman in central Argentina. Midwifery 24:490–502

Martínez Crovetto RN (2014) Algunos datos sobre etnobotánica mocoví. Bonplandia 23(2):119–131

Martínez GJ, Jiménez-Escobar D (2017) Plants of veterinary interest in the peasant culture of Sierra de Ancasti (Catamarca, Argentina). Bol Latinoam Caribe Plant Med Aromat 16(4):329–346

Martínez GJ, Luján MC (2011) Medicinal plants used for traditional veterinary in the Sierras de Córdoba (Argentina): an ethnobotanical comparison with human medicinal uses. J Ethnobiol Ethnomed 7:23. http://www.ethnobiomed.com/content/7/1/23

Martínez MR, Pochettino ML, Cortella AR (2004) Environment and illness in the Calchaquí Valley (Salta, Argentina): phytotherapy for osteo-articular and cardio-circulatory diseases. J Ethnopharmacol 95:317–327

Molares S, Ladio AH (2015) Edible and medicinal plant complexes in Patagonia Argentina: components and possible associated processes. Bol Latinoam Caribe Plant Med Aromat 14(3):237–250

Molares S, Rovero A (2016) Plantas medicinales, comestibles y aromáticas en cercos vivos de una ciudad patagónica de Argentina: características y potencialidades de un recurso poco explorado. Bol Latinoam Caribe Plant Med Aromat 15(1):41–52

Montani MC (2013) Medicinal use of plants among Huarpes descendants in the community of Lagunas del Rosario (Mendoza, Argentinia) In: Etnobotánica en zonas áridas y semiáridas del Cono Sur de Sudamérica. Ed. CEFYBO-CONICET, pp 251–270

Montenegro P (1945)[1710]. Materia Médica Misionera. Biblioteca Nacional, Buenos Aires. http://www.herbotecnia.com.ar/c-public-017.html

Muiño WA (2011) La etnobotánica médica del área de transición pampeano cuyana. Bonplandia 20(2):353–369

Muiño WA (2012) Estudio etnobotánico de plantas usadas en la alimentación de los campesions del noroeste de La Pampa Argentina. Chungara, Rev Antrop Chil 44(3):389–400

Novoa MC, Vizcaíno CE, Colares MN (1998) Anatomía y Etnobotánica de las Especies Medicinales de Monocotiledóneas de la Estepa Pampeana de Argentina: Orchidaceae y Pontederiaceae. Acta Farm Bonaer 17(3):165–172

Ochoa JJ, Ladio AH, Lozada M (2010) Uso de recursos herbolarios entre mapuches y criollos de la comunidad campesina de Arroyo Las Minas (Río Negro, Patagonia Argentina). Bol Latinoam Caribe Plant Med Aromat 9(4):269–276

Oliszewski N (2005) Estado actual de las investigaciones arqueobotánicas en sociedades agroalfareras del àrea valliserrana del noroeste argentino (0–600 d.C.) In: Relaciones de la Sociedad Argentina de Antropología XXIX 211–227
Oliva MV (2016) Usos y costumbres sobre hierbas medicinales en barrios periurbanos del sudesste de la ciudad de Salta, Argentina. Arc Med Gen Fam 13(1):18–26
OMS, UICN y WWF (1993) Directrices sobre la conservación de las plantas medicinales. Media Natura. Londres, pp 58
Pake CV (1987) Medicinal ethnobotany among refugees in Thailand. J Ethnobiol 7:13–26
Paucke F ([s/f] 2010) Hacia allá y para acá. Una estadía entre los indios Mocobíes, 1749–1767. Ministerio de Innovación y Cultura de la Provincia de Santa Fe, Santa Fe. http://www.espacio-santafesino.gob.ar/uploads/archivos/ediciones_catalogo/paucke-memorias.pdf
Pelto GH, Pelto PJ, Messer E (eds) (1989) Research methods in nutritional anthropology. United Nations University, Tokio
Perkins de Piacentino AM (2007) Misiones Jesuíticas: drogas autóctonas americanas encontradas en la botica jesuítica de la ciudad de Santa María de los Buenos Aires. 38th International Congress for the History of Pharmacy. Sevilla, September 19-22-2007
Pirondo A, Keller HA (2012) Raúl N. Martínez Crovetto: los albores de la etnobotánica em la Argentina. Introducción a *Estudios etnobotánicos V*. Bonplandia 21(2):101–107
Pirondo A, Coulleri JP, Keller HÁ, Ferrucci MS (2011) External factors influence the marketing of medicinal plants in an urban environment: the case of creoles and indigenous vendors in Corrientes, Argentina. Bol Latinoam Caribe Plant Med Aromat 10(6):553–569
Pochettino ML, Arenas P, Sánchez D, Correa R (2008) Conocimiento botánico tradicional, circulación comercial y consumo de plantas medicinales en un área urbana de Argentina. Bol Latinoam Caribe Plant Med Aromat 7(3):141–148
Puentes JP, Hurrel JA (2015) Plantas andinas y sus productos comercializados con fines medicinales y alimentarios en el Área Metropolitana Buenos Aires-La Plata, Argentina. Bol Latinoam Caribe Plant Med Aromat 14(3):206–236
Reid WV, Berkes F, Wilbanks T, Capistriano C (2006) Bridging scales and local knowledge in assessments. Island Press, Washington DC
Remorini C, Sy A (2002) Las sendas de la imperfección (tape rupa reko achy). Una aproximación etnográfica a las nociones de salud y enfermedad en comunidades Mbya. Scripta Ethnologica (XXIV) 133–147
Riat P, Pochettino ML (2015) Los remedios del monte: vigencia del conocimiento fitoterapéutico local en Los Juríes (Santiago del Estero, Argentina). Bol Latinoam Caribe Plant Med Aromat 14(2):67–82
Richeri M, Ladio AH, Beeskow AM (2013) Traditional knowledge and self-sufficiency: rural herbal medicine in Central Plain of Chubut (Argentine). Bol Latinoam Caribe Plant Med Aromat 12(1):44–58
Rodríguez MF, Rúgolo de Agrasar ZE, Aschero CA (2006) Use of plants in domestic units at the archeological site of Punta de la Peña 4, a Puna in Southern Argentina. Chungara, Rev Antrop Chil 38(2):257–271
Romeo RA (2015) Plantas empleadas en medicina popular en la provincia de Jujuy. Departamento Capital y alrededores. Dominguezia 31(2):5–10
Rondina RVD, Bandoni AL, Coussio JD (2010) Argentine medicinal plants with potential antifungal activity. Dominguezia 26(1):31–39
Rosso CN (2013) La etnobotánica histórica: el caso mocoví en la reducción de San Javier en el siglo XVIII. Etnobiología 11(3):54–65
Rosso CN, Cargnel JG (2012) Historians and etnographers: Jesuits writing in the XVIII century. The cases of Lozano and Paucke. An Esc Hist Virt 3(3):62–77
Rosso CN, Medrano C (2013) Alimentación de los grupos mocovíes asentados en la reducción de San Javier (Chaco Meridional, siglo XVIII). Cuad INAPL- Ser Esp 1(1):46–59
Rosso CN, Scarpa GF (2017) Etnobotánica de la alimentación entre los indígenas moqoit actuales de la provincia del Chaco (Argentina) y comparación con fuentes históricas de los siglos XVIII y XX. Bol Soc Argent Bot 52(4):827–840

Salazar J (2015) In: Berberián EE (ed) Condiciones de posibilidad de la reproducción social en sociedades prehispánicas y coloniales tempranas en las Sierras Pampeanas (República Argentina), 1ª ed. Centro de Estudios Históricos Prof Carlos A A Segreti, Córdoba, pp 380

Scarpa GF (2004) Medicinal plants used by the Criollos of Northwestern Argentine Chaco. J Ethnopharmacol 91:115–135

Scarpa GF (2009) Etnobotánica médica de los indígenas chorote y su comparación con la de los criollos del Chaco semiárido (Argentina). Darwiniana 47(1):92–107

Scarpa GF (2017) Etnobotánica histórica de grupos criollos de Argentina II: Puesta en valor, adscripción cultural y análisis de los usos no medicinales presentados por el gobierno argentino en la Exposición Universal de París de 1889. Bonplandia 26(2):77

Scarpa GF, Anconatani LM (2017) Etnobotánica histórica de las misiones franciscanas del este de Formosa II: Identificación y análisis de datos inéditos y reelaboración integral de fuentes ya publicadas a partir de hallazgos documentales. Dominguezia 33(2):37–79

Scarpa GF, Arenas P (2004) Vegetation units of the Argentine Semiarid Chaco: the Toba-Pilagá perception. Phytocoenologia 34(1):133–161

Scarpa GF, Montani MC (2011) Etnobotánica médica de las "ligas" (*Loranthaceae* sensu lato) entre indígenas y criollos de Argentina. Dominguezia 27(2):5–19

Scarpa GF, Pacor P (2015) Etnobotánica y religiosidad indígena chorote: el árbol "eje del mundo" y "de la regeneración inagotable". AIBR 10(1):75–98

Scarpa GF, Pacor P (2017) ¿Por qué ya no recolectan los recolectores? Procesos de estigmatización del consumo de plantas silvestres entre los indígenas chorote del Chaco salteño. Runa 38(1):5–21

Scarpa GF, Rosso CN (2014) La etnobotánica moqoit inédita de Raúl Martínez Crovetto I: descripción, actualización y análisis de la nomenclatura indígena. Bol Soc Argent Bot 49(4):623–647

Scarpa GF, Rosso CN (2018) Etnobotánica histórica de grupos criollos de Argentina III: identificación taxonómica y análisis de datos no-medicinales del Chaco húmedo provenientes de la encuesta nacional de folklore de 1921. Bonplandia 27(1):31–57

Scarpa GF, Rosso CN (2019) Etnobotánica histórica de grupos Criollos de Argentina IV: Identificación taxonómica de las plantas y análisis de datos medicinales del Chaco Húmedo provenientes de la Encuesta Nacional de Folklore de 1921. Bonplandia 28(1):5–42

Scarpa GF, Rosso CN, Anconatani L (2016) Etnobotánica médica de grupos criollos de Argentina: reconocimiento, análisis y puesta en valor de los datos presentados por el gobierno argentino em la exposición universal de París de 1889. Darwiniana, Nueva Ser 4(2):291–315

Sharry S, Abedini M, Basiglio Cordal MA, Briones V, Roussy L, Stevani R, Galarco S, Adema M (2011) Food and medicinal value of some forest species from Buenos Aires (Argentina). Emir J Food Agric 23(3):222–236

Stampella PC, Hilgert NI, Pochettino ML (2018) Unos medicinales de los cítricos (*Citrus* L., Rutaceae) entre los criollos del sur de Misiones (Argentina). Gaia Sci 21(1):90–108

Steibel PE (1997) Nombres y usos de las plantas aplicados por los Indios Ranqueles de La Pampa (Argentina). Rev Fac Agronomía-UNLpam 9(2):1–40

Suárez ME (2011) Wichi phytonymy of herbs and climbing plants of the Semiarid Chaco, Salta province, Argentina. Bonplandia 20(2):185–202

Suárez ME (2014) Etnobotánica wichí del bosque xerófito en el Chaco semiárido salteño. Editorial Autores de Argentina, pp 522

Trillo C (2016) Traditional management practices of plant resources in units of cultural landscapes on the west of the province of Cordoba, Argentina. Zonas Áridas 16(1):86–111

Trillo C, Arias Toledo B, Colantonio S (2011) Revisión de la Etnomedicina en Argentina: construcción de la disciplina y prespectiva para el futuro. Bonplandia 20(2):405–417

Vercelli N, Entraigas I, Scaramuzzino R, Migueltorena V, D' Alfonso C (2013) Medicinal plants of alkaline wetlands in the Azul creek basin (Buenos Aires province, Argentina). Rev Fca UNCUYO 45(2):285–298

Zamudio F, Kujawska M, Hilgert NI (2010) Honey as medicinal and food resource. Comparison between Polish and multiethnic settlements of the Atlantic Forest, Misiones, Argentina. Open Complement Med J 2:58–73

Chapter 3
Cuyo

3.1 Introduction

3.1.1 The Cuyo Region

Cuyo is a geographical region located in the western center of Argentina; it covers the provinces of Mendoza (MEN) and San Juan (SJU) and the occidental side of San Luis (SLU). It is flanked to the South by the Patagonia, to the East by the *Sierras Pampeanas* and the Pampeanas Plain, and to the West by Chile. The geographical characteristics of the region are diverse as there are cordilleras (high mountain ranges), mountains, and valleys with sparse vegetation and a desert climate. To the West of Cuyo is located the Andes Mountain range (*Cordillera de los Andes*) including the Aconcagua, the highest mountain of America (6959 meters high). Parallel to that are the *Cordillera Frontal* that has several independent cords and the Andean foothills of MEN and SJU. By the East, there are vast plains, the *Travesías*, which ended in the Pampa region. There are also several isolated mountain ranges belonging to the *Sierras Pampeanas*; in the lower zones of the plains, there are salt flats (e.g., Salinas del Diamante, a national protected area). In the South of MEN are plateaus and volcanoes of the *La Payunia* Provincial Reserve. In Cuyo prevails an arid climate, cold in winter and hot in summer; hence, it could have both snowstorms and frost or heat waves according to the season. In the Andes as in the desert territory, the weather is extremely hot during the day and very cold at night. The Zonda (foehn wind) is a northerly dry, hot, and suffocating mountain wind typical from this region, which appears usually between May and November but especially in winter. Cuyo is an area of seismic risk, particularly in the provinces of SJU and MEN. Also, there are chances of tornadoes, thunderstorms, landslide, alluvium, and wildfires. The wine industry is the prevailing commercial activity, being this region one of the main wine producers from South America. Dry weather and cold winters favor the development of the vines; despite the heat during summer, the height where the crops are located favors the quality of the internationally

M. A. Alvarez, *Pharmacological Properties of Native Plants from Argentina*,
https://doi.org/10.1007/978-3-030-20198-2_3

valued wines. The most common varieties are Malbec, Bonarda, Cabernet Sauvignon, Pedro Giménez, Syrah, Merlot, Tempranillo, and Chardonnay. The agricultural production includes olives, plums, damasks, peach, nuts, corn, sorghum, soy, oats, and horticultural and ornamental species. Also, there is production of honey, mineral water, fruit juices, beer, cider, and by-products of the wine industry (e.g., vinegar). Livestock production includes cattle, sheep, caprine, poultry, chinchilla, and ñandúes. As for mining, there is exploitation of salt, onyx, tungsten, wolfram, uranium, gold, silver, dolomites, bentonites, quartz, marble, and feldspar, among others (mainly in the provinces of SLU and SJU). Also, there is one of the main oils and gas basins of the country (Luján de Cuyo, Tupungato, Rivadavia, Malargüe, Maipú, and Junín departments).

The fauna is diverse; there are pumas, foxes, wild cats, guanacos (*Lama guanicoe*), red lizards, ferrets, maras or Patagonia hare (*Dolichotis patagonum*), and cuises (Southern mountain cavy). As for birds, there are ñandúes (*Rhea*), spot-winged falconet (*Spiziapteryx circumcincta*), elegant-crested tinamou (*Eudromia elegans*), and burrowing parrot (*Cyanoliseus patagonus*), among others. In some lakes (e.g., Lancanelo lake), there are flamingos, plovers, ducks, rails (*Rallidae* spp.), and swans.

The vegetation is characteristic of an arid region, with predominance of xerophilous plants. Among the flora predominate those from the Monte phytogeographical region with *Prosopis alba* and *P. nigra* (algarrobos); a variety of shrubs such as *Condalia microphylla* (piquillín), *Larrea nitida* (jarilla), *Opuntia quimilo* (quimil), *Celtis tala* (tala), and *Geoffroea decorticans* (chañar); as well as cactus (Fig. 3.1). Some of the medicinal plant species reported in Cuyo are *Acacia caven* (Molina) Molina (Fabaceae), *Acantholippia seriphioides* (A. Gray) Moldenke (Verbenaceae), *Achyrocline satureioides* (Lam.) D.C. (Asteraceae), *Acmella decumbens* (Sm.) R.K. Jansen (Asteraceae), *Acaena magellanica* (Lam.) Vahl (Rosaceae), *Adiantum raddianum* Presl. (Adiantaceae), *Aloysia gratissima* (Gillies at Hook, ex Hook) Tronc. (Verbenaceae), *Amaranthus muricatus* (Moq.) Hieron. (Amaranthaceae), *Ambrosia tenuifolia* Spreng. (Asteraceae), *Anemia tomentosa* (Savigny) Sw. var. australis Mickel (Anemiaceae), *Apium sellowianum* Wolff (Apiaceae), *Argemone subfusiformis* G.B. Ownbey (Papaveraceae), *Aristolochia argentina* Griseb. (Aristolochiaceae), *Asclepias campestris* Decne (Asclepiadaceae), *Aspidosperma quebracho-blanco* Schldl. (Apocynaceae), *Atriplex crenatifolia* Chod. & Wilczk (Chenopodiaceae), *Bacharis articulata* (Lam.) Pes. (Asteraceae), *Bauhinia forficata* Link (Fabaceae), *Bidens pilosa* L. (Asteraceae), *Buddleja araucana* Phil. (Buddlejaceae), *Buddleja mendozensis* Benth (Buddlejaceae), *Caesalpinia gilliesii* (Wal. Ex Hook.) D. Dietr. (Fabaceae), *Capparis atamisquea* Kuntze (Capparaceae), *Capsicum chacoense* A.T. Hunz. var. chacoense (Solanaceae), *Cardionema ramosissimun* (Weinm.) Nelson & Macbr. (Caryophyllaceae), *Celtis tala* Planchon (Ulmaceae), *Cestrum parqui* L'Her. (Solanaceae), *Chaptalia nutans* (L.) Pol. (Asteraceae), *Chenopodium ambrosioides* L. (Chenopodiaceae), *Colletia spinosissima* Gmelin (Rhamnaceae), *Conyza bonariensis* (L.) Cronquist. (Asteraceae), *Coronopus didymus* (L.) Smith (Brassicaceae), *Cortaderia selloana* (Schultes) Asch. & Graeb. (Poaceae), *Cucurbitella asperata* (Gill.) Walpers (Cucurbitaceae), *Cuphea glutinosa* Cham. & Schldl. (Lythraceae), *Cuscuta indecora* Choisy

Fig. 3.1 The Cuyo region. Landscapes from Mendoza (**a**), San Juan (**b**), and San Luis (**c**) provinces

(Cuscutaceae), *Fabiana imbricata* Ruiz et Pavon (Solanaceae), *Gentianella achalensis* (Hieron. Ex Gilg.) Fabris (Gentianaceae), *Geoffroea decorticans* (Gill. Ex Hook. et Arn.) Burkart (Fabaceae), *Gomphrena pulchella* Mart. (Amaranthaceae), *Grindelia pulchella* Dunal. (Asteraceae), *Hedeoma multiflorum* Benth. (Lamiaceae), *Heliotropium amplexicaule* Vahl. (Boraginaceae), *Hypericum connatum* Lam. (Hypericaceae), *Ipomoea purpura* (L.) Roth (Convolvulaceae), *Jodina rhombifolia* (Hook & Arn.) Reissek (Cervantesiacaceae), *Larrea cuneifolia* Cav. (Zygophyllaceae), *Larrea divaricata* Cav. (Zygophyllaceae), *Lepidium bonariense* L. (Brassicaceae), *Ligaria cuneifolia* (Ruiz et Pav.) Tiegh. (Loranthaceae), *Lippia integrifolia* (Griseb.) Hieron. (Verbenaceae), *Lippia turbinata* Griseb. F. (Verbenaceae), *Lithraea molleoides* (Vell.) Engl. (Anacardiaceae), *Margyricarpus pinnatus* (Lam.) Kuntze (Rosaceae), *Minthostachys mollis* (Griseb.) Epling (Lamiaceae), *Mirabilis jalapa* L. (Nyctaginaceae), *Passiflora caerulea* L. (Passifloraceae), *Perezia multiflora* (Humb. & Bonpl.) Less spp. *multiflora* (Asteraceae), *Petunia axillaris* (Lam.) Britton subsp. Parodii (Steere) Cabrera (Solanaceae), *Physalis mendocina* Philippi (Solanaceae), *Physalis viscosa* L. (Solanaceae), *Plantago tomentosa* Lam. ssp. Tomentosa (Plantaginaceae), *Pluchea sagittalis* (Lam.) Cabrera (Asteraceae), *Prosopanche bonacinae* Speg. (Hydnoraceae), *Prosopis alba* Griseb. (Fabaceae), *Salix humboldtiana* Wild. (Salicaceae), *Salpichroa origanifolia* (Lam.) Thell. (Solanaceae), *Satureja parvifolia* (Phil.) Epling (Lamiaceae), *Schinus areira* L (Anacardiaceae), *Schinus fasciculatus* (Griseb.) I.M. Johnst (Anacardiaceae), *Senna corymbosa* (Lam.) Irwin & Barneby (Fabeaceae), *Sida rhombifolia* L. (Malvaceae), *Sisyrinchium macrocarpum* Hieron. (Iridaceae), *Solanum eleagnifolium* Cav. (Solanaceae), *Solanum sisymbrifolium* Lam. (Solanaceae), *Solanum argentinum* Bitter et Lillo (Solanaceae), *Solidago chilensis* Meyen (Asteraceae), *Tagetes argentina* Cav. (Asteraceae), *Tagetes filifolia* Lag. (Asteraceae), *Tagetes mendocina* Phil. (Asteraceae), *Tagetes minuta* L. (Asteraceae), *Talinum paniculatum* (Jacq.) Gaertn. (Portulacaceae), *Tessaria absinthioides* (Hook. & Arn.) DC. (Asteraceae), *Teucrium grisebachii* Hieron. (Lamiaceae), *Tillandsia myosura* Gris. & Baker (Bromeliaceae), *Tillandsia pedicellata* (Mez) Castell. (Bromeliaceae), *Trixis cacalioides* (Kunth) D. Don (Asteraceae), *Typha dominguensis* Persoon (Typhaceae), *Urtica dioica* L. var. mollis (Steudel) Wedd. (Urticaceae), *Usnea densirostra* Taylor (Usneaceae), *Valeriana ferax* (Griseb.) Höck (Valerianaceae), *Verbesina encelioides (Cav.)* Benth & Hook. Ex Gray (Asteraceae), *Wedelia glauca* (Ortega) Eifert (Asteraceae), *Xanthium spinosum* L. (Asteraceae), and *Zuccagnia punctata* Cav. (Fabaceae), among other species (Bustos et al. 1996; Del Vitto et al. 1997; Barboza et al. 2009; Martínez Carretero 2013; Agüero 2017).

3.2 *Aloysia gratissima* (Gillies & Hook. ex Hook.) Tronc. var. gratissima (Verbenaceae)

Synonymy *Aloysia lycioides* Cham., *Lippia lycioides* (Cham.) Steud., *Lippia ligustrina* (Lag.) Kuntze var. paraguariensis Briq., *Verbena gratissima* Gillies & Hook. Ex Hook.

Vernacular name Arrayán, arrayán del campo, azahar del campo, cedrón, cedrón del monte, azar del monte, favorita, palo, muña del monte, palo amarillo, usillo, niño rupá.

General information *A. gratissima* grows in xerophilous environments and dry forests from BAI, CAT, COR, COS, ERI, FOR, LPA, MEN, MIS, SDE, SFE, SJU, SLU, and TUC. It is also found in Paraguay and Uruguay. It is an aromatic species with a lemony scent. Also, it is a melliferous and ornamental species ideal for urban gardens in temperate or temperate-warm climate (Haene 2007; Cané 2016).

Botanical characteristics *A. gratissima* is an aromatic suffriticose or shrubby plant, highly branched, 1–3 m in height, and 4–5 m in diameter. Stems are four-angled when young and rounded in maturity. The leaves are simple, deciduous, oblong-obovate, 5–35 mm in length, entire or 3- to 5-lobed by the base, opposite subsessile or briefly petiolate, densely pubescent at the inferior lamina, and glandulous and glabrous by the central vein. The foliage is grayish, persistent, and with a very fine texture. Inflorescences are spicate racemes or paniculate, with white, zygomorphic fragrant flowers. The calix is gamosepale, tubular, four-lobed with a larger anterior lobule, externally glabre, and 5 mm in length. The flowers have four stamens, ovary with two locules, and one ovule in each locule. The fruits are dry schizocarps and exceptionally drupaceous that in maturity separates in two mericarps. The seeds do not have endosperm (Domínguez 1928; Cabrera and Zardini 1978; Ragonese and Milano 1984; Rotman and Múlgura de Romero 1999; Alonso and Desmarchelier 2015; Cané 2016; O'Leary et al. 2016).

Ethnobotanical information *A. gratissima* is very used in homemade medicine. Hieronymus (1882) has already reported the use of the infusion made of *A. gratissima* cuttings to treat cold and stomach problems. Currently, the infusion is mainly used as digestive, anti-catarrhal, sedative, or sudorific. It has also been commonly used to alleviate symptoms associated with headache, bronchitis, and nervous system disorders. Besides, the infusions made with the aerial parts are used to treat varicose veins and dizziness. The infusion made from stems is reported as a treatment for hypotension; those made from leaves, to sooth menstrual pains and against hemorrhoids; and those combining flowers and leaves, as cardiotonic, stomachic, carminative, and diaphoretic and to relieve stomachache (Toursarkissian 1980; Andersen et al. 1997; Del Vitto et al. 1997; Rotman and Múlgura de Romero 1999; Múlgura de Romero et al. 2002; Barboza et al. 2009; Sérsic et al. 2010; Zeni et al. 2014; Bernasconi Salazar et al. 2017).

Chemical data The compounds found in the essential oil (EO) from the aerial parts were α- and β-pinene, sabinene, α-terpinene, β-phellandrene, α- and β-thujone, α- and β-caryophyllene, α-bisabolene, β-elemene, bicyclogermacrene, camphene, cadinol, caryophyllene oxide, citral, copaenone, copaenol, chiysanthenyl acetate, globulol, spathulenol, limonene oxide, limonene, linalool, viridiflorol, 10-epicubebol, (E)-nerolidol, germacrene D, eucalyptol, pulegone, terpinyl acetate, eugenol, myrcene, γ-terpinen-1-al- *trans*-pinocarveol, *cis*-pinocarveol, γ- and δ-elemene, geranyl acetate, β-cubebene, globulol, α-humulene, alloaromadendrene, γ-gurjunene, cubebol, elemol, germacrene B, and carotol. Other components that have been found in the aerial parts were the sesquiterpenoids guaiol and bisabolol; the triterpenoids amyrin, betulinic acid, oleanolic acid, and ursolic acid; the flavonoids apigenin-5-hydroxy-7,40-dimethyl ether, genkwanin, and luteolin-7,3-O,4-O-trimethylether; the glycoside flavonol rutin; and the phenylethanoids verbascoside and arenarioside (Soler et al. 1986; Barboza et al. 2009; Trovati et al. 2009; Benovit et al. 2015) (Fig. 3.2).

Biological activities The decoction, alcoholic extract, and hydroalcoholic extract from the aerial parts tested against *Enterococcus faecalis* (19433), *Escherichia coli* (11229), *Rhodococcus equi* (6939), *Salmonella enteritidis* (ATCC 11076), and *Staphylococcus aureus* (ATCC 25923) showed inactivation capacity and/or

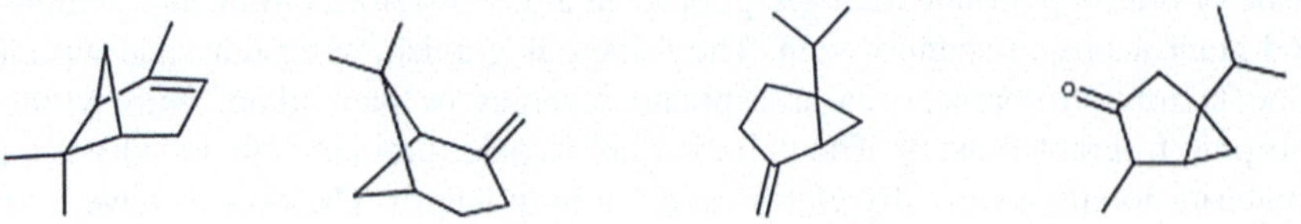

α-pinene β-pinene sabinene α-thujone

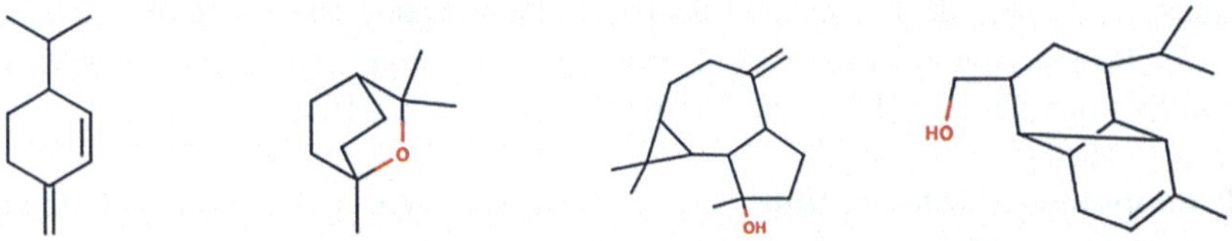

β-phellandrene eucaliptol spathulenol copaenol

Fig. 3.2 Some chemicals found in essential oils (EOs) from aerial parts of *Aloysia gratissima* (Gillies & Hook. ex Hook.) Tronc. var. gratissima

selective inhibition. The best results were attained with the alcoholic extract from the fresh plant. Among the bacteria, *Pasteurella multocida* demonstrated more sensibility to the three extracts, while *S. aureus* and *E. faecalis* presented higher resistance (Souza and Wiest 2005). It was also reported that the hydroalcoholic extract (30:70) has activity against *B. subtilis* (Duarte et al. 2004) and the alcoholic extract against *B. cereus* (Demo et al. 1998) and *S. aureus* (Oliva et al. 2001). Furthermore, the EO has activity against Junín virus (responsible of hemorrhagic fever) and herpes simplex virus type 1 (HSV-1) (García et al. 2003). It was reported that the EOs from the aerial parts have viricidal and nematicide activity (Barboza et al. 2009).

The aqueous extract and its main component, the ferulic acid (46.95 mg 100 g^{-1}), attenuated quinolinic acid-induced hippocampus cell damage in mice when they were pretreated with any of them (100 mg kg^{-1}). However, they did not prevent clonic or tonic seizure incidence or its severity. The protection produced may involve the glutamate transport modulation through an inhibitory effect on the N-methyl-D-aspartate receptor (Zeni et al. 2014). The antidepressant activity apparently involved N-methyl-D-aspartate (NMDA) receptors (Farahani et al. 2014).

Finally, an antiproliferative effect was reported. The infusion prepared with leaves of *A. gratissima* (6 and 24 g l^{-1}) collected in Rio Grande do Sul, Brazil, and the corresponding essential oil obtained by hydrodistillation (0.25% on ethanol) produced a decrease in the mitotic index on *Allium cepa* bulb cells after 24 h of treatment indicating an antiproliferative activity (Hister et al. 2009). However, it has to be considered that the essential oil composition differs according to the places the species grows (Duschatzky et al. 2004; Ricciardi et al. 2005; Franco et al. 2007).

3.3 *Lippia integrifolia* (Griseb.) Hieron. (Verbenaceae)

Synonymy *Lippia boliviana* Rusby, *Lippia boliviana* Rusby var. angusta, *Lippia boliviana* Rusby var. integrifolia Moldenke, *Lippia integrifolia* (Griseb.) Hieron. var. beckii, *Lippia turbinata* Griseb. var. integrifolia Griseb.

Vernacular name Pulco, poleo, inca-yuyo, incayuso, té del inca, manzanilla.

General information *L. integrifolia* is a native species from Argentina that is on risk of extinction. It grows in the provinces of CAT, COR, JUJ, LRI, SAL, SJU, and TUC.

Botanical characteristics *L. integrifolia* is a woody aromatic perennial shrub and 1–5 m high. Leaves are entire, green-light, 2–3 cm in length, and 4–6 mm wide, ovoid-lanceolate in adulthood. The lamina border is slightly backwards curved with a central prominent nerve in the inferior lamina. The flowers are white or pink-whitish, with a 5–6 mm corolla, originating in axillary globose-compressed clusters. The fruit has two mericarps with exalbuminate seeds (Bassols and Gurni 1996; Schiuma 1997; Rotman and Múlgura de Romero 1999; Zuloaga et al. 2008; Alonso and Desmarchelier 2015).

Ethnobotanical information In pre-Columbian America, *L. integrifolia* was considered as sacred and exclusively used by the upper Incas castes that thought that *L. integrifolia* stimulated the development of mental powers. In the case of the original settlers from the province of COR, the Diaguitas and Comechingones, they smoked *L. integrifolia* without restrictions by castes, genre, or age. Currently, its popular use is as nervous tonic, in cases of melancholy and sadness, neurasthenia and nerve pain, and in cases of stress. The infusion made with leaves and flowers is used as digestive, mostly when the disease has a nervous origin; to relieve the stomachache; to treat *empacho*; as carminative; also as emmenagogue, anti-flu, and antitussive; in bronchitis and flu; and as expectorant. The infusion prepared with the dried aerial parts is used as stomachic, emmenagogue, tonic, and diuretic. It could be added to the traditional beverage *mate*. It is also used for the elaboration of bitter nonalcoholic beverages (Ratera and Ratera 1980; Toursarkissian 1980; Martínez Crovetto 1981; Del Vitto et al. 1997; Pascual et al. 2001; Barboza et al. 2009; Agüero 2017).

Chemical data In the fresh EOs were found the compounds α-thujene, sabinene, p-cymene, limonene, terpinen-4-ol, borneol, eremophyllene, γ-elemene, β-caryophyllene, δ-elemene, α-humulene, african-1-en-6-ol, african-4,5-dione, 4,5-seco africanone, spathulenol, S (+)-*trans*-nerolidol, 2-β-,9-α-2,6,6,9-tetramethyltricyclo[6.3.0.0 (2,4-)]undec-1-(8)-ene-6,7,11-dione, 1β-,2α-,8β-2,6,6,10-tetramethylbicyclo[6.3.0]undec-5-en-2-ol(3α-hydroxy-6-asteriscane), 1α-,7β-,9-a-1-hydroxy-3,6,6,9-tetramethylbicyclo[5.4.0]undec-3-en-8-on, lippifoli1(6)-en-5-one, lippifoliane, 1,6-*cis* lippifolian-1α-ol-5-one, 1,6-*trans* lippifolian-1-α-ol-5-one, lippifolil (6)-en-4-β-ol-5-one, *trans*-africanan-1α-ol, african-5-en-1α-ol, β-caryophyllene oxide, 4,5-secoafricanan-4,5-dione, humulene epoxide II, ketone 4,1,6-*cis* and 1,6-*trans* lippifolian-1α-ol-5-one, lippifoli1(6)-en-4α-ol-5-one, integrifolian-1,5-dione, asteriscane, bicyclohumulendione, humulendione, and *trans* humul (9E)-en-2,6-dione (Catalán et al. 1983, 1991, 1992; Dartayet et al. 1984; Terblanché and Kornelius 1996; Barboza et al. 2009; Marcial et al. 2016) (Fig. 3.3).

Biological activities The choleretic and antispasmodic activity of the aqueous extracts of *L. integrifolia* (250, 500 and 750 mg kg^{-1}) was tested on rats. A dose of 500 mg kg^{-1} induced a significant and persistent increase (60.5%) in bile flow after the first 15 min from the oral administration; the dose of 750 mg kg^{-1} was also effective. A significant increase of bile acids output and the antagonism of muscle contractions induced by acetylcholine (Ach) and $CaCl_2$ on rat isolated jejunum was observed. Those effects were related to the content on mono- and dicaffeoylquinic acids (0.10% w/v expressed as chlorogenic acid) and the possible inhibition of the calcium influx into the cytoplasm (Speroni et al. 2003; Gorzalczany et al. 2008).

The aqueous extract (0.5, 1.0, 2.0 mg ml^{-1}) has anti-inflammatory effects on stomach cells (20–60%) and antiadhesive properties against *Helicobacter pylori*. Infusions (24% w/w), decoctions (25% w/w), raw polysaccharides isolated from decoctions (RPS 2.3%), and the phenolic fractions (PhF) stimulated cellular vitality of human stomach cells (AGS cells). Those results could explain the popular use of the herbal tea for digestive disorders (Marcial et al. 2014). Also, it was

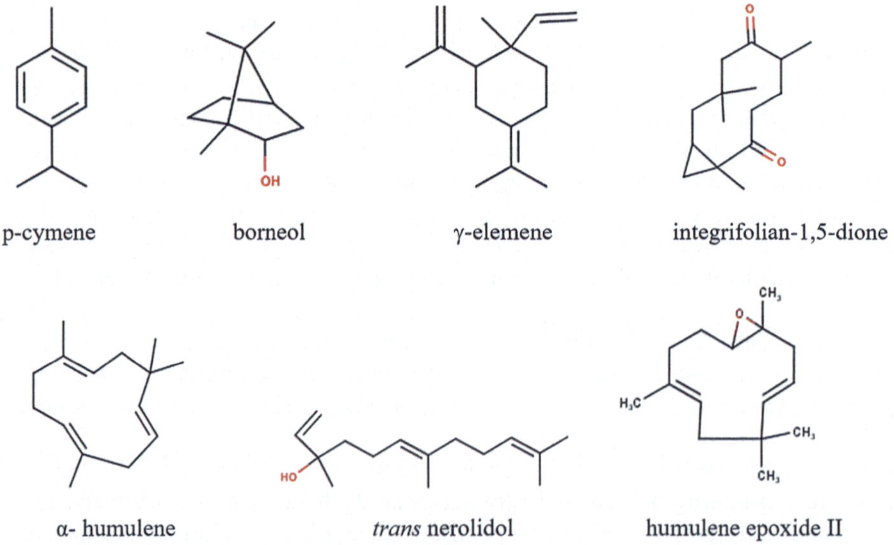

Fig. 3.3 Some chemicals found in *Lippia integrifolia* (Griseb.) Hieron

reported a stimulation of innate immunity and potential radical scavenging by *L. integrifolia* extracts (Marcial et al. 2014).

Furthermore, it was shown antimicrobial activity against Gram-positive and Gram-negative bacteria (*Mycobacterium tuberculosis*) and fungi (*Microsporum gypseum, Trichophyton metagrophytes, T. rubrum*). The EOs have antifungal activity against *Aspergillus* spp. (Bluma and Etcheverry 2008; Bluma et al. 2008a, b), and the methanolic extracts showed a limited spectrum of action and a lower potency against *Microsporum canis, M. gypseum, Epidermophyton floccosum, Trichophyton rubrum*, and *T. mentagrophytes* (MICs between 250 and 1000 μg ml^{-1}) (Muschietti et al. 2005). Also, the organic and aqueous extracts have anti-trypanocide activity against *Trypanosoma cruzi* and against *Aedes aegypti* (Gleiser et al. 2011). The EOs have a moderate activity against mosquitos (Gleiser et al. 2011).

***In vitro* culture** *L. integrifolia* was micropropagated using nodal segments and apices from a plant growing in a greenhouse as initial explants. They were transferred to vessels containing MS medium (Murashige and Skoog 1962) with 20 g l^{-1} sucrose, BAP (2.2 μM), and 7 g l^{-1} agar. The developed shoots were transferred to rooting medium composed by MS half strength media with IAA (0.28 μM). Then, the rooted plants were transferred to pots that were placed in a humidity chamber first, and finally to a greenhouse. Also, the genetic variability was studied by ISSR technique with the goal of developing a breeding program (Ianicelli et al. 2016).

Legal status The Argentina Food Code contains some tea preparations and commercial beverages containing incayuyo that are taken as aperitif (2013).

3.4 *Minthostachys mollis* (Griseb.) Epling (Lamiaceae)

Synonymy *Bystropogon kuntzeanum* Briq., *Minthostachys verticillata* (Griseb.) Epling, *M. verticillata* (Griseb.) Epling var. eupatorioides, b., *Xenopoma verticillata* Griseb., *M. mollis* Griseb.

Vernacular name Peperina, peperita, piperina, martin muña.

General information *Minthostachys verticillata* (Griseb.) Epling (Epling 1939), whose popular name is peperina, is nowadays defined in the Flora of Argentina as *Minthostachys mollis* Griseb. (Zuloaga et al. 2008; van Baren et al. 2014); however, most of the literature still refers to peperina as *M. verticillata. M. mollis* is a species with ethnobotanical, pharmacological, and commercial relevance that grows as a bush in the central and northwestern regions of Argentina, in the provinces of CAT, COR, JUJ, SAL, SDE, SLU, and TUC. The leaves and aromatic flower spikes have a distinct peppermint-like odor and are used as a condiment in foods and as a food preservative.

Botanical characteristics *M. verticillata* is an aromatic subshrub, 0.3–2.0 m in height with highly branched pubescent quadrangular stems. Leaves are ovoid, obtuse, with festooned edges, 1–5 cm in length. Flowers are white, small, and placed as clusters in the axilla of the leaf. Fruits are small dry capsules with four tiny seeds per fruit. Blooming is in summer. The fruiting period covers the months of March and April (Núñez and Cantero 2000; Cariddi 2013).

Ethnobotanical information Peperina is usually taken as an infusion for its attributed properties as digestive, or it is added, for its characteristic aroma and taste, to the traditional beverage *mate* (Bandoni et al. 2002; González Pereyra et al. 2005; Elechosa et al. 2007; van Baren et al. 2014). Also, it is added to yerba mate (*Ilex paraguariensis*) for producing *yerba mate compuesta* and to liqueurs and aperitifs (Valladares et al. 2002; Escobar et al. 2015). According to the dualistic classification from Native Americans, it is considered a hot species (Schmidt-Lebuhn 2008). Consequently, it is used in popular medicine to treat digestive diseases (*empacho*) and as an aphrodisiac. Leaves mixed with flowers are used to prepare infusions with recognized activities as stomachic, antispasmodic, antidiarrheal, antiemetic, anti-rheumatic, carminative, sedative, hemostatic, and hypotensive. The infusion made with dried leaf and stems is used as carminative, for improved fertility, and as sexual stimulant (Del Vitto et al. 1997; Barboza et al. 2009).

Chemical data Chemically, the oxygenated monoterpenes fraction represents 90% of the total *M. verticillata* EOs. The EOs from the aerial parts include pulegone, menthone, isomenthone, limonene, menthol, α-pinene and β-pinene, carvone, carvacryl acetate, γ-terpinene, linalool, piperitenone, sabinene, myrcene, (E)-β-ocimene, thymol, camphene, p-cymene, α- and γ-terpinene, isopulegol, δ-cadinene, β-caryophyllene, γ-gurjunene, and τ-cadinol (Fester and Martinuzzi 1950; Fester et al. 1960; De Feo et al. 1998; Escobar et al. 2012). The EOs from the leaves have β-caryophyllene: sesquiterpene and myrcene (monoterpene (E) β-ocimene). However, in all cases the main components are pulegone (cyclo-hexanone, 5-methyl-2-(1-methylethylidene) (61%) and menthone (cyclohexanone, 5-methyl-2-(1-methylethyl) (27%) (Valladares et al. 2002; Escobar et al. 2012; Pellegrini et al. 2017) (Fig. 3.4).

Biological activities It was reported that *M. verticillata* extracts and its EOs have antimicrobial, insecticidal, acaricidal, and antifungal properties (Primo et al. 2001; Valladares et al. 2002; Bluma et al. 2008b; Palacios et al. 2009; González and Marioli 2010; Vogt et al. 2010a; Cariddi et al. 2011; Escobar et al. 2012; Escobar et al. 2015; Pellegrini et al. 2017). Consequently, it was proposed their use in food preservation and also as insecticides. The antimicrobial activity was demonstrated against *Aspergillus* section Flavi, *Streptococcus uberis* (isolated from bovine mastitis), and *P. larvae* (responsible of American foulbrood in *Apis mellifera*). *Aspergillus* section Flavi growth rate was tested at different EOs concentrations (0, 150, 300, 500, and 700 mg kg^{-1}) being weakly inhibited by the EOs obtained by hydrodistillation at all the water activities (0.982 and 0.955 aw) tested. The lowest concentration

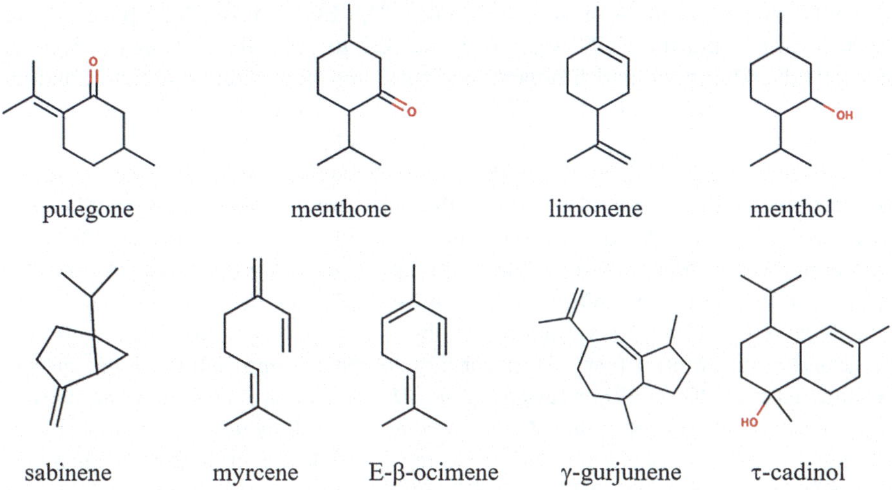

Fig. 3.4 Some of the chemical compounds found in *Minthostachys verticillata* (Griseb)

of these EOs showed growth rate stimulation in the aflatoxigenic strains, while the highest concentrations reduced sporulation and vegetative mycelia in aflatoxigenic growth zones (Bluma et al. 2008a). When the EOs and limonene were tested against *S. uberis*, the minimum inhibitory concentration (MIC) values ranged from 14.3 to 114.5 mg ml^{-1} (1.56–12.5% v/v) and the minimum bactericidal concentration (MBC) from 114.5 to 229 mg ml^{-1} (12.5–25% v/v). MICs for limonene ranged from 3.3 to 52.5 mg ml^{-1} (0.39–6.25% v/v) and MBC from 210 mg/mL (25% v/v). The aqueous extracts, namely, the water remaining after hydrodistillation (WRHD), and decoctions (D) showed the highest antibacterial activities (González and Marioli 2010). Gram-positive bacteria were more sensitive to inhibition by the EOs than the Gram-negative bacteria (Primo et al. 2001). The antimicrobial capacities of the aqueous extracts (5 mg of aerial parts per 100 ml distilled water) and the EOs determined against *P. larvae* showed a minimal inhibitory concentration (MIC) of 75 μg ml^{-1} against *P. larvae* with inhibition zone diameters in a range between 10.0 mm and 26.0 mm for EOs. On the other hand, the aqueous extracts produced 100% inhibition (100% efficacy) (Pellegrini et al. 2017). As WRHD is normally discarded after the hydrodistillation procedure, the authors proposed that the WRHD fraction could be reserved for apiaries sanitation, while EOs, for its lower biological activity, could be used for cooking or in fragrances. However, analysis of toxicity on honey bees as well as of WRHD delivery and/or application methods must be performed (Pellegrini et al. 2017). The *in vitro* antiviral activity of the EOs was tested against Suid herpesvirus type 1 (HSV-1) and Pseudorabies virus (PrV) and quantified by the plaque reduction essay with HEp-2 cells. The antiviral activity specifically affects HSV-1 and PrV multiplication at pulegone and EOs concentrations 20 to 30 times lower than the cytotoxic concentration values (CC50). Both inhibited the replication of *Suid herpesvirus* type 1 in more than 50% and exerted an inhibition over 90% at a concentration of 100 μg ml^{-1}. The EC50 were 20.25 μg ml^{-1} and 20.00 μg ml^{-1}, and the CC50 were 613 and 352 μg ml^{-1} for EOs and pulegone, respectively. Menthone and limonene had little and not selective antiviral action. The therapeutic index (TI = CC50/EC50) were 30.3 for EOs and 17.6 μg ml^{-1} for pulegone. The authors proposed the use of EOs and pulegone as therapeutic selective antiviral agents (Vogt et al. 2010a). As for its activity as insecticide, leaves of *M. verticillata* that were affected by the leaf miner *Chromatomyia platensis* (Brethes) (Diptera: Agromyzidae) and a cecidomyid gall insect (Diptera: Cecydomyiidae) showed a decrease in pulegone concentration (42.8%) and an increase in menthone concentration that almost duplicated the amounts found in no mined leaves (25.12%). No variation was observed in galled stems. The results also suggested a differential response according to the feeding habits of the insects (Valladares et al. 2002). When tested against *M. domestica*, R (+)-pulegone showed high insecticidal effects with LC50 1.7 mg l^{-1}, and menthone showed LC50 8.6 mg l^{-1}, while limonene had LC50 6.2 mg l^{-1} (Palacios et al. 2009). There are also references about the effects of D and EOs on the immunological system (enhancing or modulating the immune response). The ability of D (210 μg ml^{-1}, 104 μg ml^{-1}, 10.4 μg ml^{-1}, and 1.04 μg ml^{-1}) and EOs (0.96 μg ml^{-1}, 0.48 μg ml^{-1}, 48 μg ml^{-1}, and 4.8 μg ml^{-1}) to induce or inhibit *in vitro* basophil degranulation was tested with human basophils from dust mite non-allergic and allergic patients. The IL-13 levels were evaluated from lymphocyte cultures stimulated with only an allergen or with a combination of the main monoterpenes found in the EOs (pulegone, menthone, and limonene). IL-13 synthesis was quantified in supernatants of lymphocyte cultures from allergic patients stimulated with allergen (10,000 PNU ml^{-1}) alone or combined with EOs (6 μg ml^{-1}), pulegone (62 μg ml^{-1}), menthone (60 μg ml^{-}1), limonene (55 μg ml^{-1}), or the combination of the three monoterpenes. That combination and limonene alone were the most active to reduce IL-13 and to stimulate cell proliferation. On the other hand, EOs and monoterpenes reduced the μ-hexosaminidase release from basophils of allergic patients maybe acting as a membrane stabilizer. The greatest inhibitory effect was for EOs (10 μg ml^{-1}), pulegone (40 μg ml^{-1}), menthone (40 μg ml^{-1}), and limonene (20 μg ml^{-1}). The inhibitory effect with the EOs was highest than the one obtained with the single monoterpenes or their combination. Furthermore, the anti-allergic effect was reported to be higher than with desloratadine (Cariddi et al. 2011). On the other hand, lymphocyte-proliferating activity of D and EOs was tested and compared with cellular expansion induced by phytohemagglutinin (PHA), Pokeweed mitogen (PWM), and bacillus Calmette-Guérin (BCG) with a non-stimulated culture as control reference. The cytomorphological analysis showed mitogenic activity and induced lymphocyte cluster and colony formation similar to PHA and PWM and higher than CGB ($p < 0.02$). A 40% of the cells that proliferated were LT CD8 (+); also the basophils from allergic individual resisted up to a concentration of 104 mg ml^{-1}of D and 0.48 mg ml^{-1} of EO (González Pereyra et al. 2005).

Toxicity Cytotoxicity and genotoxicity were tested against Vero cells, human peripheral blood mononuclear cells (PBMCs), and mice bone marrow cells. No cytotoxicity was observed in Vero cells (10–1000 μg ml^{-1}). The viability of PBMCs was

72% with the highest concentration of EOs (1000 μg ml^{-1}), and no signs of apoptosis were found at the concentrations tested. EOs did not induce cytotoxicity in Vero cells and human PBMCs. In addition, EOs (100–1000 μg ml^{-1}) do not induced apoptotic effects on human PBMCs. No clinical signs of toxicity, mortality, or changes in body weights or food consumption were reported. There is no any toxicological effect in the liver, kidney, or intestine of Wistar rats up to an EOs concentration of 7 g kg^{-1} feed for 90 days. The *in vivo* assay showed that EOs (25–500 mg kg^{-1}) do not increase the frequency of micronucleus in bone marrow cells of mice. Based on those results where no cyto-genotoxic effect of EOs was found *in vitro* or, the authors conclude that EOs are safe as therapeutic agent (Escobar et al. 2012).

Legal status *M. verticillata* is among the nontoxic vegetal drugs authorized for their use with medicinal purposes. It was accepted as an official drug in the FNA 6th edition.

3.5 *Acantholippia seriphioides* (A. Gray) Moldenke (Verbenaceae)

Synonymy *Lippia seriphioides* A. Gray, *Lippia foliolosa* Phil.

Vernacular name Tomillo, tomillo andino, tomillo de la sierra, tomillo del campo, tomillo mendocino, tomillo macho, tomillo de olor, tomillo hembra, ñiancuñán, tomatillo silvestre.

General information It grows in Chile, Bolivia, and in Argentina in the provinces of BAI, CHU, LPA, MEN, NEU, RNE, SCR, SJU, and SLU in arid environments characterized by stony or saline soils, from sea levels up to 1000 m.a.s.l. (Van Baren et al. 2015). The intense overexploitation of medicinal and aromatic plants has placed this species at risk of extinction (Delucchi 2006; Elechosa et al. 2009). It is used as flavor of foods and as a condiment in replacement of the authentic thyme (*Thymus vulgaris* L.) for its typical high content of thymol and carvacrol (Ruiz Leal 1972; Roig 2001; Gastaldi et al. 2016). Its propagation could be achieved uneasily through seeds or cuttings, but the most common way in nature seems to be the spontaneous layering (Strasser et al. 2002; Strasser and Fernández 2006).

Botanical characteristics *A. seriphioides* is an aromatic bush 30–60 cm in height, branched from the basis, with 2–6 mm sessile leaves, 3–5 lobed in the apex, narrow towards the base, and with reddish spherical glands more abundant on the underside. Leaves are pubescent or subglabrous, opposite, sessile, and with a fasciculate appearance due to its axillar short braquiblast (van Baren et al. 2015). The white inflorescences are clustered at the end of the branches. The calix is tubular with two-toothed or entire lobules and 2.5–3.0 mm in length with white simple hairs. The corolla has five uneven lobules with a cylindric tube 3.0–3.5 mm in length. The fruit is formed by two fused achenes, with mericarps, and around 2 mm in length (Caro 1982).

Ethnobotanical information The infusions prepared with the aerial parts are used as a stomachic, gastrointestinal, and diaphoretic and against the influenza. Infusions made with leaves and stem are taken as aphrodisiac (Del Vitto et al. 1997; Ratera and Ratera 1980; Toursarkissian 1980; González and Molares 2004).

Chemical data Hydrodistillation from aerial parts yielded between 0.57 and 2.46% (v/w) EOs that have thymol, p-cymene, carvacrol, γ-terpinene, citral, and piperitone (Fig. 3.5). On the basis of the 13 main compounds of the EOs determined by GC-FID-MS, 4 chemotypes were determined: carvacrol, thymol, *cis*- and *trans*-dihydrocarvone, and linalool-geraniol. The co-occurrence of different chemotypes in a same population was found indicative of a pattern of variation throughout the natural distribution area of this species (Van Baren et al. 2015).

Biological activities The infusion, tinctures of dry plant material, and the EOs have antioxidant activity (determined by the DPPH assay) attributed to its content on thymol and carvacrol (Gastaldi et al. 2016). The plant EOs have antioxidant and antimicrobial activity (Ruffinengo et al. 2005; Fuselli et al. 2007; Gillij et al. 2008; Lima et al. 2011). The *A. seriphioides* EO showed a high acaricidal effect, causing high mite mortality at low oil concentration, but at the same time it had a low selectivity ratio (LD50 mite = 1.2 μl per cage, ratio selection 1.3 for 48 h of treatment). The main compounds of this EO are thymol, carvacrol, and orthocymene. Thymol and carvacrol are effective in control of *Verroea destructor* (Ruffinengo et al. 2005).

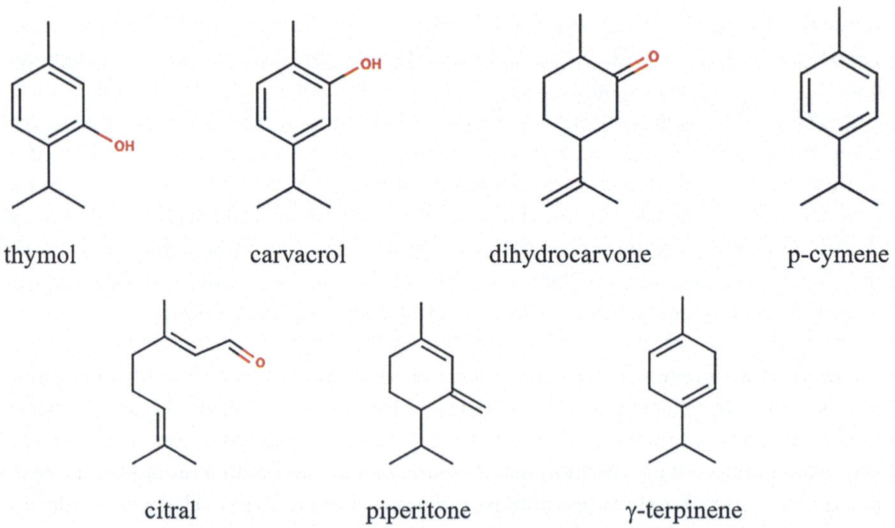

Fig. 3.5 Some of the chemical compounds found in the essential oil from *Acantholippia seriphioides* (A. Gray) Moldenke

3.6 *Achyrocline satureioides* (Lam.) D.C. (Asteraceae)

Synonymy *Achyrocline vargasiana* DC., *Gnaphalium satureioides* Lam. candicans (Kunth.).

Vernacular name Macela, marcela hembra, marcela, marcela del campo, marcelita, planta de la abeja, vira-vira, vira-vira guazú, yateí caá.

General information *A. satureioides* is one of the most spread medicinal species from Argentina; its popularity comes from its use for the treatment of a variety of disorders, principally gastrointestinal and respiratory diseases (Toursarkissian 1980; Ratera and Ratera 1980; Gattuso et al. 2007). The species grows mostly in rangelands and modified soils in the provinces of BAI, CAT, CHA, COR, COS, ERI, JUJ, LPA, MIS, SAL, SDE, SFE, SLU, and TUC (Barboza et al. 2009). Also, it is an attractant to butterflies which participate in its propagation. *A. satureioides* is also much appreciated for their delicate perfume with a light trace of anise, although the flavor is astringent. The stems, leaves, and flowers contain resins, essential oils, tannins, and other bitter principles (Ratera and Ratera 1980). For all of that, *marcela* is useful both for the pharmaceutical and for the cosmetics and bitter beverages industries (Del Vitto et al. 1997; Gattuso et al. 2007; Barboza et al. 2009; Retta 2014).

Botanical characteristics *A. satureioides* is a perennial herbaceous species and about 0.30–1.20 cm in height. It has stiffened stems covered by short white hairs. Leaves are grayish, sessile, alternate, linear, elliptic, or lanceolate, with an entire margin, and densely tomentose on the underside, 3–5 cm in length. The inflorescences are small capitula distributed in terminal glomerulus with 3–6 marginal feminine filiform flowers and 1–3 central tubular hermaphrodite flowers. Anthers are sagittate; style is short and truncated by the apex. Fruits are small brown achenes with a white pappus with faint hairs and a naked receptacle. Blooming is by the end of summer (Cabrera and Zardini 1978; Gattuso et al. 2006, 2008; Alonso and Desmarchelier 2015).

Ethnobotanical information *A. satureioides* was very popular among native communities that had been taken it as a digestive tisane. Nowadays, the aerial parts have a popular use in Argentina, Uruguay, Brazil, and Paraguay as a decoction or infusion for treating miscellaneous diseases. It is considered beneficial in cases of gastrointestinal disorders, as antidiabetic, digestive, antibacterial, antispasmodic, anti-inflammatory, antitumor, and antidiarrheal, to treat asthenia, and as tonic and stimulant. The mixture of dried inflorescences and stems is used as sedative, antiseptic, antispasmodic, for inflammation, and to aid and promote menstruation. The infusion prepared with the leaves is administrated as antitussive in bronchitis and the mixture of leaves and flowers as digestive, carminative, antispasmodic, sedative, anti-asthmatic, antitussive, and hypoglycemic. The infusion prepared with the flowers is used to treat asthma and as analgesic, antispasmodic, and anti-inflammatory.

As an external prescription, leaves are applied as anti-inflammatory in compresses and washes (Ratera and Ratera 1980; Toursarkissian 1980; Alonso and Desmarchelier 2015; Agüero 2017).

Chemical data This species has a high concentration of polyphenols and flavonoids (Fig. 3.6). In the dried entire plant, it was identified achyrofuran, quercetin, caffeic acid, chlorogenic and isochlorogenic acids, galangin, 3-methoxy galangin, quercetin-3-methylether, 2, 4′, 4-trihydroxy-6-methoxychalcone, and 7, 4′-dihydroxy-5-methoxyflavone. Quercetin, luteolin, 3O-methyl quercetin, and 3,7-dimethoxy-5,8-dihydroxyflavone were found in the dried inflorescence (Braga et al. 2006; Barboza et al. 2009; Grassi-Zampieron et al. 2009a; Carini et al. 2014; Retta 2014; Carini et al. 2015; Oliveira de Souza et al. 2018).

The inflorescences EOs contain monoterpenes (<7.3%) and sesquiterpenes (73.8%) (Fig. 3.7). Their main components were caryophyllene (39.7%), δ-canadiene

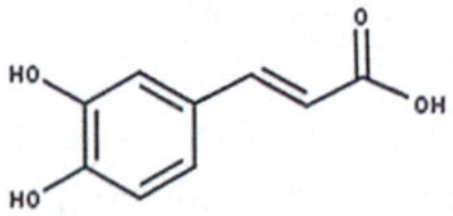
caffeic acid

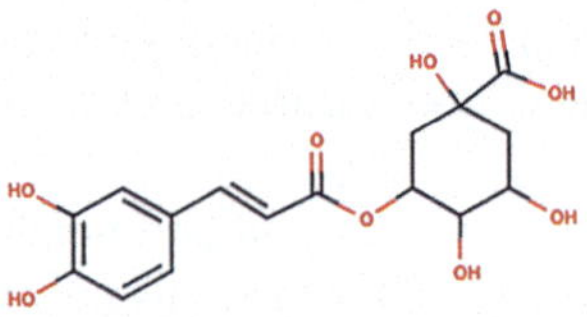
chlorogenic acid

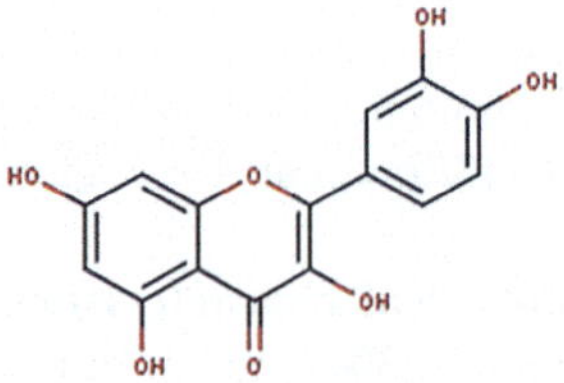
quercetin

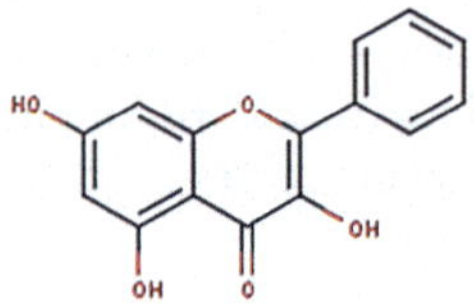
galangin

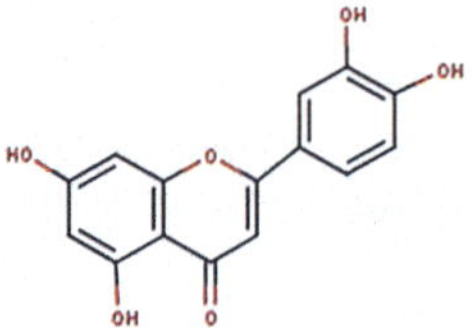
luteolin

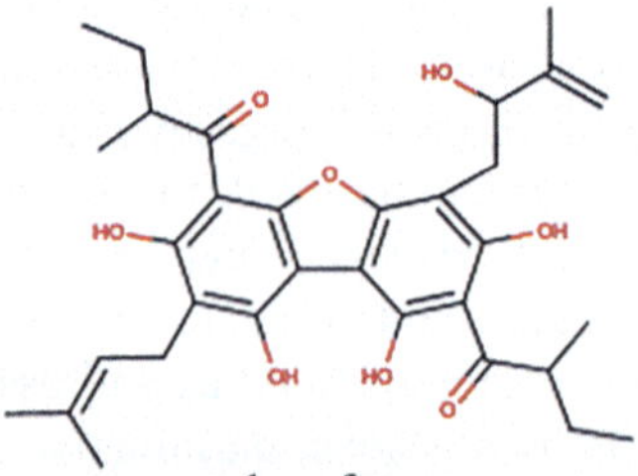
achyrofuran

Fig. 3.6 Some polyphenols and flavonoids detected in *Achyrocline satureioides* (Lam.) DC

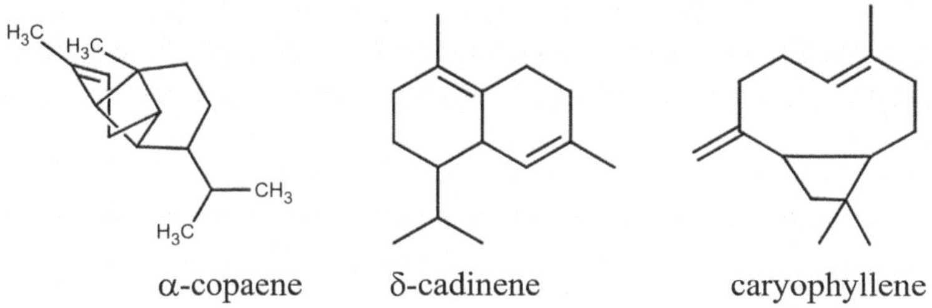

Fig. 3.7 Main components of the essential oils from *Achyrocline satureioides* (Lam.) DC

(9.4%), and α-copaene (14.2%). On the contrary, in Brazil the main components were monoterpenes (>56.4%) (Labuckas et al. 1999).

Biological activities It was reported the antioxidant and free radical scavenging activities of the *A. satureioides* aqueous and methanolic extracts (Desmarchelier et al. 1998; De Souza et al. 2002; Grassi-Zampieron et al. 2009b). The antioxidant activity was confirmed by *in silico* evaluation (PASS analysis) (Salgueiro et al. 2016). The ability of *A. satureioides* major compounds (such as the flavonoids luteolin, quercetin, 3-O-methylquercetin, and achyrobichalcone) to protect rat brain homogenates from lipid damage was reported (Rivera Megret et al. 2013). The content on those compounds was related to the hepatoprotective, immune-stimulant, cytotoxic, vasodilatory, antiherpetic, antispasmodic, relaxant, anti-ulcerative, antimicrobial, antioxidant, molluscicidal, antiviral, anti-inflammatory, and mutagenic activities attributed to the species.

The effect on skin lipoperoxidation was tested using a topical nanoemulsion made with *A. satureioides* extract or with pure quercetin on porcine skin from the back part of the ear. The nanoemulsion was prepared with *A. satureioides* inflorescences (7.5 w/v) in ethanol (80%) at a total amount of 100 μg of quercetin. The level of quercetin retained in the skin was higher with the *A. satureioides* extract-containing nanoemulsion than with the quercetin-containing nanoemulsion. Also, the thiobarbituric acid-reactive species (TBA-RS) assay showed a better result with the extract-containing nanoemulsion. The higher effectivity to avoid lipoperoxidation from the *A. satureioides* extract-containing nanoemulsion was attributed to the synergic effect of the compounds in the extract (Zorzi et al. 2015). An antioxidant hydrogel containing *A. satureioides* extract-loaded nanoemulsion (total flavonoids content, 1 mg ml^{-1}) was formulated to prevent oxidative stress on the skin (Balestrini et al. 2016).

The immunosuppressant, antioxidant, and anti-inflammatory activities of the dried inflorescences have also been attributed to the presence of flavonoids, mainly quercetin and its derivatives (Desmarchelier et al. 1998; Rivera Megret et al. 2013). However, when flower extracts were obtained by supercritical fluid extraction, its antioxidant activity did not show dependence on the phenolics concentration, suggesting the presence of other non-phenolic substances with antioxidant properties (Peruchi et al. 2005). The maximum antioxidant activity (81.6 ± 5%) was obtained at 250 bar and 30 °C, and the maximum amount of phenolics content (85 ± 3 mg gallic acid equivalent g^{-1} of extract) was recovered at 200 bar and 30 °C (Peruchi et al. 2005). As for their mechanism of action in tissue inflammation, the infusion (0.06–0.24 μg ml^{-1} quercetin equivalent) reduced phytohemagglutinin (PHA)-induced proliferation and production of interferon (IFN)-γ and interleukin (IL)-4. It also slightly increased the spontaneous generation of reactive oxygen species (ROS) from a concentration of 0.06 μg ml^{-1} quercetin equivalent and inhibited N-formyl-Met-Leu-Phe (fMLP)-induced ROS generation as well as spontaneous and fMLP-induced IL-8 production between 0.0012–0.03 μg ml^{-1} quercetin equivalents (Cosentino et al. 2008). Additionally, it was reported that the treatment of male Wistar rats with *A. satureioides* aqueous and ethanolic extracts reduced the neutrophil influx and the secretion of leukotriene B4 (LTB4) and cytokine-induced neutrophil chemoattractant 1 (CINC-1) in the exudates and the number of rolling and adhered leukocytes in the mesentery postcapillary venules, neutrophil L-selectin, *β*2-integrin, and toll-like receptor (TLR-4) expression and the oxidative burst, but did not cause an alteration in the morphology and activities of the liver and kidney. Together, the data showed that *A. satureioides* extract inhibited neutrophil functions related to the innate response and did not cause systemic toxicity (Barioni et al. 2013).

Another interesting activity of the hydroalcoholic extract of *A. satureioides* (14.8 equivalents of caffeic acid and 3 mg ml^{-1} quercetin) is the improvement of sleep quality in patients with obstructive sleep apnea syndrome (OSAS) that have received an oral administration of the extract. As sleep apnea might enhance oxidative stress through a reduction of antioxidant capacity of the blood due to hypoxia, the positive effect could be related to the antioxidant and anti-inflammatory actions attributed to the polyphenolic content of the extract (Pedemonte et al. 2013).

To test the neuroprotective effect, a model of permanent middle cerebral artery occlusion (pMCAo) with 2,3,5-triphenylltetrazolium chloride (TTC) was assessed in rats. When a decoction of dried flowers (2% w/v) was administrated to the rats during the 21st days previous to the treatment with TTC, the functional deficit caused by the tetrazolium salt was reversed and the cerebral infarct volume diminished at a plasma level of 60.4 ± 58.2 ng ml^{-1} quercetin. That effect was partially attributed to the content in flavonoids, mainly quercetin (1.14 ± 0.19 mg ml^{-1}) and luteolin (Rivera Megret et al. 2013). An infusion of the aerial parts (2% w/v) triggered a differentiation effect in PC12 cells at 10 and 20 μg ml^{-1} of total polyphenols. In that work it was demonstrated an increase of the number of neurites per cell, the percentage of cells with neurites, the number of fusiform cells, and the percentage of differentiation, which translated into an increase in long neurites. The changes induced resembled to those produced by nerve growth factor treatment. However, the differentiation features observed would likely not only be related to flavonoids (Blasina et al. 2009).

The popular use of the aqueous extract of *A. satureioides* as a hepatoprotective and digestive agent could be mediated through its antioxidant and choleretic activities (Kadarian et al. 2002; Rahimi 2015). Additionally, extracts of *A. satureioides* whole plant material were significantly active in in-house type 2 diabetes model on genetically diabetic db/db mice (oral dose 20 mg kg^{-1} q.d). The hypoglycemic activity was attributed by the authors to the compound achyrofuran as its oral administration (20 mg kg^{-1}) significantly lowered blood glucose levels (Ruffa et al. 2002; Carney et al. 2002). It was proposed the inclusion of the extracts in food formulations to treat (or prevent) type-2 diabetes and to ease the digestion process (Peruchi et al. 2005).

The achyrobichalcone, isolated from *A. satureioides* (Holzschuh et al. 2010; Carini et al. 2012), could also represent a promising chemotherapeutic biomolecule considering its similarity with other cytotoxic bi-chalcones (Carini et al. 2014). In general, the effects (attributed to the flavonoid content) were inhibition of cell proliferation, cell cycle arrest, apoptosis, and inhibition of angiogenesis, prevention of migration/metastasis, and overcoming multidrug resistance, alone or in combination with commonly used chemotherapeutic drugs (Oliveira de Souza et al. 2018). Moreover, the flavonoid extract potentiated the activity of temozolomide, a chemotherapeutic drug. The positive results of these preclinical studies suggest a promising use of these flavonoids for the development of new anticancer drugs (Oliveira de Souza et al. 2018).

The antimicrobial effect against *Staphylococcus* strains isolated from patients with youthful acne and from clinical materials (prosthesis, etc.) was tested with a sterile decoction (5% w/v final concentration) prepared from leaves from samples collected in the province of Córdoba. The inhibitory effect was demonstrated on coagulase-negative non-hemolytic strains (95%). Consequently, the topic application of the decoction to reduce the staphylococcal population was proposed (Calvo et al. 2006). The decoction of *A. satureioides* aerial parts was also active against *S. aureus*, *E. coli*, and *A. niger* (Anesini and Pérez 1993). The synergistic antibacterial activity of different components of the ethanolic extract (quercetin, 23-methyl-6-O-desmethylauricepyrone, and 3-O-methylquercetin) was also reported (Joray et al. 2013). It was also referred an activity (14.3 mm diameter of inhibition) of the ethanolic inflorescences extract on *Streptococci viridians* strains found in the oral cavity (Massori et al. 2017). The aqueous, freeze-dried, and spray-dried extracts prepared from inflorescences were tested for their activity on pathogenic intestinal flora of male Wistar rats. The freeze-dried extract prepared from the 80% ethanol (v/v) extractive solution exhibited a broad spectrum of antimicrobial activity (100–200 mg ml^{-1}) being the antibacterial effects of the extracts higher than those of amoxicillin (used as a positive control) when tested against *Bacillus cereus* (ATCC 9634) and *Staphylococcus aureus* (ATCC 1901). Total flavonoid content was 132 mg g^{-1} for the freeze-dried extract, 129.7 mg g^{-1} for the spray-dried extract, and 54.23 mg g^{-1} for the freeze-dried aqueous extract. The reported results also suggested that 80% ethanol (v/v) is more efficient for extracting the three flavonoid aglycones and chalcone from the inflorescences than water alone (Moresco et al. 2017). The n-hexane and dichloromethane extracts of the aerial parts contained, besides the compounds already mentioned, prenylated phloroglucinol α-pyrones that showed antimicrobial activity (against *E. coli* ATCC35218, *S. aureus* VISA, *S. aureus* MSSA, *E. faecalis* ATCC29212, and yeast sp.) that was related by the authors to the 3-methyl-2-butenyl chain attached to the phloroglucinol nucleus (Casero et al. 2014).

On the other hand, an antiherpetic activity was also informed (Bidone et al. 2015). To locally treat HSV-1 infection, the ethanolic extract of *A. satureioides* was incorporated into nanoemulsions (polydispersity index <0.1, mean droplet size 200 nm) with flavonoid association efficiency of almost 100%. The antiherpetic activity of the nanoemulsion (307.79 ± 10.23 μg ml^{-1} quercetin, 194.50 ± 7.91 μg ml^{-1} luteolin, 780.35 ± 31.42 μg ml^{-1} 3-O- methyl quercetin) tested by the viral plaque number reduction assay on Vero cells monolayers was significantly higher (IC50 1.4 ± 0.88 μg ml^{-1}) than that corresponding to the ethanolic extract (IC50 14.07 ± 3.46 μg ml^{-1}) (Bidone et al. 2014; Bidone et al. 2015).

As for the anti-plasmodial effect, it was reported that the treatment with *A. satureioides* had control of *Trypanosoma evansi* parasitemia in Wistar female rats. The number of trypomastigotes forms diminished both with an oral administration dose (1.5 ml kg^{-1}) of the EO (around 33% decrease) or the nano-encapsulated EO (80% decrease). On the other hand, the nano-encapsulated essential oils showed a reduction of the intense inflammation response caused by *T. evansi* and resulted more protective than the EOs on the liver and kidney (Do Carmo et al. 2015). Furthermore, at a concentration of 250 μg ml^{-1}, the ethanolic extract showed 100% of lysis on bloodstream of *T. cruzi in vitro* (Rojas de Arias et al. 1995).

It was reported a relevant inhibition of *Macrophomina phaseolina* growth with *A. satureioides* aqueous, hexane, and chloroform extracts (1000 μg ml^{-1}). The hexane and chloroform extracts were also inhibitory against *Fusarium graminearum*, *F. verticillioides*, and *Sclerotium rolfsii*. Nevertheless, the inhibition of *F. solani* was minimal (< 20%) in the conditions reported (Vogt et al. 2010a, b).

In veterinary, it was showed that the decoction from inflorescences (75 mg ml^{-1}) has inhibitory activity on bacteria isolated from bovine mastitis after 24 h of contact. From the bacteria isolated, 20 were Gram-positive (14 *Staphylococcus aureus*, 5 *Streptococcus dysgalactiae*, 1 *Streptococcus uberis*), and 3 were Gram-negative (1 *Escherichia coli*, 1 *Klebsiella* sp., and 1 *Pseudomonas aeruginosa*) (Rocha Sperotto et al. 2012).

As for its potential interest for the food industry, hydroalcoholic extract of *A. satureioides* (0.5 and 1%) was effective in decreasing the lipid oxidation of salami during storage without interfering with the acceptance of the product at a 0.5% concentration (Campagnol et al. 2011). Other authors also reported the antimicrobial activity of *A. satureioides* decoctions and hydroalcoholic extracts against bacteria of interest in food (*S. aureus*, *E. faecalis*, *S. enteritidis*, *E. coli*) being more sensitive the Gram-positive (*Staphylococcus* and *Entherococcus*) bacteria than the Gram-negative (*Salmonella* and *Echerichia*) (Mota et al. 2011). The crude hydroalcoholic extract (5% w/v hydroalcoholic extract) deactivated all strains of *Salmonella* spp. resistant to antibiotics isolated from raw material from swine and poultry (Maciel et al. 2017).

***In vitro* cultures** *A. satureioides* can be regenerated *in vitro* in Murashige and Skoog (MS) ¾ media with benzilaminopurine (BAP, 1 mg l^{-1}) and 3% w/v sucrose with uni-nodal segments (5–10 mm) as explant. Once the plant developed, rooting was induced by adding indole acetic acid (IAA, 1 mg l^{-1}). After 84 days of culture, the yield of rooted plantlets obtained was 62.35%. After transferring to pots in greenhouse, the survival was 77% after 150 days (Gattuso et al. 2007).

Toxicity The species is widely consumed by South America population without any report of toxicity. However, several studies were performed in order to assess that information. The maximum tolerated dose, determined by a sub-chronic toxicity study in mice and rats with two administration routes (oral and intraperitoneal), was 5 g kg^{-1} without any evidence of toxicity in the liver or kidney (Rivera et al. 2004). Further studies have shown that the cold aqueous extract (10–50 μg ml^{-1}) have low toxicity; it did not induce apoptotic effects on human peripheral blood mononuclear cells and did not show *in vitro* genotoxicity against Vero cells at 10–50 μg ml^{-1}; however, at high concentrations the micronucleus assay showed cytotoxicity (Sabini et al. 2013). When the same authors tested the hot aqueous extract effect, they found that it was more toxic than the cold one; however, it did not induce apoptosis. As for the cytogenotoxic effect, it was positive on bone marrow of mice (Cariddi et al. 2015). On the other hand, a cold aqueous extract (0.5, 1.0, 2.0, 3.0, and 4.0 mg ml^{-1} of mineral water) was used to study cytotoxicity by the *Allium cepa* L test. Only the higher concentrations induced cytotoxicity, but the effect was reversible and not associated with mutagenicity (Sabini et al. 2011).

Other studies have demonstrated that the decoction showed a certain cytotoxic activity on human lymphocytes at a concentration of 5% w/v (Calvo et al. 2006) and that the methanolic extract has a toxic effect against Hep G2, a human hepatocellular carcinoma cell line (Ruffa et al. 2002). Finally, the *A. salina* bioassay showed a LD50 of 2.06 mg ml^{-1}, which is considered very low for plant extracts. When different extract concentrations (30, 60, 150, and 300 μg ml^{-1}) were tested on human lymphocytes, none of them caused damage to DNA of cells, suggesting that the extracts have low potential genotoxicity (Salgueiro et al. 2016).

Legal status *A. satureioides* is an official vegetable drug in the Brazilian Pharmacopoeia (Retta et al. 2012). A quantification of the active markers was validated in order to propose it for its pharmacopoeia quality control in Argentina (Retta et al. 2007).

References

Agüero E (2017) Plantas medicinales silvestres del centro de Argentina. Guía para su reconocimiento y uso terapéutico. Editorial Ecoval, Unquillo, Córdoba, Argentina

Alonso J, Desmarchelier C (2015) Plantas medicinales. Bases científicas para su aplicación en atención primaria de la salud. Ed. Corpus, Buenos Aires

Andersen A, Carosio C, Junqueras M (1997) Principales plantas medicinales de la Provincia de San Luis. Anales de Saipa 15:217–227

Anesini C, Pérez C (1993) Screening of plants used in Argentine folk medicine for antimicrobial activity. J Ethnopharmacol 39(2):119–128

Argentina Food Code (2013) Chapter XV, Art.1192 bis (res. SPReI & SAGyP, N° 25/2011 &N° 64/2011)

Balestrini LA, Bidone J, Bertolin RC, Moresco K, Moreira JC, Teixeira HF (2016) Protective effect of a hydrogel containing *Achyrocline satureioides* extract-loaded nano emulsion against UV-induced skin damage. J Photochem Photobiol B Biol 163:269–276

Bandoni AL, Lopez MA, Juárez MA, Elechosa MA, van Baren CM, Di Leo LP (2002) Seasonal variations in the composition of the essential oil of "peperina" (*Minthostachys mollis* (Kunth) Griseb.) from a local population of the Province of Cordoba, Argentina. Essenze Deriv Agrum 72:11–14

Barboza GE, Cantero JJ, Núñez C, Pacciaroni A, Ariza Espinar L (2009) Medicinal Plants: A general review and a phytochemical and ethnopharmacological screening of the native Argentine Flora. Kurtziana 34(12):7365

Barioni ED, Santin JR, Daufenback Machado I, Fernandes de Paula Rodrigues S, Ferraz-de-Paula V, Wagner TM, Cogliati B, Correa dos Santos M, Da Silva Machado M, Faloni de Andrade S, Niero R, Poliselli Farsky H (2013) *Achyrocline satureioides* (Lam.) D.C. Hydroalcoholic extract inhibits neutrophil functions related to innate host defense. Evid Based Complement Alternat Med 2013, Article ID 787916, 12 pages. https://doi.org/10.1155/2013/787916

Bassols GB, Gurni AA (1996) *Lippia* species employed in Latin America folk medicine. Dominguezia 13(1):7–25

Benovit SC, Silva LL, Salbego J, Loro VL, Mallmann CA, Baldisserotto B, Flores EMM, Heinzmann BM (2015) Anesthetic activity and bio-guided fractionation of the essential oil of *Aloysia gratissima* (Gillies & Hook.) Tronc. in silver cat-fish Rhamdia quelen. An Acad Bras Cienc 87(3):1675–1689. https://doi.org/10.1590/0001-3765201520140223

Bernasconi Salazar JR, Karlin M, Arnulphi AS, Karlin UO, Accietto RH, Cora A (2017) Plantas del centro de la República Argentina: guia de campo, 1ª edn. Ecoval Editorial. Córdoba

Bidone J, Bica VC, Petrovick PR, Simões KLS, Bassani VL, Teixeira HF (2014) Simultaneous quantification of flavonoids from *Achyrocline satureioides* by a polar-reversed phase LC method — application to skin permeation/retention studies. Pharmazie 69:5–9

Bidone J, Fretes Argenta D, Kratz J, Ferreira Pettenuzzo L, Horn AP, Scherer Koester L, Linck Bassani V, Oliveira Simões CM, Ferreira TH (2015) Anti-herpes activity and skin/Mucosa distribution of flavonoids from *Achyrocline satureioides* extract incorporated into topical nanoemulsions. BioMed Res Int 2015:238010, 7 pages. https://doi.org/10.1155/2015/238010

Blasina MF, Vaamonde L, Morquio A, Echeverry C, Arredondo F, Dajas F (2009) Differentiation Induced by *Achyrocline satureioides* (Lam) Infusion in PC12 Cells. Phytother Res 23:1263–1269

Bluma R, Etcheverry M (2008) Application of essential oils in maizegrain: impact on *Aspergillus* section Flavi growth parameters and aflatoxin accumulation. Food Microbiol 25:324–334

Bluma R, Amaiden MR, Etcheverry M (2008a) Screening of Argentine plant extracts: Impact on growth parameters and aflatoxin B1 accumulation by *Aspergillus* section Flavi. Int J Food Microbiol 122:114–125

Bluma R, Amaiden MR, Daghero J, Etcheverry M (2008b) Control of *Aspergillus* section Flavi growth and aflatoxin accumulation by plant essential oils. J Appl Microbiol 105:203–214

Braga MEM, Perucho MP, Quispe-Condori S, Rosa PTV, Meireles MAA (2006) Phenolic Compounds Extraction from *Achyrocline satureioides*, *Curcuma longa*, *Foeniculum vulgare* and *Rosmarinus officinalis*. J Agri Food Chem 43:401–407

Bustos DA, Tapia AA, Feresin GE (1996) Ethnopharmacobotanical survey of Baichazeta district, San Juan Provine, Argentina. Fitoterapia 67(5):411–415

Cabrera AL, Zardini EM (1978) Manual de la Flora de los alrededores de Buenos Aires, 2nd edn. Editorial ACME SACI, Buenos Aires

Calvo D, Cariddi LN, Grosso M, Demo MS, Maldonado AM (2006) *Achyrocline satureioides* (Lam.) DC (Marcela): Antimicrobial activity on *Staphylococcus* spp. and immunomodulating effects on human lymphocytes. Rev Latinoam Microbiol 48(3–4):247–255

Campagnol PCB, Fries LLM, Terra NN, dos Santos BA, Furtado AS, Toneto ERL, de Campos RML (2011) The influence of *Achyrocline satureioides* ("Marcela") extract on the lipid oxidation of salami. Ciênc Tecnol Aliment Campinas 31(1):101–105

Cané L (2016) Plantas nativas. Las especies y su cultivo. Ediciones Jardín, Edit, Catapulta, Buenos Aires

Cariddi LN (2013) The role of essential oil of *Minthostachys verticillata* (Griseb) Epling, (Lamiaceae) and its active metabolites in immediate-type hypersensitivity responses. In: Govil JN, Bhattacharya S (eds) Recent progress in medicinal plants essential oils II. Studium Press, LLC, Houston, pp 487–495

Cariddi L, Escobar F, Moser M, Panero A, Alaniz F, Zygadlo J, Sabini L, Maldonado A (2011) Monoterpenes isolated from *Minthostachys verticillata* (Griseb.) Epling essential oil modulates immediate-type hypersensitivity responses *in vitro* and *in vivo*. Planta Med 77(15):1687–1694

Cariddi LN, Sabini MC, Escobar FM, Bacchetti R, Montironi I, Merckis C, Reinoso EB, Núñez Montoya S, Zanon SM, Comini LR, Sabini LI (2015) *In vitro* and *in vivo* cytogenotoxic effects of hot aqueous extract of *Achyrocline satureioides* (Lam.) DC. BioMed Research International 2015:270973, 12 pages. https://doi.org/10.1155/2015/270973

Carini JP, Kaaiser S, Ortega GG, Bassani VL (2012) Development, optimisation and validation of a stability-indicating HPLC method of achyrobichalcone quantification using experimental designs. Phytochem Anal 24(3):193–200

Carini JP, Klamt F, Linck Bassani V (2014) Flavonoids from *Achyrocline satureioides*: promising biomolecules for anticancer therapy. RSC Adv 4:3131–3144

Carini JP, Leitão GG, Schneider PH, Santos CC, Costa FN, Holzschuh MH, Klamts F, Bassani VL (2015) Isolation of achyrobichalcone from *Achyrocline satureioides* by high- speed countercurrent chromatography. Curr Pharm Biotechnol 16:66–71

Carney JR, Krenisky JM, Williamson RT, Luo J (2002) Achyrofuran, a New Antihyperglycemic Dibenzofuran from the South American Medicinal Plant *Achyrocline satureioides*. J Nat Prod 65(2):203–205

Caro JA (1982) Sistematización del género *Acantholippia grisebach* (Verbenaceae) y las especies de la Flora Argentina. Domingezia 3:1–31

Casero C, Machín F, Méndez-Álvarez S, Demo M, Ravelo AG, Pérez-Hernández N, Joseph-Nathan P, Estévez-Braun A (2014) Structure and antimicrobial activity of Phloroglucinol derivatives from Achyrocline satureioides. J Nat Prod 78(1):93–102

Catalán CA, Iglesias DI, Retamar JA, Iturraspe JB, Dartayet GH, Gros EG (1983) A sesquiterpene diketone from *Lippia integrifolia*. Phytochemistry 22(6):1507–1508

Catalán CAN, de Fenik IJS, Dartayet GH, Gros EG (1991) Integrifolian-1,5-dione and a revised structure for "africanone" biogenetically related sesquterpene ketones from *Lippia integrifolia*. Phytochemistry 30(4):1323–1326

Catalan CAN, de Fenik IJS, de Arriazu PJ, Kokke CMC (1992) 4,5-seco-african-4,5-dione from *Lippia integrifolia*. Phytochemistry 31(11):4025–4026

Cosentino M, Bombelli Rm Carcano E, Luini A, Marino F, Crema F, Dajas F, Lecchini S (2008) Immunomodulatory properties of *Achyrocline satureioides* (Lam.) D.C. infusion: a study on human leukocytes. J Ethnopharmacol 116:501–507

Dartayet GH, Catalán CA, Retamar JA, Gros EG (1984) Sesquiterpenoids from *Lippia integrifolia*-africanone, a tricyclic sesquterpene ketone. Phytochemistry 23(3):688–689

de Feo V, Ricciardi A, Biscardi D (1998) Chemical composition and antimicrobial screening of the essential oil of *Minthostachys verticillata* (Griseb.) Epl. (Lamiaceae). J Essent Oil Res 10:61–65

De Souza KCB, Schapoval EES, Bassani VL (2002) LC determination of flavonoids: separatin of quercetin, luteolin and 3-0-methylquercetin in *Achyrocline satureioides* preparations. J Pharm Biomed Anal 28(3–4):771–777

Del Vitto LA, Petenatti EM, Petenatti ME (1997) Recursos herbolarios de San Luis (República Argentina) Primera parte: plantas nativas. Multequina 6:49–66

Delucchi G (2006) Las especies vegetales amenazadas de la Provincia de Buenos Aires: Una actualización. APRONA Bol Cient 39:19–31

Demo M, Ceriatti F, Sabini L, Zanón S, Sutil S, Grosso M, Ferreyra S, Ramos B (1998) Relevamiento de plantas medicinales del sur de la provincia de Córdoba con actividad antimicrobiana. 3er. Encuentro Regional del NOA de Plantas Medicinales. 7–9 de mayo de 1998. Santiago del Estero, Argentina

Desmarchelier C, Coussio J, Ciccia G (1998) Antioxidant and free radical scavenging effects in extracts of the medicinal herb A*chyrocline satureioides* (Lam.) DC (AS). Braz J Med Biol Res 31(9):1163–1170

Do Carmo G, Baldissera MD, Vaucher RA, Rech VC, Oliveira CB, Rorato Sagrillo M, Boligon AA, Athayde ML, Alves MP, França RT, Lopes STA, Schwertz C, Mendes RE, Monteiro SG, DA Silva AS (2015) Effect of the treatment with *Achyrocline satureioides* (free and nanocapsules essential oil) and diminazene aceturate on hematological and biochemical parameters in rats infected by *Trypanosoma evansi*. Exp Parasitol 149:39–46

Domínguez J (1928) Materia Médica Argentina. Edit. Peuser, Buenos Aires

Duarte M, Figueira G, Pereira B, Magalhães PE Delarmelina C (2004) Atividade antimicrobiana do extrato hidroalcoólico de espécies da coleção de plantas medicinais do CPQBA/Unicamp. Abstract FT-23. III° Congresso Ibero-Americano de Plantas Medicinais. 20–23 de maio de 2004. Rio de Janeiro

Duschatzky CB, Martínez AN, Almeida NV, Bonivardo SL (2004) The nematicidal activity of the essential oils isolated from *Aloysia triphylla*, *A. polystachya*, *A. gratissima*, *Lippia* Argentina plants against the root-knot nematode. J Essent Oil Res 16:626–628

Elechosa MA, Molina AM, Juárez MA, van Baren CM, Di Leo LP, Bandoni AL (2007) Estudio comparativo del aceite esencial de *Minthostachys mollis* (Kunth.) Griseb., "peperina", obtenido de colectas en 21 poblaciones de las provincias de Tucumán, Cordoba, San Luis y Catamarca. Bol. Latinoam. Caribe Plant. Med. Aromaticas 6:244–245

Elechosa MA, Aguirre E, Bandoni AL, Di Leo Lira PMR, Fernández EA, Heit C, Juárez MA, López S, Martínez A, Martínez E, Marino AM, Molina AC, van Baren CM, Viturro CI (2009) Manual de recolección sustentable de plantas aromáticas nativas de la regióncentral y noroeste de la Argentina. Ed. INTA, Buenos Aires

Epling C (1939) Las Labiadas en el Noroeste de la Argentina. Lilloa 4:434–435

Escobar FM, Cariddi LN, Sabini MC, Reinoso E, Sutil SB, Torres CV, Zanon SM, Sabini LI (2012) Lack of cytotoxic and genotoxic effects of *Minthostachys verticillata* essential oil: Studies *in vitro* and *in vivo*. Food Chem Toxicol 50:3062–3067

Escobar FM, Sabini MC, Cariddi LN, Sabini MC, Mañas FJ, Cristofolini AL, Bagnis G, Gallucci MN, Cavaglieri LR (2015) Safety assessment of essential oil from *Minthostachys verticillata* (Griseb.) Epling (peperina): 90-Days oral subchronic toxicity study in rats. Regulat Toxicol Pharmacol Rtp 71(1):1–7

Farahani MS, Bahramsoltani R, Farzaei MH, Abdollahi M, Rahimi R (2014) Antidepressant plants and their mechanisms of action: a review of literature. Pharmaceutical sciences 20 (supll 1) 18 18th Iranian Pharmacy Student Seminar. http://pharm-sci.tbzmed.ac.ir

Fester GA, Martinuzzi EA (1950) La esencia de peperina. 2a comunicación. An Asoc Quím Argent 38:255–267

Fester GA, Martinuzzi EA, Retamar JA, Ricciardi AIA (1960) Estudios sobre esencias volátiles argentinas. Revista de la Facultad de Ingeniería Quimica 29:21–44

Franco ALP, Oliveira TB, Ferri PH, Bara MTF, de Paula JR (2007) Avaliação da composição química e atividade antibaceriana dos óleos essenciais de *Aloysia gratissima* (Gillies and Hook) Tronc. (alfazema), *Ocium gratissimum* L. (alfavaca-cravo) e *Curcuma longa* L. (açafrao). Rev Eletronica Farm 4:208–220

Fuselli SR, Garcia de la Rosa SB, Eguaras MJ, Fritz R, Ndagijimana M, Vannini L, Guerzoni ME (2007) Efficacy of indigenous plant essential oil Andean Thyme (*Acantholippia seriphioides* A. Gray) to control American Foulbrood (AFB) in honey bee (*Apis mellifera* L.) hives. J Essent Oil Res 19:514–519

García C, Talarico L, Almeida N, Colombres S, Duschatzky C, Damonte E (2003) Virucidal activity of essential oils from aromatic plants of San Luis, Argentina. Phytother Res 17(9):1073–1075

Gastaldi B, Assef Y, van Baren C, Di Leo LP, Retta D, Bandoni AL, González SB (2016) Actividad antioxidante en infusiones, tinturas y aceites esenciales de especies nativas de la Patagonia Argentina. Revista Cubana de Plantas Medicinales 21(1):51–62

Gattuso S, Scandizzi A, Busilacchi H, Di Sapio O, Severin C (2006) *Achyrocline satureioides* (Lam,) DC.: propagación *in vitro* a partir de segmentos nodales. Revista de Investigaciones de la Facultad de Ciencias Agrarias-UNR VII(11):45–50

Gattuso S, Scandizzini A, Busilacchi H, Di Sapio O, Severin C (2007) Achyrocline satureioides (Lam.) DC: Propagagación in vitro a partir de segmentos nodales. Rev Inv Fac Cs Agrarias 11:45–50

Gattuso M, Cortadi A, Mc Cargo J, Retta D, Bandoni A, Ferraro G, Gattuso S (2008) Estudio morfoanatómico vegetativo y floral de *Achyrocline satureioides*; *Achyrocline flaccida* y *Gnaphalium gaudichaudianum* (Asteraceae) Boletín Latinoamericano y del Caribe de Plantas Medicinales y Aromáticas BLACPMA. 7(4):193–194

Gillij YG, Gleiser RM, Zygadlo JA (2008) Mosquito repellent activity of essential oils of aromatic plants growing in Argentina. Bioresour Technol 99:2507–2515

Gleiser R, Bonino M, Zygadlo J (2011) Repellence of essential oils of aromatic plants growing in Argentina against *Aedes aegypti* (Diptera: Culicidae). Parasitol Res 108:69–78

González MJ, Marioli JM (2010) Antibacterial activity of water extracts and essential oils of various aromatic plants against *Paenibacillus larvae*, the causative agent of American Foulbrood. J Invertebr Pathol 104(3):209–213

González S, Molares S (2004) Plantas Medicinales Utilizadas en Comunidades Rurales del Chubut, Patagonia Argentina. Bol Latinoam Caribe Plant Med Aromat 3:58–62

González Pereyra ML, Cariddi LN, Ybarra F, Isola MC, Demo MS, Sabini LI, Maldonado AM (2005) Immunomodulating properties of *Minthostachys verticillata* on human lymphocytes and basophils. Rev Alergia Mexico 52(3):105–112

Gorzalczany S, Sülsen V, Redko F, Vescina C, Muschietti L, Martino V, Acevedo C (2008) Choleretic and antispasmodic effects of *Lippia integrifolia* aqueous extract. Braz J Pharmacognosy 18:16–20

Grassi-Zampieron RF, França LV, Carollo CA, do Carmo Vieira M, Oliveiros-Bastidas A, de Siqueira JM (2009a) Comparative profiles of *Achyrocline alata* (Kunth) DC. and *A. satureioides* (Lam.) DC., Asteraceae, applying HPLC-DAD-MS. Revista Brasileira de Farmacognosia Brazilian Journal of Pharmacognosy 20(4):575–579

Grassi-Zampieron RF, Vieira MC, de Siquera JA (2009b) Atividade antioxidante e captora de radicais livres dos extratos de *Achyrocline alata* (Kunth.) DC. em comparação com extratos de *Achyrocline satureioides* (Lam.) DC. Revista Brasileira de Farmacognosia Brazilian Journal of Pharmacognosy 19(2B):572–576

Haene E (2007) 100 Flores Argentinas. Ed. Albatros, Buenos Aires

Hieronymus J (1882) Plantae Diaphoricae Florae Argentinae. Edit. Peuser, Buenos Aires

Hister CAL, Laughingho HD, Da Silva CB, Do Canto Dorow TS, Tedesco SB (2009) Evaluation of the Antiproliferative effect of infusions and essential oil of *Aloysia gratissima*. Pakistan J Biol Sci 12(24):1581–1584

Holzschuh MH, Gosmann G, Schneider PH, Schapoval EES, Bassani VL (2010) Identification and stability of a new bichalcone in *Achyrocline satureioides* spray dried powder. Pharmazie 65:650–656

Ianicelli J, Pérez de la Torre M, Coviella A, Del Valle AE, Elechosa MA, van Baren CM, Pacheco MG, Escandón AS (2016) *In vitro* propagation of *Lippia integrifolia* (Griseb.) Hier. and detection of genetic instability through ISSR markers of *in vitro*-cultured plants. Rev Fac Agronom, La Plata 115(1):67–76

Joray MB, Palacios SM, Carpinella MC (2013) Understanding the interactions between metabolites isolated from *Achyrocline satureioides* in relation to its antibacterial activity. Phytomedicine 20:258–261

Kadarian C, Broussalis AM, Miño J, Lopez P, Gorzalczany S, Ferraro G, Acevedo C (2002) Hepatoprotective activity of *Achyrocline satureioides* (Lam) D. C. Pharmacol Res 45:57–61

Labuckas DO, Maestri DM, Grosso NR, Zygadlo JA (1999) Essential oils of *Achyrocline satureioides, Achyrocline alata* and *Achyrocline tomentosa*. Planta Med 65:184–186

Lima B, López S, Luna L, Aguero MB, Aragon L, Tapia A, Zacchino S, López ML, Zygadlo J, Feresin GE (2011) Essential Oils of Medicinal Plants from the Central Andes of Argentina: Chemical Composition, and Antifungal, Antibacterial, and Insect-Repellent Activities. Chem Biodivers 8:924–936

Maciel MJ, Silva da Sliva MA, Ethur E, Marchionatti Avancin CA (2017) Phytochemical indicators and antibacterial activity of the crude hydroalcoholic extract of *Achyrocline satureioides* ("macela") against antibiotic-resistant *Salmonella* spp. isolated from animal products (swine and poultry). Revista Brasileira de Higiene e Sanidade Animal 11(3):273–287

Marcial G, Sendker J, Brandt S, Perotti de Lampasona M, Catalán CAN, Font de Valdez G, Hensel A (2014) Gastroprotection as an example: Antiadhesion against *Helicobacter pylori*, anti-inflammatory and antioxidante activities of aqueous extracts from the aerial parts of *Lippia integrifolia* Hieron. J Ethnopharmacol 155:1125–1133

Marcial G, de Lampasona MP, Veja MI, Lizarraga E, Viturro CI, Slanis A, Juárez MA, Elechosa MA, Catalán CAN (2016) Intraspecific variation in essential oil composition of the medicinal plant *Lippia integrifolia* (Verbenaceae). Evid Five Chemotypes Phytochem 122:203–212

Martínez Carretero E (2013) Flora de Mendoza. Multequina 22(5):1–3

Martínez Crovetto R (1981) Fertility-regulating plants used in popular medicine in Northeastern Argentina. Parodiana 1:97–117

Massori L, Yazdani S, Rezaei F, Amraei M (2017) Phytotherapy for *Streptococcus viridians*. J Pharm Sci Res 9(7):1205–1208

Moresco KS, Silveira AK, Zeidán-Chuñiá F, Fonseca Correia AP, Oliveria RR, Borges AG, Grun L, Barbé-Tuana F, Zmozinski A, Brandelli A, Goretti Rodrigues Vale M, Pens Gelain D, Link Bassani V, Fonseca Moreira JC (2017) Effects of *Achyrocline satureioides* inflorescence extracts against pathogenic intestinal bacteria: chemical characterization, *in vitro* tests, and *in vivo* evaluation. Evid Based Complement Alter Med 2017:4874865, 10 pages. https://doi.org/10.1155/2017/4874865

Mota FM, Carvalho HH, Wiest JM (2011) Atividade antibacteriana *in vitro* de inflorescências de *Achyrocline satureioides* (Lam.) DC. – Asteraceae ("macela", "marcela") sobre agentes bacterianos de interesse em alimentos. Rev Bras Pl Med Botucatu 13(3):298–304

Múlgura de Romero ME, Martínez SG, Atkins S, Rotman AD (2002) Morfología de las inflorescencias en Verbenáceae, Verbenoideae III: Tribu Lantaneae Darwinina. Nueva Serie 40(1–4):1–15

Murashige T, Skoog F (1962) A revised medium for rapid growth and bio assays with tobacco tissue cultures. Physol Plant 15:473–497

Muschietti L, Derita M, Sülsen V, de Dios MJ, Ferraro G, Zacchino S, Martino V (2005) In vitro antifungal assay of traditional Argentine medicinal plants. J Ethnopharmacol 102(2):233–238

Núñez C, Cantero J (2000) Las plantas medicinales del sur de la provincia de Córdoba. Edit. Universidad Nacional de Río Cuarto, Argentina

O'Leary N, Lu-Irving P, Moroni P, Siedo S (2016) Taxonomic revision of *Aloysia* (Verbenaceaem Lantaneae) in South America. Ann Mo Bot Gard 101:568–609

Oliva M, Will I, Ordoñez L, García Ovando H, Demo YM. (2001). Estudio comparativo del efecto antimicrobiano de extractos crudos vegetales y quinolonas. Abstract P-115. X° Simposio Latinoamericano de Farmacobotánica. 8-11de abril de 2001. Comodoro Rivadavia, Chubut

Oliveira de Souza PO, Bianchi SE, Figueró F, Heimfarth L, Moresco KS, Mayer Goncalves R, Bender Hoppe J, Peres Klein C, Gazzana Salbego C, Pens Gelains D, Link Bassani V, Zanotto Gilho A, Fonseca Moreira JC (2018) Anticancer activity of flavonoids isolated from *Achyrocline satureioides* in gliomas cell lines. Toxicol In Vitro 51:23–33

Palacios SM, Bertoni A, Rossi Y, Santander R, Urzúa A (2009) Insecticidal activity of essential oils from native medicinal plants of Central Argentina against the housefly, *Musca domestica* (L.). Parasitol Res 106(1):207–212

Pascual ME, Slowing K, Carretero E, Sánchez Mata D, Villar A (2001) *Lippia*: traditional uses, chemistry and pharmacology: a review. J Ethnopharmacol 76:201–214

Pedemonte M, Rivera F, Testa M, Velluti R, Dajas F, Nadruz E (2013) *Achyrocline satureioides* (LAM) D.C. would improve sleep quality in patients with Obstructive Sleep Apnea Syndrome: a pilot study. Sleep Sci 6(4):135–140

Pellegrini MC, Alonso-Salces RM, Umpierrez ML, Rossini C, Fuselli SR (2017) Chemical composition, antimicrobial activity and mode of action of essential oils against *Paenibacillus larvae*, ethiological agent of American foulbrood on *Apis mellifera*. Chem Biodivers 14(4) https://doi.org/10.1002/cbdv.201600382

Peruchi MP, Braga MEM, Meireles MAA (2005) *Achyrocline satureioides* (Macela) flowers extracts obtained by supercritical fluid extraction. 10th European meeting on Supercritical Fluids: Reaction, Materials and Natural Products Processing

Primo V, Rovera M, Zanon S, Oliva M, Demo M, Daghero J, Sabini L (2001) Determination of the antibacterial and antiviral activity of the essential oil from *Minthostachys verticillata* (Griseb) Epling. Rev Argent Microbiol 33:113–117

Ragonese A, Milano V (1984) Vegetales y Sustancias Tóxicas de la Flora Argentina. En: Enciclopedia Argentina de la Agricultura y Jardinería, 2ª. Ed. Edit. Acme, Buenos Aires. pp 256–257

Rahimi M (2015) A review: anti diabetic medicinal plants used for diabetes mellitus. Bull Env Pharmacol Life Sci 4(2):163–180

Ratera EL, Ratera MO (1980) Plantas de la Flora Argentina empleadas en medicina popular. Ed. Hemisferio sur S.A.

Retta DS (2014) Determinación de calidad de "marcela" *Achyrocline satureioides* (Lam.) DC. (Asteraceae). Parámetros fitoquímicos. Dominguezia 30(2):5–17

Retta SD, López P, Gattuso M, Gattuso S, Ferraro G, Bandoni Al (2007) Normalización farmacopeica de marcela (*Achyrocline satureioides* (Lam. DC.): Validación del método de cuantificación. Boletín Latinoamericano y del Caribe de Plantas Medicinales y Aromáticas (BLACPMA) 6(6):380–381

Retta D, Dellacassa E, Villamil J, Suárez SA, Bandoni AL (2012) Marcela, a promising medicinal and aromatic plant from Latin America: a review. Ind Crop Prod 38:27–38

Ricciardi GAL, Van Baren CM, Lira PDL, Ricciardi AIA, Lorenzo D, Dellacasa E, Bandoni AL (2005) Volatile consitutents from aerial parts of *Aloysia gratissima* (Gilies and Hook) Tronc. var. gratissima growing in Corrientes, Argentina. Flavour Frag J 21:698–703

Rivera Megret FR, Tejera Correa D, AbinCarriquiri JA, Prunell dos Santos G, Martínez Busi M, Dajas Méndz F (2013) *Achyrocline stureioides* (Lam,) DC. (marcela) reduces brain damage in permanent focal ischemia in rats. Revista Cubana de Plantas Medicinales 18(3):445–460

Rivera F, Urbanavicius J, Gervaz E, Morquio A, Dajas F (2004) Some aspects of the *in vivo* neuroprotective capacity of flavonoids: bioavailability and structure-activity relationship. Neurotox Res 6(7–8):543–553

Rocha Sperotto V, Levay Murari A, Rebellato da Silva DA, Rubert Possenti CG, Wiest JM, Marchionatti Avancini CA (2012) Activity of the Decoction of *Achyrocline satureioides* D.C. (Lam.) – Asteraceae ("macela") against Standard and Isolated Bacteria from Bovine Mastitis. Acta Sci Vet 40(3):1052–1059

Roig FA (2001) Flora medicinal mendocina, las plantas medicinales y aromáticas de la provincia de Mendoza, Argentina. Ed. EDIUNC, Mendoza

Rojas de Arias A, Ferro E, Inchausti A, Ascurra M, Acosta N, Rodríguez E, Fournet A (1995) Mutagenicity, insecticidal and trypanocidal activity of some Paraguayan Asteraceae. J Ethnopharmacol 45:35–41

Rotman AD, Múlgura de Romero ME (1999) Verbenaceae. Aportes Botánicos de Salta-Seria Flora 5(11):1–83

Ruffa MJ, Ferraro G, Wagner ML, Calcagno ML, Campos RH, Cavallaro L (2002) Cytotoxic effect of Argentine medicinal plant extracts on human hepatocellular carcinoma cell line. J Ethnopharmacol 79:335–339

Ruffinengo S, Eguaras M, Floris I, Faverin C, Bailac P, Ponzi M (2005) LD50 and repellent effects of essential oils from Argentinian wild plant species on *Varroa destructor*. J Econ Entomol 98:651–655

Ruiz Leal A (1972) Flora popular Mendocina. Deserta 3:9–296

Sabini MC, Cariddi LN, Escobar FM, Bachetti RA, Sutil SB, Contigiani MS, Zanon SM, Sabini LI (2011) Evaluation of cytogenotoxic effects of cold aqueous extract from *Achyrocline satureioides* by *Allium cepa* L test. Nat Prod Commun 6(7):995–998

Sabini MC, Cariddi LN, Escobar FM, Mañas F, Comini L, Reinoso E, Sutil SB, Acosta C, Núñez Montoya S, Contigiani MS, Zanon SM, Sabini LI (2013) Evaluation of the cytotoxicity, genotoxicity and apoptotic induction of an aqueous extract of *Achyrocline satureioides* (Lam.) D C. Food Chem Toxicol 60:463–470

Salgueiro ACF, Folmer V, da Rosa HS, Costa MT, Boligon AA, Paula FR, Roos DH, Punetl GO (2016) *In vitro* and *in silico* antioxidant and toxicological activities of *Achyrocline satureioides*. J Ethnopharmacol 194:6–14

Schiuma G (1997) Las plantas aromáticas y medicinales silvestres como productos primarios. Anales de Saipa 15:13–19

Schmidt-Lebuhn AN (2008) Ethnobotany, biochemistry and pharmacology of *Minthostachys* (Lamiaceae). J Ethnopharmacol 118:343–353

Sérsic A, Cocucci A, Benítez-Vieyra S, Cosacov A, Díaz L, Glinos E, Grosso N, Lazarte C, Medina M, Moré M, Moyano M, Nattero J, Paiaro V, Trujillo C, Wiemer P (2010) Flores del Centro de Argentina. Una guía ilustrada para conocer 141 especies típicas. Academia Nacional de Ciencias, Córdoba

Soler E, Dellacasa E, Moyna M (1986) Composition of *Aloysia gratissima* flower essential oil. Planta Med 52:488–490

Souza AA, Wiest JM (2005) Atividade anti-bacteriana de *Aloysia gratissima* (Gill et Hook) Tronc. (garupá, ervasanta), usada na medicina tradicional no Rio Gran-de do Sul – Brasil. Rev Bras Pl Med Botucatu 9(3):23–29

Speroni E, Cervellati R, Govoni P, Guizzardi S, Renzulli C, Guerra M (2003) Efficacy of different *Cynara scolymus* preparations on liver complaints. J Ethnopharmacol 86:203–211

Strasser B, Fernández EA (2006) Tomillo del campo (*Acantholippia seriphioides* (A. Gray) Mold.). En: Cultivemos plantas nativas. Ed. Universidad Nacional de San Luis, San Luis

Strasser B, Fernández EA, Bandoni AL, Juárez MA, Elechosa MA (2002) Estudios para la conservación de una población de *Acantholippia seriphioides* (A. Gray) Mold. en San Luis y de las variaciones en lacomposición de su aceite esencial. Communication to the I Congreso Latinoamericano de Fitoquímica y IV Reunión de la Sociedad Latinoamericana de Fitoquímica, Buenos Aires

Terblanché FC, Kornelius G (1996) Essential oil constituents of the genus Lippia (Verbenaceae)-A Literature review. J Essent Oil Res 8:471–485

Toursarkissian M (1980) Plantas Medicinales de la Argentina. Sus nombres botánicos, vulgares, usos y distribución geográfica. Editorial hemisferio sur

Trovati G, Sanchez EA, Chierice GO (2009) TLC Separation and Identification of the Essential Oil Constituents from *Aloysia gratissima*. J Liq Chromatogr Relat Technol 32:890–895

Valladares GR, Zapata A, Zygadlo J, Banchio E (2002) Phytochemical induction by herbivores could affect quality of essential oils from aromatic plants. J Agric Food Chem 50(14):4059–4061

Van Baren CM, Di Leo LP, Elechosa MA, Molina AM, Juarez MA, Martínez A, Perelman S, Bandoni AL (2014) New insights into the chemical biodiversity of *Minthostachys mollis* in Argentina. Biochem Syst Ecol 57:374–383

Van Baren C, Elechosa MA, Di Leo LP, Retta DS, Juárez MA, Martínez AJ, Molina AM, Bandoni AL (2015) *Acantholippia seriphioides*: Chemical biodi-versity of wild populations from the Cuyo region in Argentina. Boletín Latinoamericano y del Caribe de Plantas Medicinales y Aromáticas 14(1):33–41

Vogt MV, Sutil SB, Escobar FM, Sabibi MC, Cariddi LN, Torres CV, Zanon SM, Sabini LI (2010a) *Minthostachys verticillata* essentials oil and its major components: antiherpeticselective action in HEp-2 cells. Mol Med Chem 21:117–120

Vogt V, Tonn C, Sabini L, Rosas S (2010b) Fungitoxic effects of *Achyrocline satureioides* (marcela) on plant pathogens. Mol Med Chem 21:109–112

Zeni ALB, Vandresen-Filho S, Dal-Cim T, Martins WC, Bertoldo DB, Maraschin M, Tasca CI (2014) *Aloysia gratissima* prevents cellular damage induced by glutamatergic excitotoxicity. J Pharm Pharmacol 66(9):1294–1302

Zorzi GK, Caregnato F, Fonseca Moreira JC, Ferreira Teixeira H, Santana Carvalho EL (2015) Antioxidant Effect of Nanoemulsions Containing Extract of *Achyrocline satureioides* (Lam) D.C.—Asteraceae. AAPS Pharm Sci Tech 17(4). https://doi.org/10.1208/s12249-015-0408-8

Zuloaga FO, Morrone ON, Belgrano MJ, Marticorena C, Marchesi E (eds) (2008) Catálogo de las plantas vasculares del Cono Sur. Monogr. Syst Bot Missouri Bot Gard 107(1–3):i–xcvi, 1–3348

Chapter 4
The Pampa: An Infinite Plain

4.1 Introduction: The Pampa

The Pampa plain (Fig. 4.1) is an extensive surface with undulations (from ancient dunes) that comprise the province of BAI; North East of the province of LPA; South of the provinces of COR, SFE, and ERI; and sections of the provinces of SLU, RNE, and MEN. It limits to the North with the Mesopotamian region and the CHA plain, by the West with the *Sierras Pampeanas* and the Cuyo region, by the South with the Colorado River and Patagonia, and by the East with the Atlantic Ocean, the Río de la Plata, and the Paraná River. The plain is interrupted towards the South by the systems of *Sierras de Tandilia* and *Sierra de la Ventana* (500–1300 m height). *Sierras de Tandilia* extends from the city of *Mar del Plata* to *Las Flores* stream; its highest height is *La Juanita* Hill (524 m). *Sierra de la Ventana* extends from the *Guaminí* lagoon towards the Atlantic coast; its highest height is the *Tres Picos* hill (1239 m). *Sierras of Lihuel Calel* are placed in the province of LPA.

In the Pampa plains prevails a temperate humid to subhumid climate with warm summers. The rains are distributed throughout the year, without a rainy season (600 mm at the SW and 1100 mm at the NE). By the West there is an extensive system of freshwater or saline lakes and lagoons sporadically linked to each other. The characteristic winds of the region are *Pampero* and *Sudestada*. *Pampero* wind is cold and dry, comes mainly in summer from the Southwest bypassing the *cordillera de los Andes*, and produces rain, temperature fall, and large clouds of dust. *Sudestada* wind is cold and humid and comes from the Atlantic Ocean and Río de la Plata producing floods on the right margin of Río de la Plata and in the south of the littoral region. The Pampa is one of the most fertile areas in the world; the soils are rich with a high content of organics and nutrients. The farming regions are particularly susceptible to flooding during the thunderstorms.

The Pampa plain is an important agro-industry region (agriculture and livestock production). The agricultural capacity is remarkable; the agricultural economy is based on the cultivation of wheat, corn, flax, oats, barley, rye, sunflower, potatoes,

M. A. Alvarez, *Pharmacological Properties of Native Plants from Argentina*,
https://doi.org/10.1007/978-3-030-20198-2_4

Fig. 4.1 A landscape from the Pampa plain. PG Alvarez personal collection

peanut, and soybean. More recently, vineyards established in the Buenos Aires province. There is also horticultural production.

Much of the area is also used for cattle breeding for meat (Charolais, Hereford, and Aberdeen Angus) and dairy (Holando-Argentino breed). There is also intense pig (Duroc Jersey, Poland China, and Berkshire), equine, and sheep cattle and avian production.

Fishing exploitation extends from the San Antonio cape to the South. Hake, from the Atlantic Ocean, and shad, from the rivers, are the main fished species. As for mining there are exploitations of granite and limestone basins. The industrial activity of the region is outstanding, ranging from petrochemical, automotive to textile.

Among the fauna, the great animals are today threatened (puma, wildcat, pampa gray fox, pampas deer, skunk and ferret), there are other mammals (plain vizcacha, southern mountain cavy, coypu, armadillos, weasels), birds (nandú or rhea, southern screamer, partridges, elegant crested tinamou, birds of prey, burrowing owl), and birds associated with aquatic environments (herons, Rallidae spp., white faced ibis, white stork, neotropic cormorant, etc.). Some of the wetlands from the region are important centers for migratory birds from the Northern hemisphere. The humid Pampa ecosystem is one of the richest grazing areas in the world.

As for the flora, the original formations are temperate rangelands mainly of Gramineae (genus *Nassella*, *Piptochaetium*, and *Andropogon* like *Bromus*, *Stipa*, *Poa*, *Paspalum*, *Eragrostis*, etc.), halophile rangeland, dune grasslands, forest communities restricted to ravines, and cords of shells *(cordones de conchillas)* (Fig. 4.2). Today, the ceibo (*Erythrina crista-galli*) and the willow (*Salix humboldtiana*), part of the primitive vegetation of the region, only grow in the riparian zones. Other species, like the quebradillo (*Acanthosyris spinescens*) and the jarilla (*Larrea* spp.), grow in the steppe regions. The Pampas are bounded by the drier Argentine Espinal grasslands, which form a semicircle around the north, west, and south of the Humid Pampas. In the Pampas from ERI, the grasslands associate arboreal elements such as *Prosopis alba*, *P. affinis*, *Celtis tala*, and *Phytolacca dioica* (Cabrera 1976; Oyarzábal et al. 2018).

Fig. 4.2 Plant communities from the Pampa eco-region. PG Alvarez personal collection

4.2 *Erythrina crista-galli* L. var. crista-galli (Fabaceae)

Synonymy *Coladendron crista-galli* (L) Kuntze, *Erythrina fasciculate* Benth., *Erythrina lauriflora* Jacq., *Erythrina pulcherrima* Tod., *Erythrina speciosa* Tod., *Micropteryx crista-galli* (L.) Walp., *Micropteryx lauriflora* (Jacq.) Walp., *Micropteryx fasciculata* (Benth.) Walp.

Vernacular name Seibo, ceibo, ceibo macho, seibo entrerriano, choipo, cockspur coral tree (Eng.).

General information The scientific name of the species describes the color and aspect of its flower; the word *erythros* comes from the Greek and means red color; *crista-galli* meaning is "the red crest of roosters." In 1942, *E. crista-galli* flower was named the national flower from Argentina (national decree 138474). It has also been chosen as the national flower from Uruguay. The native name of this species is *sui-nandi* (*Guaraní* language) that means parrot's food. *E. crista-galli* flowers are visited for its nectar by hummingbirds, *Icterus pyrrhopterus* (*boyerito alférez*), and other birds. The beauty of the flowers promotes its use to decorate streets and parks (Fig. 4.3). It is usual to see epiphytes such as ferns, cacti, orchids, and the Bromeliaceous *Tillandsia aeranthos* growing on its branches (Chebez and Masariche 2010). *E. crista-galli* grows in the provinces of BUE, CHA, COS, ERI, FOR, JUJ, LPA, MIS, SAL, SDE, SFE, and TUC. Other two *Erythrina* species that grow in Argentina are *E. falcata* (*ceibo de Jujuy*) and *E. dominguezii* (*ceibo rosado*). *E. falcata* has red flowers lighter than those from *E. crista-galli* and grows in the woods of TUC, SAL, and JUJ and in the MIS forest. *E. dominguezii* has pink or salmon-red flowers and grows in the woods from CHA and forests from SAL (Haene 2007; Lozano and Zapater 2010; Cané 2016). It also grows in Bolivia, Brazil, Chile, Paraguay, and Uruguay.

Besides its medicinal uses, the wood is used to make bows for percussion instruments, horse saddles, and orthopedic devices (Abeddini et al. 2000), the pulp is used

Fig. 4.3 *Erythrina crista-galli* L. var. crista-galli (Fabaceae). Branches with leaves and flowers (**a**), detail of the flowers (**b**), fruits (**c**), flowers and fruits (**d**). MA Alvarez personal collection

to paper production for its color and lightness, and the extract of the flowers is used as a fabric dye (Demaio et al. 2015). Finally, recent research (Enciso et al. 2017) has demonstrated that a mixture of anthocyanins extracted from *ceibo* flowers can be used as a sensitizer to produce low-cost dye-sensitized solar cells.

Botanical characteristics *E. crista-galli* (cockspur coral tree) is a deciduous tree, 4–10 m high, with a trunk diameter by the base of around 1 m, with a gray and rough bark. The trunk is short and reduced; the crown is wide and irregular. Branches are arched, glabrous, and with stings. Leaves are deciduous, alternate, pinnate-trifoliate, 3–11 cm in length, and 1.5–5.5 cm in width. The leaflets have entire margins. Stems and petiole could have stings and thorns. Flowers are zygomorphic, hermaphrodite, butterfly-shaped, fleshy with a red bell calix, and a corolla with five petals. One of the petals is the longest, elliptic, and 3–4 cm in length (the standard), the lateral two are smaller (the wings), and the other two are fused (the keel or *carina*) wrapping ten stamens 3.5–5.0 cm in length. That architecture gives to the flowers an aspect from which derives the name of the species, *crista-galli*. Flowers can be single or gathered in foliose branches. Fruits are curved legumes, 10–20 cm in length and 1.0–1.5 in width, dehiscent, and brown or black with lighter spots. This species has 1–6 kidney-shaped seeds per legume, 10–19 mm in length. The trees gather and grow in groups called *ceibales* in the banks of rivers and streams; they are highly tolerant to waterlogging (Larré et al. 2013). Seeds are, in those cases, transported by water. Blooming is from late spring up to mid-summer; fruit ripening starts in August (Cabrera and Zardini 1978; Lahitte et al. 1999; Díaz Cillio 2008; Sharry et al. 2011).

Ethnobotanical information Leaves, flowers, and bark are used in folk medicine. In the eighteenth century, the Jesuit monk Florian Paucke reported the use of the *ceibo* bark decoction by the *Mocovíes*, native people from SFE, to heal puma bites

(Demaio et al. 2015). Other uses are in dermatosis (to heal wounds), asthma, hemorrhoids, and also as antiseptic and narcotic. The decoction of bark is prescribed to treat urinary, vaginal, and respiratory tract infections. The decoction of leaves is taken for their attributed hypnotic, sedative, anti-hemorrhoidal, and astringent activities (Martínez Crovetto 1964; Tourkarkissian 1980; Maier et al. 1999; Etcheverry et al. 2003). An emollient syrup is made with the flowers for treating cough, colds, catarrh, bronchitis, etc. (Ratera and Ratera 1980; Barboza et al. 2009; Sharry et al. 2011; Alonso and Desmarchelier 2015).

Chemical data Erythrina isoquinoline alkaloids are distributed in all the plants (Fig. 4.4). Erythraline was determined as the major alkaloid in leaves, while in fruits was found erythraline and erythrinine. Fruit wall tissue appears to be the major site of the spirocyclic alkaloid biosynthesis (Maier et al. 1999). Seeds produced erysodine as main constituent. Also, they have erythramine, erysonine, erysopine, erysovine, and hypaphorine (Deulofeu et al. 1947; Orekhov 1955).

The essential oil (EO) obtained by hydrodistillation of fresh flowers was mainly made up of linalool (53.9%), spathulenol (9.6%), and α-terpineol (8.4%) (Fig. 4.5) (Lamarque et al. 1998).

The leaves have the isoquinolines, crystadine and crystamidine (Fig. 4.6), and oleanolic acid, erythrinine, erysodine, and erythrosine (Ito et al. 1973).

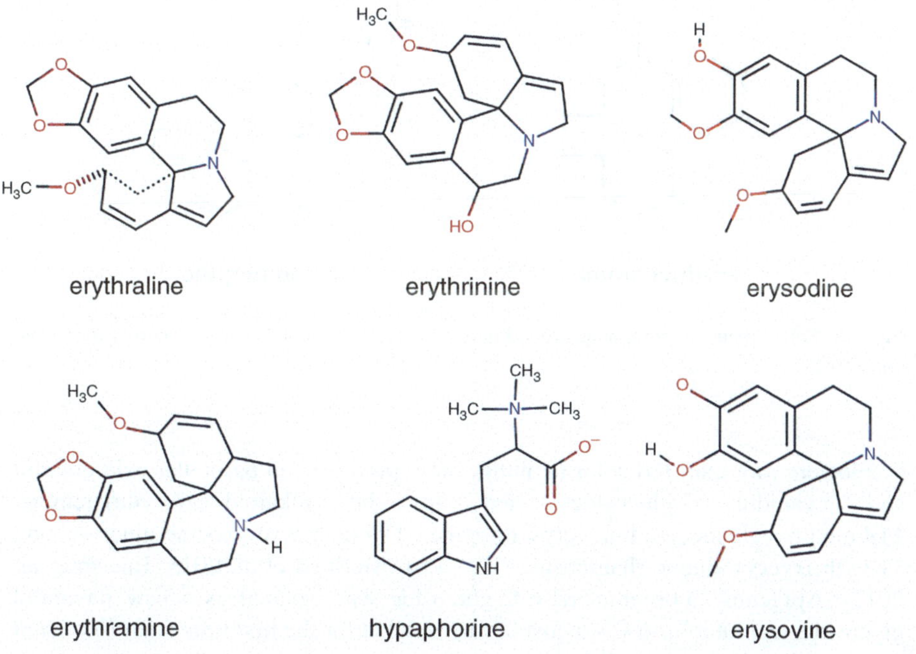

Fig. 4.4 Some isoquinoline alkaloids from *Erythrina crista-galli* L. var. crista-galli

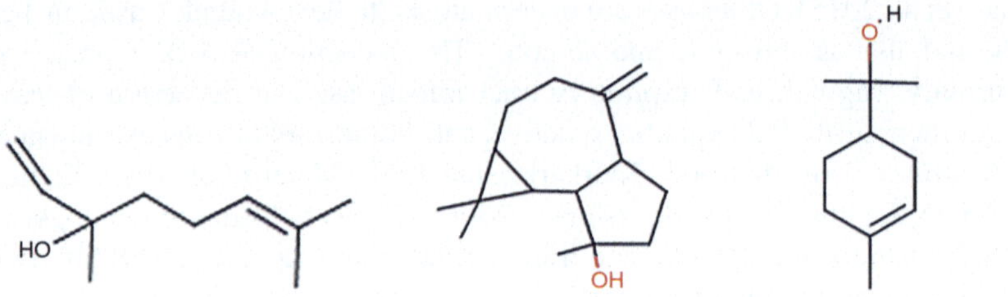

linalool spathulenol α-terpineol

Fig. 4.5 Some compounds found in the *Erythrina crista-galli* L. var. crista-galli essential oils

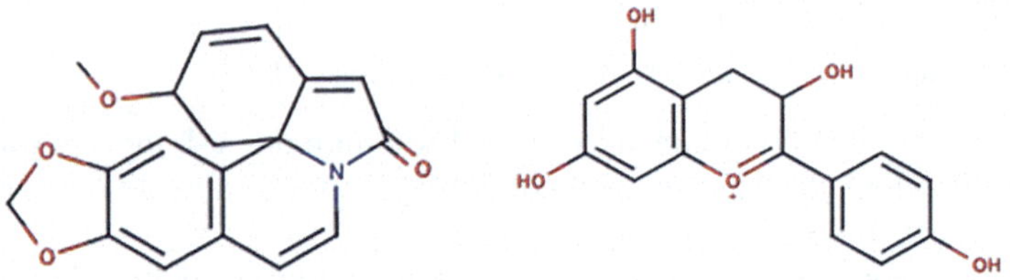

crystamidine pelargonidin

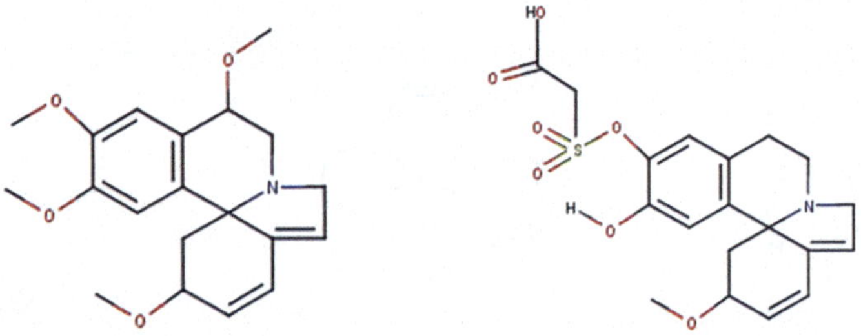

erythristemine erysothiopine

Fig. 4.6 Other chemical compounds found in leaves and flowers of *Erythrina crista-galli* L. var. crista-galli

Flowers owe their red color to anthocyanin pigments, in particular pelargonidin and cyanidin 3-glucosides and to the alkaloids erythristemine, 11*β*-methoxyglucoerysodine, erysothiopine, 11*β*-hydroxyerysodine-glucose, and 11-hydroxyerysotinone-rhamnoside (Fig. 4.6) (Merlugo et al. 2015; Enciso et al. 2017). Apigenin-7-O-rhamnosyl-6-C-glucoside was isolated as a new flavonoid glycoside, and luteolin-6-C-glucoside was isolated for the first time from *Erythrina crista-galli* (Ashmawy et al. 2013).

On the other hand, seeds have arginine, alanine, and aspartic acid and *E. crista-galli* lectins (ECA) that bind selectively to glycolipids with Galβ4GlcNAc and Fucα2Galβ4GlcNAc terminal sequences (Teneberg et al. 1994).

Biological activities The extracts from the dried aerial parts have been reported to have antibacterial, analgesic, and anti-inflammatory activities. Also, dried leaf extracts have antibacterial, psychotropic, and CNS depressor activity (Barboza et al. 2009; Alonso and Desmarchelier 2015).

The methanolic and ethanolic bark extracts have antibacterial activity against *S. aureus* and *B. subtilis* and antifungal activity against *Candida albicans* and *C. krussel*. On the other hand, extracts from the aerial part extracts were active against *B. subtilis* (Mitscher et al. 1988; Vivot et al. 2006, 2007, 2012).

As for its activity on the central nervous system (CNS), the extracts and purified fractions from leaves have a depressor effect in CNS and on the spontaneous motor activity in rats (Etcheverry et al. 2003). A sedative effect could be attributed to the isoquinoline alkaloids (Ito et al. 1973).

The aqueous leaf extracts had analgesic effect in mouse with a effect was at a dose of 1000 mg kg^{-1} and an antinociceptive effect of 53%. It was also described a curarizing effect of the alkaloids from seeds and bark (Deulofeu 1959).

The dichloromethane extract from leaves showed significant anti-inflammatory effect (51% inhibition) in mouse 3 h. post administration; also, the test of edema in mouse ears showed a 72% inhibition (Miño et al. 2002). The methanol crude extracts from leaves protected from hemolysis induced by 2,2′-azobis (2-amidinopropane) (AAPH) through inhibition of erythrocyte lipid peroxidation (Victol et al. 2015).

The *in vitro* antiparasitic activity of the ethanolic extracts from bark and leaves of *E. crista-galli* L., *E. dominguezii* Hassl., and *E. falcate* Benth. was evaluated on *Trypanosoma cruzi* trypomastigotes forms and *Leishmania amazonensis*, *L. infantum*, and *L. braziliensis* promastigotes (Yaluff et al. 2015). Against *Leishmania* strains, only *E. falcata* Benth. leaf extracts showed activity; however, bark extracts of the tree species have anti-leishmania activity. On the other hand, only *E. falcate* Benth. leaf and bark extracts showed inhibitory activity against both strains of *T. cruzi*. Another study reported antimalarial activity from three phenolic compounds (phaseollidin, sandwicensin, and lonchocarpol A) from the stem bark of *E. crista-galli* (Tjahjandarie et al. 2014).

A matrix containing *E. crista-galli* lectin (ECA) supported the undifferentiated growth of human embryonic stem cells (hESC) and human-induced pluripotent stem cells (hiPSC). Those results suggested a potential application of ECA as a matrix for culture and differentiation of human pluripotent stem cells (Mikkola et al. 2013).

The aqueous and aqueous methanol extracts and its fractions showed significant estrogenic activities evaluated by *Arabidopsis thaliana* pER8: GUS reporter assay and by their effect on the proliferation of the estrogen-responsive cell line MCF-7 (Ashmawy et al. 2013).

The aqueous extract of *Erythrina* showed cytotoxic activity against breast and lung cancer cell lines (RathiSre et al. 2015). The superior cytotoxic activity of the aqueous extract over the aqueous methanol extract is attributed to the presence of low amounts of erythrinan and benzylisoquinoline alkaloids in higher amounts (Ashmawy et al. 2013).

Toxicity Oral toxicity of the leaf infusion and the methanolic extract from the alkaline infusion in rats was negative at a dose of 2000 mg kg^{-1} (Etcheberry et al. 2004). Seeds are toxic probably as a result of its alkaloid content.

4.3 *Phytolacca dioica* L. (Phytolaccaceae)

Synonymy *Phytolacca arborea* Moq., nom.nud, *P. populifolia* Salisb., *Pircunia dioica* (L.) Moq., nom. illeg., *Phytolacca dioica* L. var. ovalifolia Chodat.

Vernacular name Ombú, umbú, calamaic, morolawén, zumbí.

General information Its massive enlarged base may grow to 4 m in diameter, spreading above the ground, making the tree looking as standing on a dune. The base and the trunk have water storage tissues accumulating up to 80% of water, which could be an excellent fire-resistant adaptation to the intense grassland fires that are common in this region. Since *P. dioica* derives from herbaceous ancestors, its trunk has an unusual secondary growth rather than true wood. As a result of its fast growth, its wood is soft and spongy. Since the sap is poisonous, it is not grazed by cattle, and it is immune to locust and other pests. *P. dioica* grows in south of Brazil, Paraguay, Uruguay, and Argentina. In Argentina it is found in the provinces of BAI, CHA, COS, DFE, ERI, FOR, MIS, SAL, SFE, SLU, and TUC. Although it is a species native from the provinces of FOR, COS, and MIS, it is characteristic from the Pampa plain. It is an ideal tree for shade, but it has to consider its huge diameter at its base (>3 m). It is used in parks and squares (Díaz Cillio 2008). The flowers are visited by Diptera and Hymenoptera during the day and by Lepidoptera in the night. In Spain the ombú is called *Bella sombra* and umbra tree in English-speaking countries.

Botanical characteristics *P. dioica* is an 8–12 m in height tree with a huge crown, a thick trunk, and large roots. As it is not ligneous, it cannot be considered as a true tree but rather a huge herbaceous plant (Chebez and Masariche 2010). Leaves are simple, fleshy, petiolate, alternate, caducous, glabrous, elliptic, or ovate, 5–13 mm in length and 2–8 cm in width, with a yellow entire margin and a pinkish petiole. Flowers are dioecious, apetalous, white or yellow, distributed in pendulous branches, 7–12 cm diameter. Male flowers have 20–30 stamens. Feminine flowers have 8–14 carpels, free in the upside during blooming. The fruit is a lobulated green-yellowish berry with one black seed (Fig. 4.7). Blooming is from November to February; fructification is from November to May (Cabrera and Zardini 1978; Lahitte et al. 1999, Alonso and Desmarchelier 2015).

Fig. 4.7 *Phytolacca dioica* L., Phytolaccaceae. Tree (**a**), leaves and fruits (**b**), pendulous flower clusters (**c**). MA Alvarez personal collection

Ethnobotanical information The vernacular name ombú is a derivation from the Guaraní name *ïmboú* that means "tree that attracts the rain." The dried entire plant is used as febrifuge, vulnerary, vermifuge, and antiseptic. The infusion made with leaves and root bark is taken as emetic, antirheumatic, antiseptic, astringent, and vulnerary. The dried leaf is considered as a purgative. Leaves are used in case of cephalgia, as wound healer and coagulant. The bark ashes are used to produce soap for their saponin content (Tourkarkissian 1980; Martínez Crovetto 1981; Quiroga et al. 2001).

Chemical data Ribosome-inactivating proteins (RIPs) are RNA N-β-glycosidases that are classified into three types (1, 2, and 3) on the basis of the number and type of their constituting polypeptide chains. *P. dioica* leaves and seeds contain at least four type 1 RIPs, namely, PD-L1, PD-L2, PD-L3, and PD-L4. The expression of RIPs in leaves has a seasonal and age regulation (Di Maro et al. 2012). Leaves also have the flavonol ombuoside (Fig. 4.8); fresh flowers have fructose. The dried seeds contain the RIP PD-S2 with a probable role as defense protein and/or as a supplier of essential amino acids during seedling development (Parente et al. 1993; Di Maro et al. 1999, 2012). Autumn leaves also have a secretory class III peroxidase belonging to the plant heme-dependent peroxidase family (Guida et al. 2011).

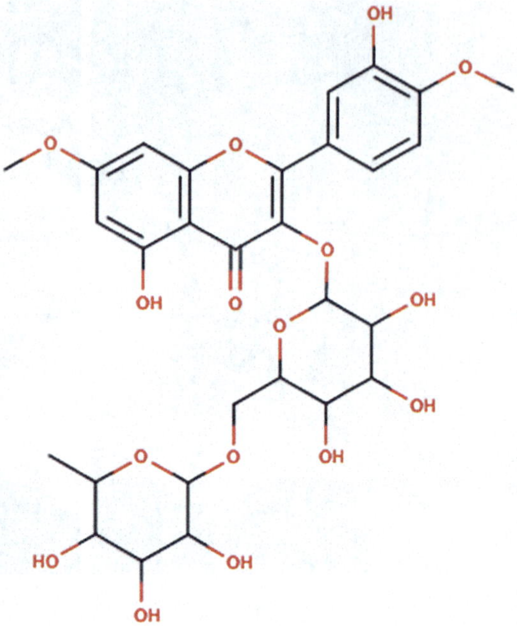

Fig. 4.8 Ombuoside from *Phytolacca dioica* L. leaves

Biological activities The extracts of the entire plant have molluscicidal, anti-inflammatory, antifungal, and antibacterial effects (Blanco et al. 1998). Also, the RIP D2 has demonstrated an *in vivo* antiviral activity in two pathogen systems, tobacco mosaic virus (TMV), *Nicotiana tabacum*, and tobacco necrosis virus (TNV), *Phaseolus vulgaris*. It was presumed that the antiviral activity is expressed only when D2 and the virus are in the same cell. Moreover, the oxidative burst in D2-treated leaves could result from the acceleration of the hypersensitive response. On the other hand, D2 cannot translocate into the phloem or induce systemic acquired resistance (SAR) signals (Faoro et al. 2009).

The acid hydrolysate from *P. dioica* berries containing triterpenoid sapogenins, such as phytolaccagenins (PG), has antifungal potency when tested against *Candida albicans* ATCC 10231, *Cryptococcus neoformans* ATCC 32264, and clinical isolates of *C. albicans* and *C. neoformans* (Di Liberto et al. 2010). PD-L4 is a type 1 RIP isolated from summer leaves; a recombinant version (rPD-L4) was tested against *E. coli*, *P. aeruginosa*, *S. aureus*, and the plant pathogen *A. tumefaciens*. Both the rPD-L4 as its denaturized version demonstrated antimicrobial activity at a concentration of 3 μM. The PD-L4 peptides obtained by chemical fragmentation were also tested. A cryptic peptide, involving residues 40–65 of full-length protein ($PDL_{440–465}$ peptide), showed a significant antimicrobial activity on the different bacterial strains tested. Since cationic antimicrobial peptides (CAMPS) are of interest, the authors tested the activity of the $PDL_{440–465}$ peptide towards HaCat and CaCo-2 human cells lines. $PDL_{440–465}$ triggered an anti-inflammatory response in lipopolysaccharide

(LPS)-treated CaCo-2 cells. By *in silico* analysis, homologous cryptic peptides in other type 1 RIPs from Phytolaccaceae were identified with a possible key role during seedling development, a process strongly susceptible to attack by pathogens (Pizzo et al. 2015). Therefore, the *Phytolacca* RIPs have a potential use in crop plant biotechnology to increase resistance to fungal and virus pathogens (Di Maro et al. 2009).

Toxicity When aqueous leaf and berry extracts (50, 100, 200 mg kg^{-1}) were administered to Wistar rats for 14 days, a progressive reduction in the serum levels of platelets, white blood cells, neutrophils, and monocytes was observed concomitantly with a dose-dependent increase of lymphocytes and a reduction of eosinophils. As for enzymes, a decrease in serum activity of alkaline phosphatase, alanine transaminase, and aspartate transaminase was reported. Those alterations were accompanied by a significant reduction in body weight, absolute organ weight, and organ-body weight ratio. All the abovementioned alterations could be an indication of toxicity; hence, their use as an oral medicine would not be safe (Ashafa et al. 2010; Ashafa et al. 2011).

4.4 *Salix humboldtiana* Willd. var. humboldtiana (Salicaceae)

Synonymy *Salix chilensis* Molina var. crispa, *Salix magellanica* Lam.

Vernacular name Sauce colorado, sauce criollo, sauce chileno, sauce isleño, sauce amargo, treique, cheique, reique, huayao.

General information *S. humboldtiana* is the only *Salix* species native from South America. It is a marsh tree widely distributed in Argentina that grows along rivers, lakes, and lagoons in the provinces of BAI, CAT, CHA, CHU, COR, COS, DFE, ERI, FOR, JUJ, LPA, MIS, NEU, RNE, SAL, SDE, SFE, SJU, and TUC. The wood is light and soft. It is used for rural constructions to make boxes, packaging, toys, oars, matches, poles, and fences (Ratera and Ratera 1980; Barboza et al. 2009; Sharry et al. 2011).

Botanical characteristics It is a dioecious species, 15–25 m in height and 40–80 cm in diameter. Its foliage is deciduous; the crown is conic or rounded with fallen foliage and pendulous branches (Fig. 4.9). The bark is grayish-brown, rough, cracking in thick and irregular pieces. Leaves are simple, light green, alternate, 3–10 cm in length, and 0.6–0.8 cm in width; they are lanceolate, glabrescent, with a toothed-serrate margin. Petiole is 0.5–1.0 cm in length; leaves and petiole are glabrescent. Inflorescences are pendulous terminal aments, 4–10 cm in length. The masculine flowers are green-yellowish; the 5 mm in length calix has six unequal stamens sustained by lanceolate-ovate bracts. The feminine flowers are green; the calix is 2–3 mm in length, with a superior unilocular ovary, with multiple ovules.

Fig. 4.9 *Salix humboldtiana* Willd. var. humboldtiana. Tree (**a**) and leaves (**b**). MA Alvarez personal collection

Fruits are small brown-greenish ovoid capsules, dehiscent by 2–4 valves, with many seeds surrounded by copious white cottony hairs (Di Sapio and Gattuso 1992; Cané 2016; Agüero 2017).

Ethnobotanical information The vernacular name *huayao* derives from *wayaw* in *Quechua* language. The bark is used as a substitute of quinine. Also, it is referred the use of the bark stem against influenza and dermatosis and as antirheumatic, antispasmodic, sedative, febrifuge, astringent, and tonic (Tourkarkissian 1980; Di Sapio and Gattuso 1992; Líate and Hurrell 1994; Barboza et al. 2009; Sharry et al. 2011). A preparation of boiled leaves and pieces of bark is administered to treat diarrhea (Muñoz et al. 2000; Bourdy et al. 2004). The ashes from the bark are used to treat dermic affections. As a shampoo, it is used as a hair revitalizing (Dimitri et al. 1997; Lahitte 1997).

Chemical data The leaves have flavonoids, coumarin, tannins, salicylic acid, and the alkaloid salicine (Fig. 4.10) (Líate and Hurrell 1994).

Biological activities The antimicrobial activity of the aqueous extract of aerial parts against *Brucella canis* and *Staphylococcus aureus* was reported (Toso et al. 2006). *S. humboldtiana* leaf extract showed stronger activity against *Plasmodium vinckei in vitro* (99% inhibition at 10 mg ml^{-1}) than *in vivo* (32% at 1000 mg kg^{-1}) (Muñoz et al. 2000). The bitter bark has febrifuge and tonic, sedative, and antispasmodic activities. Salicine from bark has antirheumatic properties (Demaio et al. 2015, Medina et al. 2015). The well-known salicylate derivatives responsible for the febrifuge and anti-rheumatism activity contained in the genus *Salix* justify the traditional uses of this species.

Fig. 4.10 Chemical structure of the alkaloid salicine from *Salix humboldtiana* Willd. var. humboldtiana

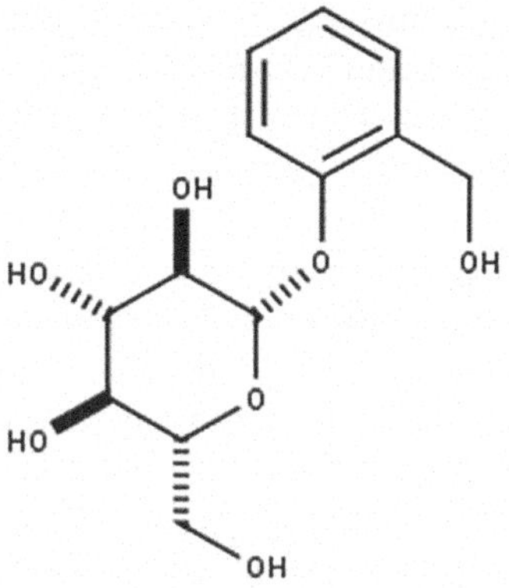

4.5 *Grindelia pulchella* Dunal var. discoidea (Hook. & Arn.) A. Bartoli & Tortosa (Asteraceae)

Synonymy *Grindelia anomala* DC., *Grindelia discoidea* Hook. & Arn.

Vernacular name Quilcha amarilla, botoncillo, pichana amarilla.

General information *G. pulchella* produces a crude resin similar to that obtained with *G. camphorum* in the Arizona desert (Tortosa et al. 2000); henceforth, the use of *Grindelia* resins as alternatives to nonrenewable petroleum-based substances was suggested (Timmermann and Hoffmann 1985). *G. pulchella* grows in the provinces of BAI, CHU, COR, DFE, ERI, LRI, MEN, SDE, SLU, and TUC (Barboza et al. 2009).

Botanical characteristics *G. pulchella* is a shrub, 25–70 cm high, with lanceolate or oblong alternate leaves 1.5–5.5 cm long, glabrous, toothed, with long petioles. The apical capitula (heads) have dimorphic yellow flowers, the marginal flowers are feminine, ligulate (sometimes absent), and the central flowers are tubular and hermaphrodite. The receptacle is glabrous; fruits are thick smooth or slightly striated glabrous achenes. *G. pulchella* has a characteristic secreting system composed of glandular trichomes in the epidermis of stems, leaves, and involucral bracts (Cabrera and Zardini 1978; Bartoli et al. 2011).

Ethnobotanical information Most of the *Grindelia* species are used in folk medicine as antispasmodic and diuretic. The aerial parts are used as febrifuge and antirheumatic (Barboza et al. 2009).

Chemical data The essential oils from the aerial parts contain the sesquiterpenes aromadendrene, α-bisabolene, and γ-cadinene; the monoterpenes bornyl acetate, camphene, and carvone; the diterpene cordobic acid; and the flavonol pachypodol (Barboza et al. 2009). Also *G. pulchella* produces grindelic acid and resins composed of a mixture of labdane diterpenoids (Figs. 4.11 and 4.12) (Timmermann and Hoffmann 1985).

Biological activities Grindelic acid possesses antimicrobial activity towards several phytopathogenic fungal and bacterial strains (Hernández et al. 2005). Grindelic acid hydroxylated derivatives, the acids 6-β-hydroxygrindelic, 6-α- hydroxygrindelic, and 3-β-hydroxygrindelic, were successfully obtained by fungal bioconversion from grindelic acid (Hernández et al. 1997, 2002; Orden et al. 2005) and showed bioactive properties towards *Tenebrio molitor* larvae species and phytopathogen fungi and bacteria. Also, the ethanolic and hydro-methanolic extracts from aerial parts showed feed-deterrent effects (200 μg cm^{-2}) on *Epilachna paenulata* Germ (Coleopteran-Coccinellidae) larvae (IA 23.1 and 80%, respectively). The aqueous, methanolic, and hexane extracts inhibited *Avena sativa* germination and to a lesser extent *Raphanus sativus* germination. The aqueous extract (2 mg ml^{-1}) has antibacterial activity (MIC > 4) against *E. coli* ATCC 25922 (Palacios et al. 2007).

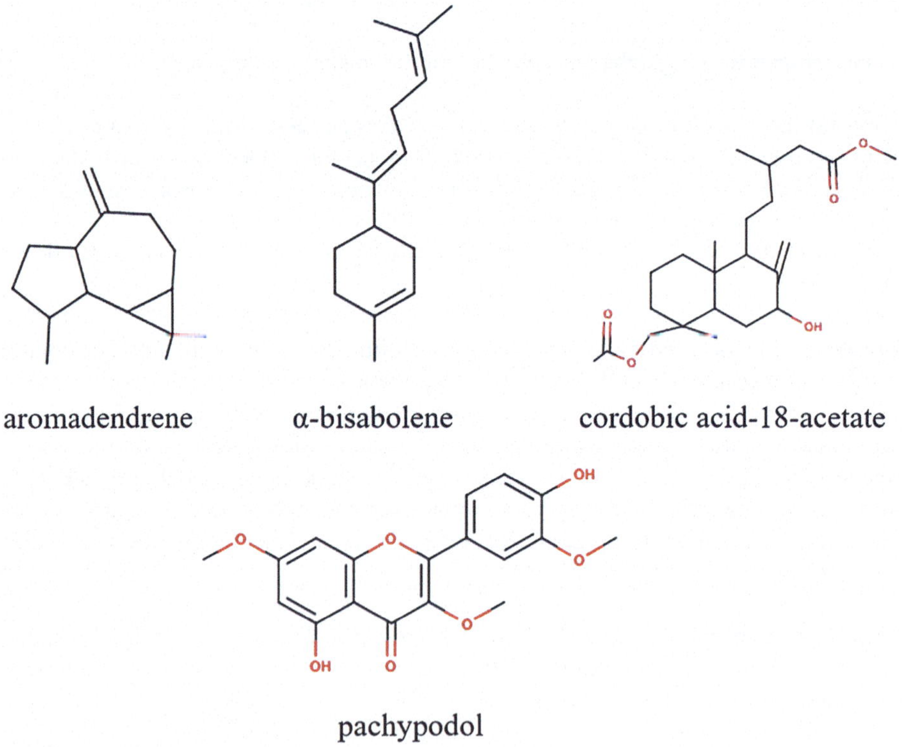

Fig. 4.11 Chemical compounds from *Grindelia pulchella* Dunal var. discoidea (Hook. & Arn.) A. Bartoli & Tortosa essential oils

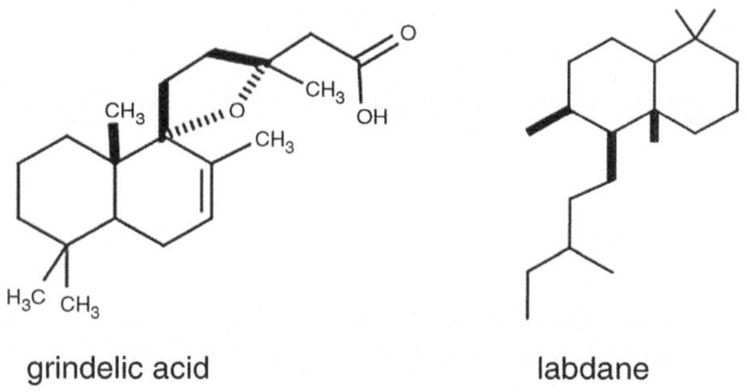

Fig. 4.12 Grindelic acid and labdane from *Grindelia pulchella* Dunal var. discoidea (Hook. & Arn.) A. Bartoli & Tortosa

***In vitro* cultures** Calli cultures were established from leaves of *G. pulchella* seedlings in MS media (Murashige and Skoog 1962) with indolbutiric acid (IBA) and benzylamine purine (BAP) at different concentration relationship (Hernández et al. 2005). Grindelane diterpenes were produced in all the callus lines established, while grindelic acid was only produced in the lines growing in MS media with 20.0 μM IBA and 18.0 μM BAP as plant growth regulators. Cell suspension cultures were initiated from friable calli and maintained in the same conditions on rotary shaker. However, they only produced grindelic acid that was released to the culture media. When elicitation was performed with $CuSO_4$ (1.0–2.0 mM) and DMSO (1 μl ml^{-1} culture), the terpenoid compound production increased at expense of cell growth (Hernández et al. 2000, 2005).

4.6 *Larrea* spp. (Zygophyllaceae)

Larrea spp. are widespread distributed along the country (Pampa plains, Monte, Chaco, Patagonia) being generally known as *jarrillas* (Morello 1958; Ezcurra et al. 1991). Larrea species inhabit arid ecosystems, with low temperature, or temperature fluctuations, low absolute humidity, and high solar radiation (Alonso and Desmarchelier 2015). Moreover, isolated populations of *L. nitida* and *L. divaricata* occur in arid and semiarid areas of Chile, Bolivia, and Peru (Hunziker et al. 1972). The infusions prepared with *jarrillas* have a strong bitter taste and a typical odor.

Botanical Characteristics

Larrea species rarely exceed 2–5 m height and can be differentiated by their leaf characters. *Larrea nitida* has compound leaves, while *L. divaricata* and *L. cuneifolia* have only two leaflets with a small filiform mucro. Flowers are solitary and axillary, with five partially fused green sepals and five imbricated-free yellow petals. Filaments are partially fused with a basal scale, forming a cup-shaped receptacle where nectar is accumulated. The penta-carpellate, globose, and pubescent ovary is superior, with a basal nectariferous tissue. The style and the stigma are simple in all the species; the presence of an early female stage (protogyny) has been reported (Simpson et al. 1977). The fruit is a schizocarp dehiscent into five one-seeded mericarps (Rossi et al. 1999).

4.6.1 Larrea cuneifolia *Cav.*

Synonymy *Neoschroetera cuneifolia* (Cav.) Briq., *Schroeterella cuneifolia* (Cav.) Briq.

Vernacular name Jarilla, jarilla macho, jarilla crespa, jarilla nortesur.

General information It grows in the provinces of BAI, CAT, CHU, COR, LPA, LRI, MEN, NEU, RNE, SAL, SDE, SJU, SLU, and TUC. Stems are used to roof construction and also as fuel. Leaves are used as tinctorial (Ratera and Ratera 1980; Ladio and Lozada 2009).

Ethnobotanical information It is widely used as anti-inflammatory; for externally treating wounds; as diaphoretic, oxytocic, and emmenagogue; and for footbath and rheumatism (Ratera and Ratera 1980; Barboza et al. 2009; Ladio and Lozada 2009).

Chemical data The dried leaves have 11 flavonoid aglycones (Fig. 4.13): quercetin 3,7,3′,4′-tetramethyl ether (retusine), quercetin 3,7-dimethyl ether, quercetin 3,3′-dimethyl ether, quercetin 7, 3′-dimethyl ether (rhamnazin), quercetin 3′-methyl ether (isorhamnetin), kaempferol 3,7-dimethyl ether (kumatakenin), kaempferol 3-methyl ether (isokaempferide), kaempferol ‘I-methyl ether (rhamnocitrin), kaempferol, luteolin 7,3′-dimethyl ether (velutin), luteolin 3′-methyl ether (chrysoeriol), apigenin 7-methyl ether (genkwanin), apigenin, and dihydromyricetin 3′,5′-dimethyl ether (dihydrosyringetin) (Valesi et al. 1972). Also, the flavonols quercetin 3,7,3′-trimethyl ether and quercetin 7,3′,4′-trimethyl ether were determined (Sakakibara et al. 1976). Also, the leaves have the protein asparagine (Valesi et al. 1972).

Biological activities *L. cuneifolia* Cav. methanol and chloroform extracts have larvicidal activity against *Culex quinquefasciatus* larvae (Batallán et al. 2013). The ethanol extract has antibacterial activity against Gram-positive (Amani et al. 1998; Quiroga et al. 2001) and Gram-negative bacteria (Zampini et al. 2007). The aqueous extracts have antioxidant properties (Carabajal et al. 2017). *L cuneifolia* Cav. infusion (0.1 g ml^{-1}), decoction (0.1 g ml^{-1}), and tincture (0.1 g ml^{-1} alcohol 96°) were tested against *Fusarium subglutinans*, *F. thapsinum*, *F. bothii*, and *F. graminearum* sensu stricto. All the strains showed a greater susceptibility to tinctures, except in the case of *F. thapsinum* that was mostly inhibited by the infusion. Decoctions were more active than infusions against *F. subglutinans*, *F. bothii*, and *F. graminearum* sensu stricto (Jiménez and Sgariglia 2010). Ethanolic extracts of leaves and bark were tested against *Salmonella enteritidis* (Gram-negative), *Enterococcus faecalis* (Gram-positive), and two strains of agricultural relevance, *Xanthomonas axonopodis* pv. *phaseoli* (Gram-negative) and *Clavibacter michiganensis* (Gram-positive).

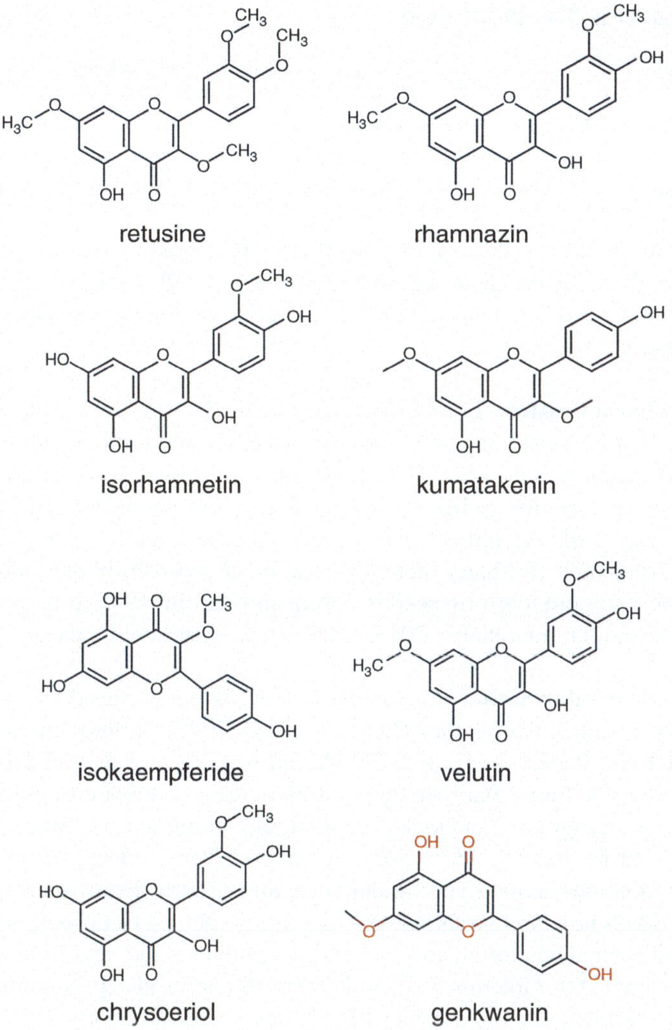

Fig. 4.13 Some of the flavonoids found in *Larrea divaricata* Cav. leaves

S. enteritidis was inhibited with all the extracts (MIC 106 to 220 μg ml^{-1}). Leaf extracts were active against *C. michiganensis*, *E. faecalis*, and *S. enteritidis*. The phytopathogenic bacteria were more resistant (MIC 216–228 μg ml^{-1}) (Lorenzo et al. 2018). A patent (KR101348471B1) was published with a formulation based on *L. cuneifolia* Cav. for preventing and treating estrogen-related menopausal symptoms (hot flashes, osteoporosis, venous thrombosis, and any of the symptoms of atrophic vaginitis).

4.6.2 Larrea divaricata *Cav.*

Synonymy *Schroeterella divaricata* (Cav.) Briq., *Neoschroetera divaricata* (Cav.) Briq., *Larrea monticelli* Perrone & Caro.

Common name Jarilla, jarilla hembra, chamanilla, jarilla del cerro, yarilla.

General information *L. divaricata* is widespread distributed in savannas along the country, mostly in the provinces of BAI, CAT, CHU, COR, LPA, LRI, MEN, NEU, RNE, SAL, SDE, SJU, SLU, and TUC. It is a nectar-producing xenophile bush that attracts bees.

Botanical characteristics *L. divaricata* is a bush, 2–3 m high, with woody and flexible stems. Leaves are opposite, waxy, coriaceous, and bifoliate, with the knife-shaped leaflets united at the base. The small yellow solitary flowers are located in the foliar axils. They have five pubescent sepals, five yellow petals, ten reddish stamens distributed in two whorls, dithecous dehiscent anthers, and a five-carpel ovary. The spheroidal fruits have five hairy mericarps that break at maturity into units containing each one a smooth reniform seed (Cabrera and Zardini 1978; Ratera and Ratera 1980; Alonso and Desmarchelier 2015; Cané 2016; Bernasconi Salazar et al. 2017).

Ethnobotanical information *L. divaricata* Cav. is widely used in Argentine in folk therapy as anti-inflammatory (Del Vitto et al. 1997) or antirheumatic (Ratera and Ratera 1980; Pedernera et al. 2006). Aerial parts (leaves and tender branches) are taken as a tea as anti-inflammatory; for treating external wounds; as diaphoretic, febrifuge, emmenagogue, oxytocic, odontalgic, antitussive, antirheumatic, and hypotensive; or for treating and preventing varicose veins. Also, it is used as rubefacient, for footbaths, as antihyperlipidemic, against gastric disturbances, and peptic ulcer. For its balsamic properties is taken to reduce mucus. The decoction with addition of salt is a good deodorant and foot antiperspirant (Mabry and Bohnstedt 1997; Tourkarkissian 1980; Goleniowski et al. 2006; Stege et al. 2006; Barboza et al. 2009; Ladio and Lozada 2009; Agüero 2017; Sen and Chakraborty 2017).

Chemical data It was reported the presence of the nontriterpenoid larreagenine, nordihydroguaiaretic acid (NDGA), α-thujene, myrcene, limonene, E-ocimenone, thymyl acetate, γ-elemene, spathulenol, β-oplopenone, α-cadinol, eugenol, carvacrol, limonene, and thymoquinone (Waller and Gisvold 1945; Boiteux et al. 2018). Also, apigenine-7-methylether, 3,4′-dihydroxy-3′,4-dimethoxy-6,7′-cyclolignan and aromadendrene oxide were identified (Vogt et al. 2013). The resin that covers the leaves contains flavonoid aglycones, several lignans, and NDGA (Saigler et al. 1974; Barboza et al. 2006) (Fig. 4.14).

Biological activities Extracts of *L. divaricate* display immunomodulatory (Franchi-Micheli et al. 1986; Schreck et al. 1992; Anesini et al. 1996a), antimicro-

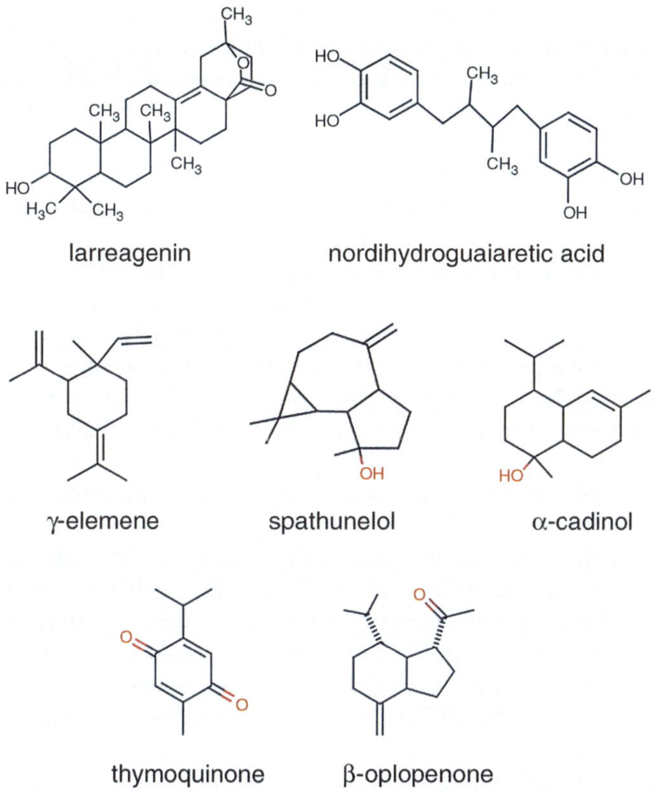

Fig. 4.14 Chemical compounds from *Larrea divaricata* Cav.

bial (Quiroga et al. 2001; Stege et al. 2006), antitumoral (Anesini et al. 1996b, 1999), and anti-ulcerogenic activities (Gisvold and Thaker 1974; Anesini and Perez 1993; Amani et al. 1998; Quiroga et al. 2001, 2004; Pedernera et al. 2006; Davicino et al. 2007).

The aqueous extract has tumorstatic capacity (Anesini et al. 1997); the methanol and dichloromethane extracts have *in vitro* cytotoxicity (Bongiovanni et al. 2007).

The anti-inflammatory activity, attributed to NDGA, was displayed in cotton pellet granuloma and adjuvant induced arthritis models in rats. The methanolic extract has anti-ulcerogenic and anti-inflammatory effect in ethanol- and HCl-induced ulcers in rats. In the first case, the inhibition was dose dependent. Moreover, the cold aqueous extract, infusion, and decoction had inhibitory activity at 0.04–0.1 mg l^{-1} against clarithromycin and metronidazole susceptible and resistant *Helicobacter pylori* strains, which was attributed to flavonoid activity (Stege et al. 2006). Those results validate the popular use of *L. divaricata* Cav. as an anti-inflammatory agent with gastric anti-ulcerogenic properties (Franchi-Micheli et al. 1986; Schreck et al. 1992; Pedernera et al. 2006).

The leaf aqueous extract was found to have both proliferative (at low concentrations) and antiproliferative action on murine lymphoma cell line (BW 5147). The proliferative effect was related to leukotriene modulation, while the antiproliferative effect was related to the modulation of both hydrogen peroxide and nitric oxide. The same extract induced proliferation in a murine lymphocytic leukemia cell line (El-4) by the presence of nordihydroguaiaretic acid (NDGA) (Davicino et al. 2011a, b). Hence, a fraction with low NDGA quantity was proposed as a potential therapy for lymphoma and leukemia treatment (Martino et al. 2013).

The antimicrobial and antimitogenic activity of the aqueous extracts was explained by interference on the arachidonic acid metabolism via lipoxygenase (Anesini and Perez 1993; Anesini et al. 1999). The extracts *in vitro* inhibited the transcription and/or replication of the viruses HIV-1, several herpes types, and *Junín* (Garcia et al. 2003; Konigheim et al. 2004; Bernasconi Salazar et al. 2017). On the other hand, immunization with *L. divaricata* crude extracts triggers the production of antibodies that cross-react with proteins from Gram-negative bacteria including *Pseudomonas aeruginosa* favoring its opsonophagocytosis by murine macrophages, which was found promissory for developing a potential vaccine against *P. aeruginosa* (Canale et al. 2018). An association of *L. divaricata* Cav. and *Coffea arabica* extracts is commercialized for hair growth. It was reported that the spry formulation improved the overall hair volume and appearance, increased hair thickness, induced hair growth, and decreased hair loss without adverse local reactions in patients with no cicatricial alopecia after a 3-month treatment. The amount of dandruff decreased, but it was attributed to the inhibitory effect of *C. arabica* on *Malassezia furfur* (Alonso and Anesini 2017).

The ethanolic and chloroformic extracts of aerial parts showed *in vitro* antifungal activity against *P. notatum*, *Lenzites elegans*, *Schizophyllum commune*, *Trichoderma* spp., *Ganoderma applanatum*, *Pycnoporus sanguineus*, *Fusarium oxysporum*, and *A. niger* and the yeasts *S. carlsbergensis* and *Rhodotorula* spp. (Quiroga et al. 2001). The chloroformic extract showed the highest antifungal activity, attributed to apigenine-7-methylether, NDGA, and 3,4′-dihydroxy-3′,4-dimethoxy-6,7′-cyclolignan being this last compound the most active both *in vitro* (*Fusarium graminearum* MIC 15.6 $\mu g\ ml^{-1}$) as *in vivo*. The authors proposed the management of wheat seedling blight and crown rot by using *L. divaricata* Cav. formulations (Vogt et al. 2013). Finally, *L. divaricata* extracts have also shown activity against the phytopathogens *Monilinia fructicola*, the most destructive pre- and postharvest pathogen in stone fruit worldwide (Boiteux et al. 2018), *Botrytis cinerea* (Hapon et al. 2017), *Fusarium graminearum*, *F. solani*, *F. verticillioides*, and *Macrophomina phaseolina* (Vogt et al. 2013).

***In vitro* culture** *In vitro* cultures of *L. divaricata* were established (Palacio et al. 2006). The addition of precursors (l-phenylalanine, cinnamic acid, ferulic acid, and sinapic acid) affected the yields of the compounds of medicinal interest (NDGA, p-coumaric acid, ferulic acid, and sinapyl alcohol). Feeding l-phenylalanine at 0.5, 1.0 and 3.0 mM concentration resulted in an increase of NDGA of up to 301.35 ± 1.19, 285.23 ± 28.44, and 190.53 ± 19.50 $\mu g\ g^{-1}$ DW, respectively. At a concentration of 3 mM, l-phenylalanine increased p-coumaric acid concentration from 47.43 ± 9.01 to 1157.28 ± 47.79 μg g^{-1} DW. The addition of 0.5 μM cinnamic acid enhanced growth but not NDGA yields. The addition of cinnamic acid (1 and 1.5 μM), ferulic acid (0.1, 0.5, and 1 mM), and sinapic acid (0.1, 0.5, and 1 mM) resulted toxic for the cells and the production of NDGA and the phenylpropanoids were suppressed. The sinapyl alcohol content was not affected by any of the precursors tested (Palacio et al. 2011). Calli, both with and without organogenesis, produced NDGA and quercetin, as did plantlet and wild plants. Cell suspension cultures produced NDGA, p-coumaric acid, ferulic acid, and sinapyl alcohol. When undifferentiated cultures were induced to produce adventitious shoots, the phenolic production increased significantly. The capacity of undifferentiated tissues to produce secondary metabolites is very limited, but when the undifferentiated cultures underwent organogenesis, the yields usually increased (Alvarez 2014). However, plantlets regenerated from adventitious shoots of *L. divaricate* calli did not show the same phenolic pattern as wild plants, with 3.6- and 5.9-fold lower levels of NDGA and quercetin, respectively (Palacio et al. 2012).

4.6.3 Larrea nitida *Cav.*

Synonymy *Larrea balsamica* (Molina) I.M. Johnst., *Covillea nitida* (Cav.) Vail, *Mimosa balsamica* Molina.

Common name Jarilla crespa, jarilla de la sierra, jarilla fina.

General information *L. nitida* mainly grows in the pre-Andes mountainous area from the province of SAL to the North to the province of CHU to the South. It could be found in the provinces of BAI, CAT, CHU, LPA, LRI, MEN, NEU, RNE, SAL, and SJU (Kiesling 2003; Agüero et al. 2011).

Ethnobotanical information The infusion of the dried leaf and stem is used as emmenagogue. A not specified part is referred to as anti-inflammatory, choleretic, antiperiodic, emmenagogue, sudorific, stimulant, balsamic, and vulnerary (Tourkarkissian 1980; Ladio and Lozada 2009; Barboza et al. 2009).

Chemical data In the aerial parts—2′4′-dihydroxychalcone, quercetin-3,3′,7-trimethyl ether, NDGA, 3′ methyl-nordihydroguaiaretic acid (MNDGA), 4-[4-(4-hydroxy-phenyl)-2,3-dimethyl-butyl]-benzene-1,2-diol, norisoguaiacine, and ferulic acid (Fig. 4.15). Two epoxylignanes were isolated: compound 4 meso-(7S,8S,7′R,8′R)-3,4,3′,4′-tetrahydroxy-7,7′-epoxylignan and compound 5 (7S,8S,7′S,8′S)-3,3′,4′-trihy droxy-4-methoxy-7,7′-epoxylignan. In the leaves: asparagine. In the volatile compounds from aerial parts and propolis predominates monoterpenoid compounds, mainly o-cymene and limonene (Agüero et al. 2011).

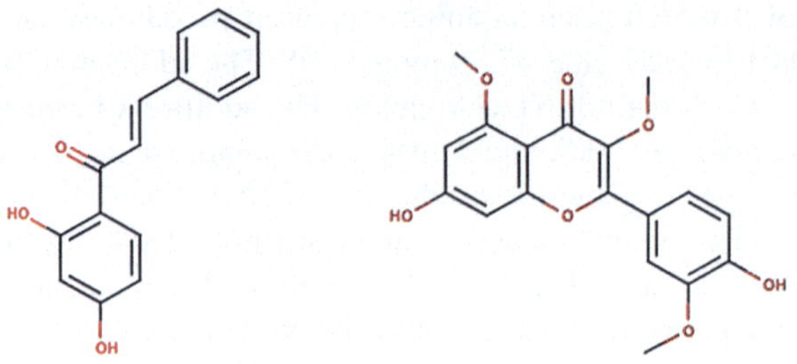

2',4'-dihydroxychalcone quercetin 3,5,3'-trimethyl ether

Fig. 4.15 Chemical compounds from *Larrea nitida* Cav. aerial parts

Biological activities Leaf and stem extracts showed antitumoral activity. Twigs and leaf resin have antioxidant activity. Propolis and extracts from aerial parts of *L. nitida* Cav. growing in the mountain valleys of Iglesia and Calingasta (SJU) have antifungal activity attributed to MNDGA, NDGA, and (4-[4-(4-hydroxy-phenyl)-2,3-dimethyl-butyl]-benzene-1,2-diol) (Agüero et al. 2011).

References

Abeddini W, Boeri P, Marinucci L, Ruscitti M, Scelzo L (2000) Biotécnicas aplicadas a especies forestales nativas. Invest Agr Sist Recur For 9(1):31–43

Agüero E (2017) Plantas medicinales silvestres del centro de Argentina. Guía para su reconocimiento y uso terapéutico. Edit. Ecoval, Buenos Aires, Argentina

Agüero MB, Svetaz L, Sánchez M, Luna L, Lima B, López ML, Zacchino S, Palermo J, Wunderlin D, Feresin GE, Tapia A (2011) Argentinean Andean propolis associated with the medicinal plant *Larrea nitida* Cav. (Zygophyllaceae). HPLC–MS and GC–MS characterization and antifungal activity. Food Chem Toxicol 49:1970–1978

Alonso MR, Anesini C (2017) Clinical evidence of increase in hair growth and decrease in hair loss without adverse reactions promoted by the commercial lotion ECOHAIR®. Skin Pharmacol Physiol 30:46–54

Alonso J, Desmarchelier C (2015) Plantas medicinales. Bases científicas para su aplicación en atención primaria de la salud. Ed. Corpus. Buenos Aires, Argentina

Alvarez MA (2014) *In vitro* plant cell cultures as biofactories. In: Plant Biotechnology for Health. Switzerland, Springer, pp 33–60

Amani SM, Isla MI, Vattuone MA, Poch MP, Cudmani NG, Sampietro AR (1998) Antimicrobial activities in some Argentine medicinal plants. Acta Hortic 501:115–122

Anesini C, Perez C (1993) Screening of plants used in Argentine folk medicine for antimicrobial activity. J Ethnopharmacol 39:119–128

Anesini C, Genaro A, Cremaschi G, Sterin Borda L, Cazaux C, Borda E (1996a) Immunomodulatory activity of *Larrea divaricata*. Fitoterapia 67:329–334

Anesini C, Genaro A, Cremaschi G, Zubillaga M, Boccio J (1996b) "*In vivo*" and "*in vitro*" antitumoral action of *Larrea divaricata* Cav. Acta Physiol Pharmacol Ther Latinoam 46:33–40

Anesini C, Boccio J, Cremaschi G, Genaro A, Zubillaga M, Borda LS, Borda ES (1997) *In vivo* antitumoural activity and acute toxicity study of *Larrea divaricata* Cav. extract. Phytother Res 11:521–523

Anesini C, Genaro A, Cremaschi G, SterinBorda L, Borda E (1999) Antimitogenic effect of *Larrea divaricata* Cav.: participation in arachidonate metabolism. Comp Biochem Physiol C Pharmacol Toxicol Endocrinol 122:245–252

Ashafa AOT, Sunmonu TO, Afolayan AJ (2010) Toxicological evaluation of aqueous leaf and berry extracts of *Phytolacca dioica* L. in male Wistar rats. Food Chem Toxicol 48:1886–1889

Ashafa AOT, Sunmonu TO, Afolayan AJ (2011) Effects of leaf and berry extracts of *Phytolacca dioica* L. on haematological and weight parameters of Wistar rats. Afr J Pharm Pharmacol 5(2):150–154

Ashmawy NS, Ashour ML, Wink M, El-Shazly M, Chang FR, Swilam N, Ab-del-Naim AB, Ayoub N (2013) Polyphenols from *Erythrina crista-galli*: structures, molecular docking and phytoestrogenic activity. Molecules 21:726. https://doi.org/10.3390/molecules21060726

Barboza GE, Cantero JJ, Núñez CO, Ariza Espinar L (2006) Flora medicinal de la Provincia de Córdoba (Argentina). Museo de Botánica, UNC, Córdoba, pp 1168–1171

Barboza GE, Cantero JJ, Núñez C, Pacciaroni A, Ariza Espinar L (2009) Medicinal plants: a general review and a phytochemical and ethnopharmacological screening of the native Argentine Flora. Kurtziana 34(12):7365

Bartoli A, Galati BG, Tortosa RD (2011) Anatomical studies of the secretory structures: glandular trichomes and ducts, in *Grindelia pulchella* Dunal (Astereae, Asteraceae). Flora 206:1063–1068

Batallán G, Torre R, Flores F, Konigheim B, Ludueña-Almeida F, Tonn C, Contigiani M, Almirón W (2013) Larvicidal activity of crude extracts from *Larrea cuneifolia* (Zygophyllaceae) and of its metabolite nordihydroguaiaretic acid against the vector *Culex quinquefasciatus* (Diptera: Culicidae). Rev Soc Bras Med Trop 46:84–87

Bernasconi Salazar JR, Karlin M, Arnulphi AS, Karlin UO, Accietto RH, Cora A (2017) Plantas del centro de la República Argentina: guia de campo, 1ª edn. Ecoval editorial, Córdoba

Blanco FDV, Cafaro V, Di Maro A, Scognamiglio R, Siniscalco G, Parente A, Di Donato A (1998) A recombinant ribosome-inactivating protein from the plant *Phytolacca dioica* L. produced from a synthetic gene. FEBS Lett 437:241–245

Boiteux J, Monardez C, Fernández MA, Esoino M, Pizzuolo P, Silva MF (2018) *Larrea divaricata* volatilome and antimicrobial activity against *Monilinia Fructicola*. Microchem J 142:1–8

Bongiovanni G, Cantero J, Eynard A, Goleniowski M (2007) Organic extracts of *Larrea divaricata* Cav. induced apoptosis on tumoral MCF7 cells with a higher cytotoxicity than nordihydroguaiaretic acid or paclitaxel. J Exp Ther Oncol 7:1–7

Bourdy G, Chávez de Michel LR, Roca-Coulthard AR (2004) Pharmacopoeia in a shamanistic society: the Izoceño-Guaraní (Bolivian Chaco). J Ethnopharmacol 91:189–208

Cabrera A L (1976) Regiones fitogeográficas argentinas. In: W F Kugler (ed.). Enciclopedia Argentina de Agricultura y Jardinería, 2nd edition. Acme, Buenos Aires 2 (1): 1–85

Cabrera AL, Zardini EM (1978) Manual de la flora de los alrededores de Buenos Aires, 2nd edn. Editorial ACME SACI, Buenos Aires

Canale FP, Valle Dávila S, Sasso CV, Pellarín NW, Mattar Domínguez MA (2018) Immunization with *Larrea divaricata* Cav. proteins elicits opsonic antibodies against Pseudomonas aeruginosa and induces phagocytic activity of murine macrophages. Microb Pathog 118:257–267

Cané L (2016) Plantas nativas. Las especies y su cultivo. Ediciones Jardín, Edit. Catapulta, Buenos Aires

Carabajal MPA, Isla MI, Zampini IC (2017) Evaluation of antioxidant and anti-mutagenic activity of herbal teas from native plants used in traditional medicine in Argentina. S Afr J Bot 110:258–265

Chebez JC, Masariche M (2010) Nuestros árboles- De norte a sur, descubriendo los árboles de la Argentina. Ed. Albatros, Buenos Aires

Davicino R, Mattar MA, Casali YA, Correa SG, Pettenati EM, Micalizzi B (2007) Actividad antifungica de extractos de plantas usadas en medicina popular en Argentina. Rev Peru Biol 14:247–251

Davicino R, Genaro AM, Cremaschi G, Anesini C (2011a) Leukotrienes antagonize the antiproliferative effect of *Larrea divaricata* Cav. on a lymphoma cell line interfering with cAMP intracellular level and PKC activity. Cancer Investig 29(1):29–36

Davicino R, Alonso R, Anesini C (2011b) *In vivo* and *in vitro* activity of *Larrea divaricata* Cav. on EL-4 cells. Hum Exp Toxicol 30:965–971

Del Vitto LA, Petenatti EM, Petenatti ME (1997) Herbal resource of San Luis (Argentina) First part: native plants. Multequina 6:49–66

Demaio P, Karlin UO, Medina M (2015) Árboles nativos de Argentina. vol 1: Centro y Cuyo. Ecoval Edit. Córdoba, Argentina

Deulofeu V (1959) The curarizing alkaloids of Erythrina species. In: Bovet D, Bovet Nittland F, Marini Bettolo GB (eds) Curare and Curare-like agents. Elsevier, Amsterdam

Deulofeu V, Labriola R, Hug E, Fondovila M, Kaufmann A (1947) Studies on argentine plants. VIII. The alkaloids of *Erythrina crista-galli*. Chromatographic separation of erythrinine and erysodine. J Org Chem 12:486–489

Di Liberto M, Svetaz L, Furlán RLE, Zacchino AS, Delporte C, Novoa MA, Asencio M, Cassels BK (2010) Antifungal activity of saponin-rich extracts of *Phytolacca dioica* and of the sapogenins obtained through hydrolysis. Nat Prod Commun 5(7):1013–1018

Di Maro A, Valbonesi P, Bolognesi A, Stirpe F, De Luca P, Gigliano GS, Gaudio L, DelliBovi P, Ferranti P, Malorni A, Parente A (1999) Isolation and characterization of four type-1ribosome-inactivating proteins, with polynucleotide: adenosine glycosidase activity, from leaves of *Phytolacca dioica* L. Planta 208:125–131

Di Maro A, Chambery A, Carafa V, Constantini S, Colonna G, Parente A (2009) Structural characterization and comparative modeling of PD-Ls 1e3, type 1 ribosome-inactivating proteins from summer leaves of *Phytolacca dioica* L. Biochimie 91:352–363

Di Maro A, Berisio R, Ruggiero A, Tamburino R, Severino V, Zacchia E, Parente A (2012) Structural and enzymatic properties of an in vivo proteolytic form of PD-S2, type 1 ribosome-inactivating protein from seeds of *Phytolacca dioica* L. Biochem Biophys Res Commun 421:514–520

Dimitri M J, Leonardis R F, Biloni JS (1997) El Nuevo libro del árbol: especies forestales de la Argentina oriental. El Ateneo Bs As 5(2):184–185

Di Sapio OA, Gattuso MA (1992) Caracteres morfoanatómicos de diferentes cortezas de la medicina folklórica argentina. I. *Salix humboldtiana* Willd (Salicaceae) Dominguezia 10(1):18–26

Díaz Cillio LM (2008) Guía de plantas argentinas del jardín botánico de la Ciudad de Buenos Aires. Díaz Cillio L editor. Argentina

Enciso P, Decoppet JD, Grätzel M, Wörner M, Cabrerizo FM, Cerdá MF (2017) A cockspur for the DSS cells: *Erythrina crista-galli* sensitizers. Spectrochim Acta A Mol Biomol Spectrosc 176:91–98

Etcheverry SR, Fernández MA, Rates SK, Parrillo S, Vázquez A, Heinzen HA (2003) Pharmacological activity and phytochemical studies of *Erythrina crista-galli* extracts. Mol Med Chem 1:8–12

Etcheberry S, Vázquez A, Heinzen H (2004) Validación del uso medicinal de Erythrina crista-galli L. VIII Simposio Argentino y XI Simposio Latinoamericano de Farmacobotánica. Buenos Aires, Argentina

Ezcurra E, Montaña C, Arizaga S (1991) Architecture, light interception, and distribution of Larrea species in the Monte desert, Argentina. Ecology 72:23–34

Faoro F, Conforto B, Di Maro A, Parente A, Iriti M (2009) Activation of plant defence response contributes to the antiviral activity of Diocin 2 from *Phytolacca dioica*. Induced resistance in plants against insects and diseases IOBC/wprs Bull 44:53–57

Franchi-Micheli S, Luzzi S, Ciuffi M, Zilletti L (1986) The effects of lipoxygenase inhibitors and leukotriene antagonists on anaphylaxis. Agents Actions 18:242–244

Garcia CC, Talarico L, Almeida N, Colombres S, Duschatzky C, Damonte EB (2003) Virucidal activity of essential oils from aromatic plants of San Luis, Argentina. Phytother Res 17:1073–1075

Gisvold O, Thaker E (1974) Lignans from *Larrea divaricata*. J Pharm Sci 63:1905–1907

Goleniowski ME, Bongiovanni GA, Palacio L, Núñez C, Cantero JJ (2006) Medicinal plants from "Sierra de los Comechingones". Argentina J Ethnopharmacol 107:324–341

Guida V, Criscuolo G, Tamburino R, Malorni L, Parente A, Di Maro A (2011) Purification and enzymatic properties of a peroxidase from leaves of *Phytolacca dioica* L. (Ombú tree). BMB Rep 4(1):64–69. https://doi.org/10.5483/BMBRep.2011.44.1.64

Haene E (2007) 100 flores argentinas. Edit Albatros, Buenos Aires

Hapon MV, Boiteux JJ, Fernández MA, Lucero G, Silva MF, Pizzuolo PH (2017) Effect of phenolic compounds present in Argentinian plant extracts on mycelial growth of the plant pathogen *Botrytis cinerea* pers. Phyton 86:270–277

Hernández XE, Kurina M, Giordano OS (1997) Production of 6-β-hydroxygrindelic acid from grindelic acid by *Alternaria alternata*. Biotechnol Lett 19(12):1223–1225

Hernández XE, Kurina Sanz MB, Giordano OS (2000) Grindelic production in *Grindelia pulchella* cell suspension cultures elicited with CuSO4. Molecules 5(3):614–615

Hernández XE, Carrizo RA, Giordano OS, Kurina M (2002) Hydroxylation of grindelic acid by filamentous fungus. J Argent Chem Soc 90(1–3):27–30

Hernández XE, Orden AA, Giordano OS, Kurina M (2005) Effects of elicitor and copper sulfate on grindelic acid production in submerged cultures of *Grindelia pulchella*. Electron J Biotechnol 8(3):276–283

Hunziker JH, Palacios RA, Valesi AG, Poggio L (1972) Species disjunction in *Larrea*: evidence from morphology, cytogenetics, phenolic compounds, and seed albumins. Ann Mo Bot Gard 59:224–233

Ito K, Furukawa H, Haruna M, Ito M (1973) Studies on the Erythrina alkaloids. V. Alkaloid of *Erythrina crista-galli* (L.) cv Maruba Deiko. Yakugaku Zasshi 93:1674–1678

Jiménez CM, Sgariglia MA (2010) Evaluación de actividad antifúngica de *Larrea divaricata* Cav. y *Larrea cuneifolia* Cav. sobre especies del género Fusarium. Cuartas Jornadas de Jóvenes Investigadores UTN-Conicet. Tucumán 22, 23 y 24 de Junio de 2010

Kiesling R (2003) Zygophyllaceae. In: Sigma E (ed) Flora de San Juan. Estudio Sigma, República Argentina, vol II, Buenos Aires, pp 39–46

Konigheim BS, Goleniowski ME, Contigiani MS (2004) Cytotoxity and antiviral activity of one ligand of *Larrea divaricata*. Drug Design Rev 2:81–83

Ladio AH, Lozada M (2009) Human ecology, ethnobotany and traditional practices in rural populations inhabiting the Monte region: resilience and ecological knowledge. J Arid Environ 73:222–227

Lahitte HB, Hurrell JA, Belgrano MJ, Jankowski LS, Mehltreter K et al (1997) Plantas de la Costa. Las plantas nativas y naturalizadas más comunes de las costas del Delta del Paraná, Isla Martín García y Ribera Platense. Literature of Latin America (L.O.L.A.). Buenos Aires, Argentina

Lahitte HB, Hurrell JA, Valla JJ, Jankowski LS, Bazzano D, Hernández AJ (1999) Árboles urbanos. Biota Rioplatense IV. Inventario de la biota de la región del Delta del Paraná, Isla Martín García y Ribera Platense. Literature of Latin America (L.O.L.A.). Buenos Aires. 320 p

Lamarque A, Maestri D, Zygadlo JA, Grosso NR (1998) Volatile constituents from flowers of *Acacia caven* (Mol.) Mol. var. caven, *Acacia aroma* Gill. ex Hook., *Erythrina crista-galli* L. and *Calliandra tweedii* Benth. Flavour Fragr J 13:266–268

Larré CF, Fernando JA, Marini P, Bacarin MA, Peters JA (2013) Growth and chlorophyll a fluorescence in *Erythrina crista-galli* L. plants under flooding conditions. Acta Physiol Plant 35:1463–1471

Líate HB, Hurrell JA (1994) Flora arbórea de la Isla Martín García nativa y naturalizada. Reserva Natural y Cultural. Provincia de Buenos Aires. República Argentina. Directores: ISSN 0325-1225. Serie informe (47): 227-229

Lorenzo ME, Gómez PE, Segovia AF, Quiroga A, Figueroa LC, Werning ML, Baroni MV (2018) Contenido fenólico y actividad antibacteriana de extractos etanólicos de *Larrea divaricata* Cav. y *Larrea cuneifolia* Cav. de la provincia de Catamarca (Departamento Capital). 1° Jornadas de divulgación científica y técnica. Facultad de Ciencias Agrarias. Universidad Nacional de Catamarca

Lozano EC, Zapater MA (2010) El género *Erythrina* (Leguminosae) en Argentina. Darwin 48(2):179–200

Mabry TJ, Bohnstedt CF (1997) Larrea: a chemical resource. In: Campos Lopez E, Mabry TJ, Fernandez Tavizon S (eds) Centro de Investigación en Química Aplicada, Mexico, Saltillo, Coah, pp. 217–235

Maier UH, Rödl W, Deus-Neumann B, Zenk MH (1999) Biosynthesis of Erythrina alkaloids in *Erythrina crista-galli*. Phytochemistry 52:373–382

Martínez Crovetto RN (1964) Estudios Etnobotánicos I. Nombres de plantas y su utilidad según los indios tobas del este del Chaco. Bonplandia 1(4):279–333

Martínez Crovetto RN (1981) Las plantas utilizadas en medicina popular en el Noroeste de Corrientes (Argentina). Miscelánea 69:1–140. Inst. M. Lillo, Tucumán

Martino R, Sülsen V, Alonso R, Anesini C (2013) A fraction rich in phenyl propanoids from *L. divaricata* aqueous extract is capable of inducing apoptosis, in relation to H2O2 modulation, on a murine lymphoma cell line. Leuk Res 37:1137–1143

Medina M, Demaio P, Karlin UO, (2015) Árboles nativos de Argentina. Vol. 1: Centro y Cuyo. Ecoval Edit. Córdoba

Merlugo L, Santos MC, Sant'Anna LS, Cordeiro EWF, Batista LAC, Miotto STS, Garcia CV, Moreira CM, Mendez SL (2015) Alkaloids in *Erythrina* by UPLC-ESI-MS and *in vivo* hypotensive potential of extractive preparations. Evid Based Complement Alternat Med 2015:959081, 12 pages. https://doi.org/10.1155/2015/959081

Mikkola M, Toivonen S, Tamminen K, Alfthan K, Tuuri T, Satornaa T, Natunen J, Saarinen J, Tiittanen M, Lampinen M, Valmu L, Partanen J, Otonkoski T (2013) Lectin from *Erythrina crista-galli* supports undifferentiated growth and differentiation of human pluripotent stem cells. Stem Cells Dev 22(5):1–10

Miño J, Gorzalczany S, Moscatelli V, Ferraro G, Acevedo C, Hnatyszyn O (2002) Actividad antinociceptiva y antinflamatoria de *Erythrina crista-galli* (Ceibo). Acta Farm Bonaer 21:93–98

Mitscher LA, Gollapudi RSR, Gerlach DC, Drake SD, Veliz EA, Ward JA (1988) Erycristin, a new antimicrobial petrocarpan from *Erythrina crista-galli*. Phytochemistry 27:381–385

Morello J (1958) La provincia fitogeográfica del Monte. Opera Lilloana 2:5–155

Muñoz V, Sauvain M, Bourdy G, Callapa J, Rojas I, Vargas L, Tae A, Deharo E (2000) The search for natural bioactive compounds through a multidisciplinary approach in Bolivia. Part II. Antimalarial activity of some plants used by Mosetene Indians. J Ethnopharmacol 69:139–155

Murashige T, Skoog F (1962a) A revised medium for rapid growth and bio assays with tobacco tissue cultures. Physiol Plant 15(3):473–497

Orden AA, Cifuente DA, Borkowski EJ, Tonn CE, Kurina MB (2005) Stereo and regioselective hydroxylation of grindelic acid by *Aspergillus niger*. Nat Prod Res 19(6):625–631

Orekhov A (1955) Chemistry of alkaloids. Akademia Nauk. 2 ed. Moscú

Oyarzábal M, Clavijo J, Oakley L, Biganzoli F, Tognetti P, Barberis I, Maturo HM, Aragón R, Campanello PI, Prado D, Oesterheld M, León RJC (2018) Unidades de vegetación de la Argentina. Ecol Austral 28:040–063

Palacio L, Bongiovanni G, Luchino N, Goleniowski ME (2006) Development of an *in vitro* culture of *Larrea divaricata* for biomass production. Mol Med Chem 10:14–16

Palacio L, Cantero JJ, Cusidó R, Goleniowski M (2011) Phenolic compound production by *Larrea divaricata* Cav. plant cell cultures and effect of precursor feeding. Process Biochem 46:418–422

Palacio L, Cantero JJ, Cusidó RM, Goleniowski M (2012) Phenolic compound production in relation to differentiation in cell and tissue cultures of *Larrea divaricata* (Cav.). Plant Sci 193–194:1–7

Palacios SM, Maggi ME, Bazán CM, Carpinella MC, Turco M, Muñoz A, Alonso RA, Núñez C, Cantero JJ, Defago MT, Ferrayoli CG, Valladares GR (2007) Screening of Argentinian plants for pesticide activity. Fitoterapia 78:580–584

Parente A, De Luca P, Bolognesi A, Barbieri L, Battelli MG, Abbondanza A, Sande MJW, Gigliano GS, Tazzari PL, Stirpe F (1993) Purification and partial characterization of single-chain ribosome-inactivating proteins from the seeds of *Phytolacca dioica* L. Biochim Biophys Acta 1216:43–49

Pedernera AM, Guardia T, Guardia Calderón C, Rotelli AE, de la Rocha NE, Di Genaro S, Pelzer LE (2006) Anti-ulcerogenic and anti-inflammatory activity of the methanolic extract of *Larrea divaricata* Cav. in rat. J Ethnopharmacol 10:415–420

Pizzo E, Zanfardino A, Di Giuseppe AMA, Bosso A, Landi N, Ragussi S, Varcamonti M, Notomista E, Di Maro A (2015) A new active antimicrobial peptide from PD-L4, a type 1 ribosome inactivating protein of *Phytolacca dioica* L.: a new function of RIPs for plant defence? FEBS Lett 589:2812–2818

Quiroga EN, Sampietro AR, Vattuone MA (2001) Screening antifungal activities of selected medicinal plants. J Ethnopharmacol 74:89–96

Quiroga EN, Sampietro AR, Vattuone MA (2004) *In vitro* fungitoxic activity of *Larrea divaricata* Cav. Extracts. Lett Appl Microbiol 39:7–12

Ratera EL, Ratera MO (1980) Plantas de la flora argentina empleadas en medicina popular. Edit. Hemisferio sur S.A. Buenos Aires

RathiSre PR, Reka M, Poovazhagi R, Arul Kumar M, Murugesan K (2015) Antibacterial and cytotoxic effect of biologically synthesized silver nanoparticles using aqueous root extract of *Erythrina indica* Lam. Spectrochim Acta A Mol Biomol Spectrosc 135:1137–1144

Rossi BE, Debandi GO, Peralta IE, Martínez Palle E (1999) Comparative phenology and floral patterns in *Larrea* species (Zygophyllaceae) in the Monte desert (Mendoza, Argentina). J Arid Environ 43:213–226

Saigler DS, Jakupcak J, Mabry TJ (1974) Wax esters from *Larrea divaricata*. Phytochemistry 13:983–986

Sakakibara M, Di Feo D Jr, Nakatani N, Timmermann B, Mabry TJ (1976) Flavonoid methylethers on the external leaf surface of *Larrea tridentata* and *L. divaricata*. Phytochemistry 15:727–731

Schreck R, Albermann K, Bauerle PA (1992) Nuclear factor kB: an oxidative stress-responsive transcription factor of eukaryotic cells (a review). Free Radic Res Commun 17:221–237

Sen S, Chakraborty R (2017) Chapter 19 Herbs, gastrointestinal protection, and oxidative stress. In: Gastrointestinal tissue. Oxidative stress and dietary antioxidants. Section I II Antioxidants and Gastroenterology Jordi Gracia-Sancho and Josepa Salvadó eds. Elsevier Inc. pp. 259–274

Sharry S, Abedini MA, Basiglio Cordal V, Briones V, Roussy L, Stevani R, Galarco S, Adema M (2011) Food and medicinal value of some forest species from Buenos Aires (Argentina). Emir J Food Agric 23(3):222–236

Simpson BB, Neff JL, Moldenke AR (1977) Reproductive systems of Larrea. In: Mabry TJ, Hunziker J, Di Feo DR Jr (eds) Creosote bush: biology and chemistry of *Larrea* in New World Deserts. Dowden, Hutchinson and Ross, Stroudsburg, pp 92–114. 284 pp

Stege P, Davicino R, Vega A, Casali Y, Correa S, Micalizzi B (2006) Antimicrobial activity of aqueous extracts of *Larrea divaricata* Cav. (Jarilla) against *Helicobacter pylori*. Phytomedicine 13(9–10):724–727

Teneberg S, Angström J, Jovall P, Karlsson KA (1994) Characterization of Binding of Galj34GlcNAc-specific Lectins from *Erythrina crista-galli* and *Erythrina corallodendron* to Glycosphingolipids. J Biol Chem 269(11):8554–8563

Timmermann BN, Hoffmann J (1985) Resins from *Grindelia*: a model for renewable resources in arid environments. In: Wickens GE, Goodwin JR, Field DV (eds) Plants for arid lands. George Allen and Unwin, London, pp 357–368

Tjahjandarie TS, Pudjiastuti P, Saputri RD, Tanjung M (2014) Antimalarial and antioxidant activity of phenolic compounds isolated from *Erythrina crista-galli* L. J Chem Pharm Res 6(4):786–790

Tortosa RD, Bertoli A, Bartoloni N, Puga E (2000) Optimum plant density for *Grindelia pulchella* Dunal (Asteraceae): a potential resin-producing crop in Arginine. Agrociencia 34(2):247–250

Toso R, Steibel PE, Troiani HO, Oriani DS, Ardoino S, Toribio M, Boeris MA (2006) Búsqueda sistematizada de plantas com actividad farmacológica utilizando el banco de extractos vegetales de plantas nativas y naturalizadas de la província de La Pampa. Rev bras ciênc vet 8(1):25–30

Tourkarkissian M (1980) Plantas medicinales de la Argentina. Sus nombres botánicos, vulgares, usos y distribución geográfica. Editorial hemisferio sur. Argentina

Valesi AG, Rodriguez E, Vaner Velde G, Mabry TJ (1972) Methylated flavonols in *Larrea cuneifolia*. Phytochemistry 11:2821–2826
Victol NS, Costa HMG, Takegava VF, Catelan TBS, Gebara KS, Silva CAH, Formagio AS, Tirloni CAS (2015) Anti-hemolytic protection and activity in lipid peroxidation of *Erythrina crista-galli* L. 23rd congress of the International Union for Biochemistry and Molecular Biology. 44th annual meeting of the Brazilian Society for Biochemistry and Molecular Biology Foz do Iguaçu, PR, Brazil, August 24th to 28th, 2015
Vivot E, Muñoz J, Herrero I, Dragán A, Sequin C (2006) Antimicrobial activity of dichloromethane extracts of eleven plants of the flora argentina. Pharmacol Online 3:845–849
Vivot E, Herrero I, Sánchez C, Sequin C (2007) Detección de actividad antifúngicas in vitro de extractos vegetales en candidiasis. XVI Congreso Ítalo-Latinoamericano de Etnomedicina. La Plata, Argentina
Vivot EP, Sánchez C, Cacik F, Sequin C (2012) Actividad antibacteriana en plantas medicinales de la flora de Entre Ríos (Argentina). Ciencia, Docencia y Tecnología 45:165–185
Vogt V, Cifuente D, Tonn C, Sabini L, Rosas S (2013) Antifungal activity *in vitro* and *in vivo* of extracts and lignans isolated from *Larrea divaricata* Cav. against phytopathogenic fungus. Ind Crop Prod 42:583–586
Waller CW, Gisvold OJ (1945) A phytochemical investigation of *Larrea divaricata* Cav. J Am Pharm Assoc 34:78–81
Yaluff G, Maldonado M, Ferreria ME, Benitez B, Lopez T, Vera N (2015) Actividad antiparasitaria *in vitro* de extractos etanólicós de *Erythrina crista-galii* L., *Erythrina falcata* Benth y *Erythrina dominguezii* Hassl. Mem Inst Investig Cienc Salud 13(2):60–66
Zampini IC, Cudmani N, Isla MI (2007) Antimicrobial activity of Argentine medicinal plants on antibiotic-resistant bacteria. Acta Bioquim Clin Latinoam 41:385–393

Chapter 5
Mesopotamia-Paranaense Forest

5.1 Introduction: The Mesopotamia

The Northeast region of Argentina comprises two subregions: the Chaco region and the Mesopotamia. Mesopotamia is located throughout the northeast of the country and involves the provinces of Misiones (MIS), Corrientes (COS), and Entre Rios (ERI). It is surrounded by two mighty rivers, Paraná and Uruguay, whose Northeastern tributaries are the Iguazú, San Antonio, and Pepirí-Guazú Rivers. The Iguazú Falls, declared "World Natural Heritage of Humanity," is a system of 275 waterfalls in the middle of the Paranaense Forest, located 17 kilometers from the mouth of the Iguazú River in the Paraná River, in the borderline between the province of MIS and Brazil (Fig. 5.1). The total wide of the falls is 27 km, and the average height is 60–80 m, with 1800 m^3 s^{-1} water flow; two-thirds of the falls are in Argentina (https://iguazuargentina.com/es/parque-nacional-iguazu). The Paraná and Uruguay Rivers flow downstream to the south up to the Rio de la Plata where they empty, crossing the provinces Chaco (CHA), COS, ERI, and Santa Fe (SFE). The Delta of Paraná marks the boundary between SFE and ERI and occupies the southern coasts of these provinces. The Uruguay River is the border between Argentina (the eastern border of the provinces MIS, COS, and ERI) and Uruguay.

Both subregions, Mesopotamia and Chaco, have rich and varied flora and fauna, especially in the subtropical protected forest sectors of MIS and in the forests and more temperate parks of COS and ERI, such as the Iberá Wetlands natural reserve.

In the provinces of MIS and COS, the soil is brick colored for the presence of a mineral called laterite, composed of aluminum, silica, and large amounts of iron oxide, which gives different shades of red to the soil.

It has a subtropical weather with high temperatures and intense rainfalls all along the year, even in winter. The southern zone of the Mesopotamian subregion is hotter, more humid and rainier than the northern, during summer. In this subregion the climate is warm and humid to the north and temperate and dry to the west; the precipitations decreased from east to west.

M. A. Alvarez, *Pharmacological Properties of Native Plants from Argentina*,
https://doi.org/10.1007/978-3-030-20198-2_5

Fig. 5.1 The Iguazú Falls. Province of Misiones, Argentina. MA Alvarez personal collection

Agriculture and livestock are the most outstanding economy resources of the region. Crops, corn, rice, beans, cassava, citrus, cotton, sugarcane, tobacco, and especially tea and yerba mate are the most relevant. As for livestock, there are breeding of cattle, horses, and pigs. The main industries are timber and steel and the exploitation of iron deposits (MIS) and hydroelectric energy (COS: Iberá, Paraná Medio, and Yaciretá-Apipé power stations).

5.2 The Paranaense Forest

This ecoregion, also known as ecoregion of the Misiones Forest, is part of the Atlantic Forest, one of the most relevant biodiversity world hotspots that contain about 7% of the world species and a high degree of endemism (Di Bitetti et al. 2005). The climate is humid subtropical or tropical. There are plenty of rivers and streams. In the Paranaense Forest is placed the Guaraní aquifer (up to 1000 m depth) one of the world highest fresh water reservoirs.

As for the fauna, there are more than 500 bird species described for the region, mammals such as *Panthera onca* (jaguar), *Puma concolor* (puma), *Leopardus pardalis* (ocelot), *Tapirus terrestris* (tapirus), *Mazama* spp., *Nasua nasua* (*coati*), *Speothos venaticus* (vinegar fox), armadillo (*Chlamyphoridae* and *Dasypodidae*), and various species of primates. Among the reptiles, there are *Eunectes murinus* (green anaconda) and *Caiman latirostris* (*overo* yacare caiman). The region has more than 50 amphibian species.

The Paranaense Forest is an extremely rich habitat with rich soils; the main vegetation type is semi-deciduous forests. Some of the plant species from the region are *Aspidosperma polyneuron*, *Ilex paraguariensis*, *Nectandra* spp., *Euterpe edulis*, *Ocotea* spp., and *Astronium balansae* (Oyarzabal et al. 2018). Within the area is the *Parque Nacional Iguazú* that has 53 to 73 arboreal species per hectare, 85 species of orchids, and more than 3000 species of vascular plants (1/3 from the total plant vascular species from Argentina) (Placci and Giorgis 1994; Zuloaga et al. 2000; Giraudo et al. 2003). Among the plant species used in popular medicine are *Aspidosperma quebracho-blanco* Schltdl. (Apocynaceae), *Anadenanthera colubrina* (Vell.) Brenan var. cebil (Griseb.) Altschul (Fabaceae) *Aristolochia triangularis* Cham. Et Schlecht. (Aristolochiaceae), *Blechnum occidentale* L. var. occidentale (Blechnaceae), *Bulnesia sarmientoi* Lor. Ex Griseb. (Zygophyllaceae), *Baccharis articulata* (Lam.) Pers., *Baccharis trimera* (Less.) DC, *Cecropia pachystachya* Trecul. (Cecropiaceae), *Chenopodium ambrosioides* L. (Chenopodiaceae), *Heteropterys angustifolia* Griseb. (Malpighiaceae), *Ilex paraguariensis* A. St.-Hil var. paraguariensis (Aquifoliaceae), *Lepidium bonariense* L. (Brassicaceae), *Maytenus ilicifolia* Mart. ex Reissek (Celesteraceae), *Passiflora caerulea* L. (Passifloraceae), *Plantago tomentosa* Lam. (Plantaginaceae), *Phoradendron liga* (Gill) Eichl. (Viscaceae), *Schinus molle* L. (Anacardiaceae), *Sida cordifolia* L., *Sida rhombifolia* L. (Malvaceae), and *Tabebuia impetiginosa* (Mart. ex DC.) Standl. (Bignoniaceae).

5.3 The Jesuit Missions

The Jesuit Order arrived to the territory in 1608 and established settlements that were inhabited by *Guarani* natives in the borders of present-day Paraguay, Brazil, and Argentina. The missions acted as a *state within a state* in which the native people, guided by the Jesuits, remained autonomous and isolated from Spanish colonists and Spanish rule. In the missions the natives found protection from enslavement (*bandeirantes*) and the forced labor of *encomiendas*. Under the leadership of both the Jesuits and native caciques, the reductions achieved a high degree of autonomy within the Spanish colonial empire. With the use of native labor, the reductions became economically successful. In 1767, the Jesuits were expelled from the Guaraní missions and the Americas by order of the Spanish King, Charles III (Woodrow 1985; Anonimous 2017). In Argentina, the reductions settled in the province of MIS were *San Ignacio Mini*, *Nuestra Señora de Santa Ana*, *Nuestra Señora de Loreto*, *Nuestra Señora de la Candelaria*, *Santa María la Mayor*, *Corpus Christi*, *Santos Apóstoles Pedro y Pablo*, *San José*, *San Javier de Yaguaraities*, *Nuestra Señora de la Concepción*, and *Santos Mártires*, which were named World Heritage Sites by UNESCO in 1983. In the province of COS were established the *Nuestra Señora de Yapeyú*, *Santo Tomé*, *La Cruz*, and *San Carlos Caapi* reductions. In the province of Córdoba, la *Manzana Jesuítica* (Jesuit Block) and *Estancias de Córdoba* were named World Heritage Sites in 2000.

Nuestra Señora de Loreto was founded by fathers José Cataldino and Simón Masseta to the left of the Paranapanema River. In 1613 they moved it to the margin of the Yabebirí stream. It was the birthplace of the first printing press in South America, founded by father José Serrano and father Neumann (MIS).

Fig. 5.2 *San Ignacio Miní* Mission, Misiones, Argentina. MA Alvarez personal collection

San Ignacio Miní (Fig. 5.2) was also founded by father Cataldino and Masseta in the Guirá in 1610. It moved for the first time to the right margin of Yabebirí stream in 1631 and finally in 1660 to the place where its ruins are today (MIS).

Nuestra Señora de Santa Ana, founded by fathers Romero and C. de Mendoza in 1633 near Yacuiba stream, moved to the East of Paraná River in 1638 and finally in 1660 to the place where its ruins stand today (MIS).

Nuestra Señora de la Candelaria, funded in 1637 in Ibicuí by fathers Roque González and Romero, was funded again in Caazapá mini (today Brazil). In 1667 it was moved to the Paraná River coast to the Southwest of La Candelaria. It was considered the capital of the missions (MIS).

Santa María la Mayor, funded in 1628 by fathers Diego de Boroa and Claudio Ruyer in the right bank of Iguazú River, in 1637 moved south to the site where its ruins are today.

The urban layout of the missions is attributed to father Antonio Sepp. The village was divided into two equal parts. The church and the courtyard of the fathers were in the center of the *plaza*. The streets ran parallel to the sides of the square. The main street stretched from the main entrance to the *plaza* and the church. At the side of the temple was the school or residence, on the other side the cemetery and the large room (*coty guazú*). The other sides were occupied by the houses of the natives. A house for the visitors was placed far from the houses of the natives and the *cabildo* in front of the *plaza*. Without pre-established location were the jail, the school, and a hospital. The layout was symmetrical, straight with north-south or east-west orientation (Anonimous 2017).

In the schools, the curriculum was the *Ratio Studiorum*. In addition to Christian doctrine, writing, and reading, the Jesuits considered important the teaching of singing and music (lyre, violin, harp) (Anonimous 2017).

From the commercial point of view, it behaved like a communal stay, with yerba mate as the only safe sale product that gave them large profits since it replaced alcohol. Other crops were cotton, sugarcane, and wheat and groves of oranges, olives, pears, vineyards, apple and peach orchards, and walnut trees. The Jesuits introduced the silkworm to the region. They had 200,000 head of cattle in Yapeyú. The reduction manufactured glasses, tin vessels, mechanical watches, compasses, mirrors, violins, bugles, harps, guitars, and astronomical lenses with rock crystals. They discovered that the *Itacurú* stone had large amounts of iron, which allowed them to produce farming tools and bells. They have also built dams and irrigation channels. In January 1768 Carlos III from Spain banished the Jesuits from Paraguay. At the beginning of the 19th century the reductions are already abandoned ruins (Woodrow 1985).

5.4 *Cecropia pachystachya* Trecul (Cecropiaceae)

Synonymy *Cecropia adenopus* Mart. ex Miq., *Cecropia adenopus* Mart. ex Miq. var. vulgaris, *Cecropia lyratiloba* Miq., *Cecropia adenopus* Mart. ex Miq. var. lyratiloba, *Cecropia adenopus* Mart. ex Miq. var. macrophylla, *Cecropia glauca* Rojas Acosta, *Cecropia catarinensis* Cuatrec., *Cecropia cinerea* Miq.

Vernacular name Ambay, ambai, ambahi, ambaí, amba-í, ambahú, am-baiba, ambay guazú, palo lija, palo de lija, umbaúba, embaúva (Amorin and Rosow 1988; Alonso and Desmarchelier 2015).

General information It is a medium-size tree that grows in groups or isolated in temperate regions next to the riverbanks and into the Neotropical rainforest in the provinces of CHA, COS, FOR, MIS, and SFE. Also, it could be found in the south of Brazil and Paraguay (Consolini et al. 2006).

Botanical characteristics *C. pachystachya* is a dioecious tree, 10–20 m height, with knotty hollow branches, and latex in the bark. The leaves are light wide palm-shaped; lamina is peltate, palmately compound, divided in long leaflets, dark green on the upper side and silver gray and hairy on the lower side. Leaflets are entire, obtuse, sometimes acuminated, with a petiole. The primary and secondary nerves are brownish yellow, prominent on the underside. The petiole is 10–30 cm length, cylindrical, 4–11 mm diameter, rigid, and brittle (ANP 1978). The flowers are small, dioecious, distributed in digitated big fleshy receptacles (Fig. 5.3). The fruits are little, reddish, edible achenes. The species that grew in the temperate region of the country, the Espinal biogeographic region, is smaller (up to 1 m height).

Ethnobotanical information *C. pachystachya* or *ambay* has been traditionally used by Guaraní indigenous medicine (Gupta 1995). The oldest reference to *ambay* is by Dominguez and Soto (1925) describing a syrup made with bark and leaves of *ambay* that was prescribed as bequic and expectorant. Its use as expectorant, cardiotonic, to treat cough and catarrh is cited by Ratera and Ratera (1980) and Martínez Crovetto (1981). Hypoglycemic, diuretic, and sedative activity was also reported (Soraru and Bandoni 1978). There are some references to its use against eye infections. The trunk aqueous extract is used against cutaneous bores and as antidiarrheal (Barboza et al. 2009; Alonso and Desmarchelier 2015).

Chemical data The more relevant chemicals isolated from the aerial parts of the plant are orientin, isoorientin, rutin, chlorogenic acid, and pomolic acid. In dried leaves luteolin-C-glycosides, orientin, isoorientin, vitexin, isovitexin, catechin, protocatechuic acid, isoquercitrin, β-sitosterine, stigmast-4-en-3-one, α- and

Fig. 5.3 *Cecropia pachystachya* Trecul (Cecropiaceae). Parque Nacional Iguazú, Misiones, Argentina. MA Alvarez personal collection

β-amyrin, β-sitosterol, cecropine, cecropinine, and arachidonic acid; the amino acids alanine, glycine, serine, valine, isoleucine, asparagine, proline, aspartic acid, phenylalanine, glutamic acid, and arginine; the sugars glucose, fructose, and galactose; and the glucosides ambain and ambainin were found (Fig. 5.4). Also, the plant contains oxidases, mucilages, and saponins. The bark has cecropine and tannic acid. The chloroform/methanol (9:1) extraction from the bark yielded fisetinidol-3-O-β-D-xylopyranoside ($C_{20}H_{22}O_9$) or anadanthoside as the main compound (Piacente et al. 1999, Costa et al. 2011).

In roots, there are β-sitosterol and tormentic acid. In roots and stems, there are 2α-acetoxy-3β,19α-dihydroxy-11α,12α-epoxyursan-28,13β-olide,3β-acetoxy-2α,19α-dihydroxy-11α,12α-epoxyursan-28,13β-olide and 2-O-acetyleuscaphic acid.

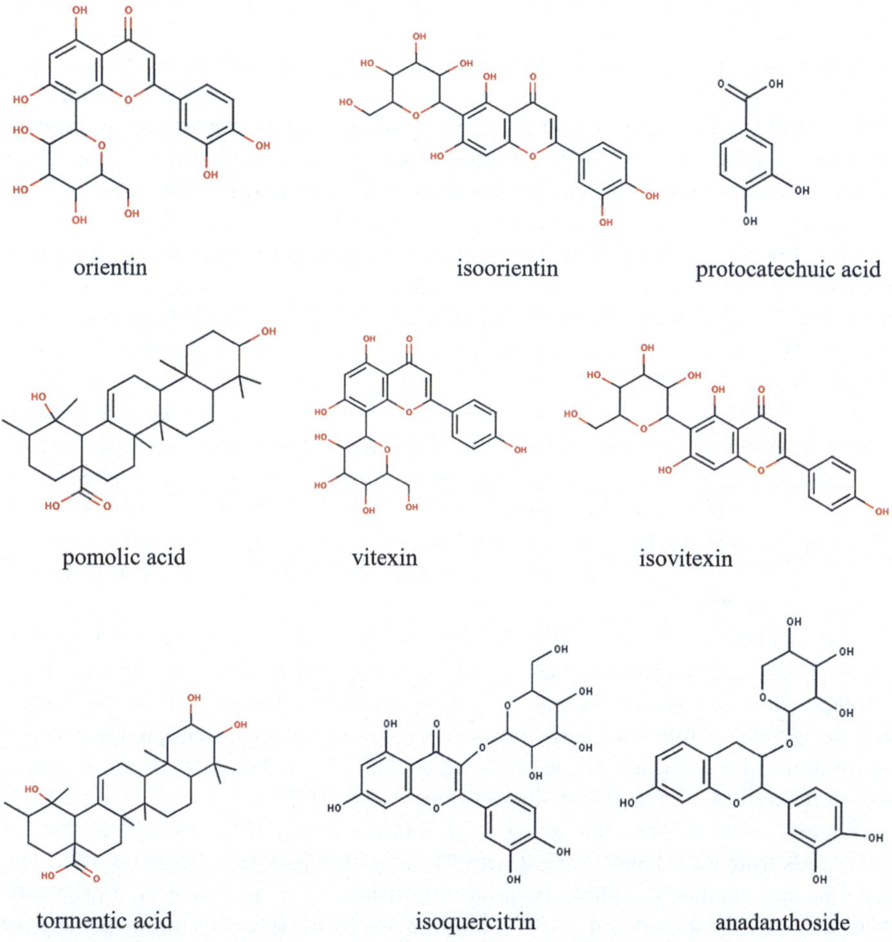

Fig. 5.4 Some chemical compounds isolated from *Cecropia pachystachya* Trecul

Biological activities Effects of *C. pachystachya* aqueous extracts as anti-inflammatory, cardiotonic, sedative, and antioxidant have been associated with their high level of antioxidant compounds (Velázquez et al. 2003; Consolini and Migliorisi 2005; Consolini et al. 2006; Schinella et al. 2008; Uchoa et al. 2010; Aragão et al. 2013). Chlorogenic acid, the main phenolic compound of the aqueous extract, plays a major role in the protective effect to H_2O_2 challenge (Mendonça et al. 2016). The leaf aqueous extract resulted beneficial against behavioral and biochemical modifications induced in mice by chronic unpredictable stress. Apparently, its phenolic compounds could modulate glutathione peroxidase involved in the prevention of stress-induced oxidative damage in mice hippocampus. Therefore, it was proposed the use of *ambay* aqueous preparations for treating depressive disorders (Gazal et al. 2014).

The effect on cardiac rhythm differs between plants from temperate regions and those from Neotropical regions; extracts from the first ones produced tachycardia, while those from the second ones did not (Consolini and Migliorisi 2005). However, both decreased rat spontaneous locomotion and exploratory behavior producing a similar sedative effect. The central inhibition attributed to flavonoids (Consolini et al. 2006) could be the reason of *ambay* efficiency to treat cough and asthma. A stronger hypotensive effect and an *in vivo* cardiotonic activity mediated by an anticholinergic effect of extracts from Neotropical specimens were also informed. The analgesic and central depressor effect was also described (Perez-Guerrero et al. 2001).

In a chronic renal failure model in rats, the aqueous extract attenuated hypertension and renal lesions by decreasing angiotensin converting enzyme activity in the renal cortex, which could also be mediated by flavonoids (Maquiaveli et al. 2014). The hexane extract from leaves showed anti-inflammatory activity (Hikawczuk et al. 1998).

The methanolic extract of leaves was tested in normal and alloxan-induced diabetic rats (dose: 80 mg kg^{-1}). Normal rats showed a better hypoglycemic activity on the oral glucose tolerance test, while diabetic rats had a significantly higher hypoglycemic effect (< 116 mg ml^{-1} glucose after 12 h, 68% of blood glucose reduction). That effect was partially attributed to chlorogenic acid and C-glycosylated flavones (Ivorra et al. 1989); however, the mechanism of action has not been proposed (Aragão et al. 2010).

The methanolic extract of dried leaves inhibited quorum sensing when the biosensors *Chromobacterium violaceum* ATCC 31532 and *E. coli* pSB403 were used (Brango-Vanegas et al. 2014), but did not inhibit bacterial growth. Thus, the extract and the isolated compounds could be used in the development of antipathogen drugs or for antifouling coatings (Brango-Vanegas et al. 2014). The extracts did not have a diuretic activity on rats (Consolini and Migliorisi 2005).

Finally, an anti-leishmanial activity (Mello Cruz et al. 2013) was attributed to the flavonoids from the ethanol leaf extract (5% w/v) from Salvador, Bahia (Brazil). The ethyl acetate-methanol soluble fraction diminished *L. amazonensis* (L.) promastigote axenic growth/survival (70% inhibition in 96 h), inhibited arginase activity (IC50 7 μg ml^{-1}), and altered the mitochondrial kinetoplast DNA (K-DNA) array. As *Leishmania* arginase participated in the maintenance of the infection in mammalian hosts, its inhibition by *C. pachystachya* extracts was of paramount relevance. Orientin (luteolin-8-C-glucoside), an arginase inhibitor (IC50 15.9 lM), was the major component of the fraction; lower concentrations of quercetin, another arginase inhibitor, were detected as well (Mello Cruz et al. 2013).

Toxicity The infusions prepared with dried leaves (9.2% of crude extract) did not show any genotoxic effect in animals. On the other hand, a neurotoxic effect was found as the aqueous extract induced DNA damage on brain tissue independently of the dose (500–2,000 mg kg^{-1}) (Mendonça et al. 2016). The extract did not show mutagenic activity in five *S. typhimurium* strains tested (TA100, TA98, TA97, TA102, and TA1535).

Interactions The extracts are synergic with digoxin (Domínguez and Soto 1925).

Legal status *C. pachystachya* was included in the Argentinian National Pharmacopeia sixth edition (1978). It is there established that "*ambay* is the dried leaf from *C. adenopus* Martius and its varieties in different states of fragmentation with no more than 2 % of foreign organic material." In 2004 it was introduced in the primary healthcare system of the province of Misiones.

5.5 *Anadenanthera colubrina* (Vell.) Brenan var. cebil (Griseb.) Altschul (Fabaceae)

Synonymy *Acacia cebil* Griseb., *Anadenanthera macrocarpa* (Benth.) Brenan, *Piptadenia cebil* (Griseb.) Griseb., *Piptadenia macrocarpa* Benth., *Piptadenia macrocarpa* Benth. var. cebil, *Piptadenia hassleriana* Chodat, *Piptadenia hassleriana* Chodat var. fruticosa, *Piptadenia macrocarpa* Benth. f. puberula, *Piptadenia macrocarpa* Benth. f. rupestris, *Piptadenia macrocarpa* Benth. f. microcarpa, *Piptadenia macrocarpa* Benth. var. vestita, *Piptochaetium macrocarpa* Benth. var. plurifoliata.

Vernacular name Cebil, cebil colorado, cebil moro, curupaih, curupay, kurupá, yopo, vilca, cohaba.

General information *A. colubrina* is a caducous tree that grows on rocky hillsides in well-drained soil, often near rivers, in the provinces of CAT, CHA, COR, COS, ERI, FOR, JUJ, MIS, SAL, SDE, and TUC. Its multiple uses include leather tanning, as fuel (firewood), to make charcoal and bleach (natural soap), and for rural construction. For its rapid growth, its condition as a nitrogen fixer, and its resistance to strong wind, it has importance for the establishment of agroforestry systems (Justiniano and Fredericksen 1998). *A. colubrina* flourishes between September and November, and the fruit matures from August to September (Lorenzi and Matos 2002).

Botanical characteristics *A. colubrina* is a tree 5–30 m tall, with an irregular crown, and a 30–80 cm diameter trunk, covered by a rough, grayish, and thorny bark. Leaves are bipinnately compound with opposite leaflets, 4–6 mm length. The white flowers are gathered in panicles with globular spikes. The fruits are flattened, rigid, glabrous, brilliant-brown dehiscent legumes containing 5–10 dark seeds (Fig. 5.5). Seeds are brown, smooth, and laterally compressed (Viglione and Vallejo 2011). Heartwood is pale brown and darkens to reddish brown with darker colored veins on exposure. Sapwood is yellow, brown, or light pink (Lorenzi and Matos 2002).

Ethnobotanical information It was used by the native communities of *Wichis*, *Comechingones* (Córdoba), *Ranqueles*, and *Mbayas*, among other native inhabitants, as hallucinogen in medical rituals for its psychoactive principles (tryptamine-derived alkaloids) (Schultes and Hofmann 1993; Pérez Gollán and Gordillo 1993; Alonso and Desmarchelier 2015).

The bark and seed powder was snuffed; semi-roasted seeds were used to make the tribe *kurupáih* or *kurupa* (a spell according to Montoya) that produced ecstasies when inhaled as snuff (*rapé*). Those *Guaraní* natives that owned the secret to prepare *kurupá* were called *kurupadyaras* (Amorín 1974). The *Mosetene* ethnic group from Bolivia used the bark to tan leather (Vargas & Quintana). Currently, it is used in folk medicine to treat dysentery, as vermifuge and antipyretic. Bark preparations are used as abortive, astringent, depurative, antitussive, and expectorant and for healing infections (Toursarkissian 1980). The gum is prescribed for respiratory

Fig. 5.5 *Anadenanthera colubrina* (Vell.) Brenan var. cebil (Griseb.) Altschul (Fabaceae). Parque Nacional Iguazú, Misiones, Argentina. MA Alvarez personal collection

problems. Seed are abortive and also used in case of heart complications and for treating the disease called *susto* (Martínez and Pochettino 1999). A syrup formulation is taken for treating cough, pertussis, and bronchitis. A bark macerate is used in cases of inflammation and leukorrhea. The alcoholic preparation is applied on external wounds, as hemostatic and cicatrizing (Matos 1997; Palmeira et al. 2010). The bark releases a resin helpful in cases of skin infections (Mors et al. 2000). The fruit is poisonous (Agra et al. 1996).

Chemical data The main active ingredient is calcium bufotenate. The dried aerial parts contain alnusenol, anadanthoflavone, lupenone, lupeol, betulinic acid, α- and β-amyrin, β-sitosterol, stigmasterol, apigenin, 4-hydroxybenzoic acid, and cinnamic acid. The heartwood contains dalbergin, 3,4,5-dimethoxy-3-dalbergione, and 5-methoxy-N-methyltryptamine. The dried bark has anadanthoside (®setinidol-3-O-b-D-xylopyranoside), gallic acid, caffeic acid, chlorogenic acid, rutin, catechin, quercetin, isoquercetin, luteolin, apigenin, and cinnamic acid. The gum has galactose, arabinose, and acidic heteropolysaccharides (Fig. 5.6). The seed contains bufotenin and N, N-dimethyltryptamine (Fig. 5.7) (Piacente et al. 1999; Barboza et al. 2009; Leite et al. 2010).

Biological information The ethanolic extracts (2024 $\mu g\ ml^{-1}$) of *A. colubrina* are a natural source of antibacterial compounds (flavonoids, tannins, and terpenes) (Reginatto et al. 2001). The extract was effective against the broad-spectrum resistant bacteria *E. coli* 27 (isolated from surgical wounds), *E. coli* ATCC10536,

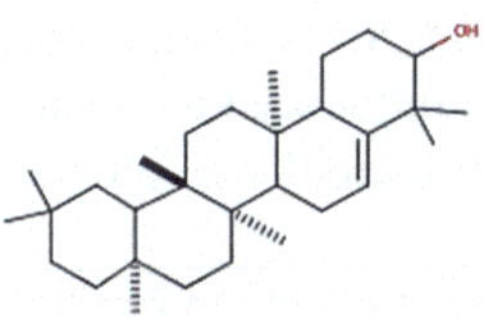

alnusenol

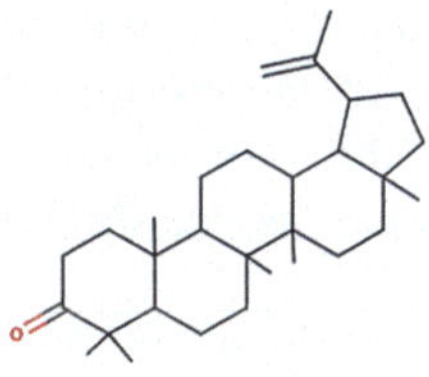

lupenone

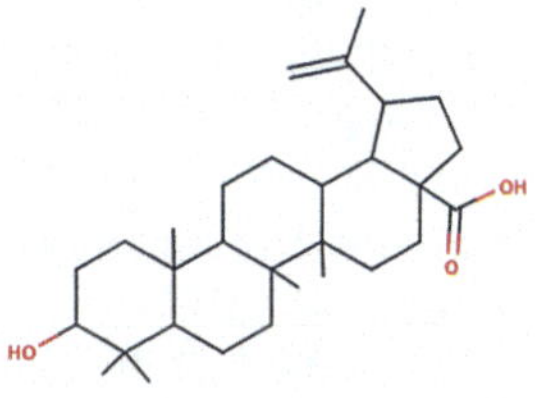

betulinic acid

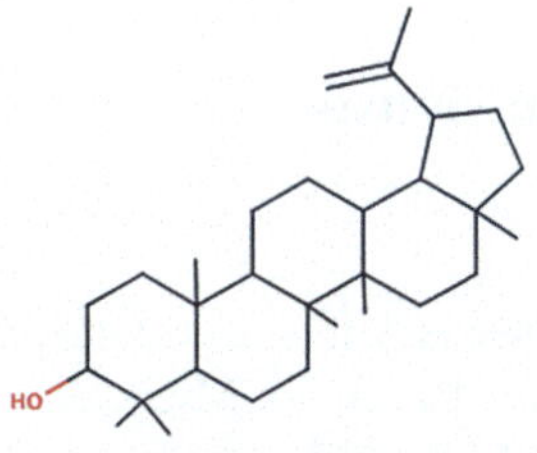

lupeol

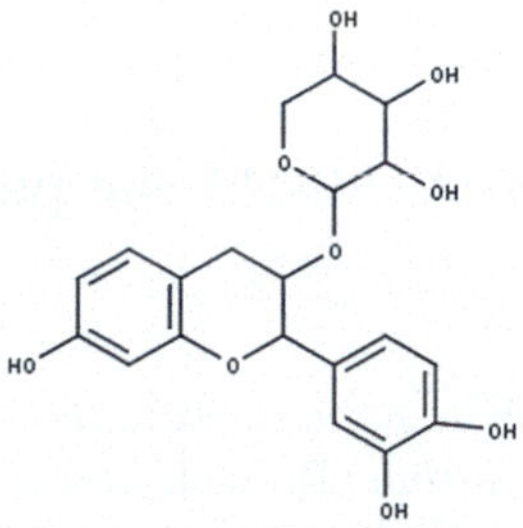

anadanthoside

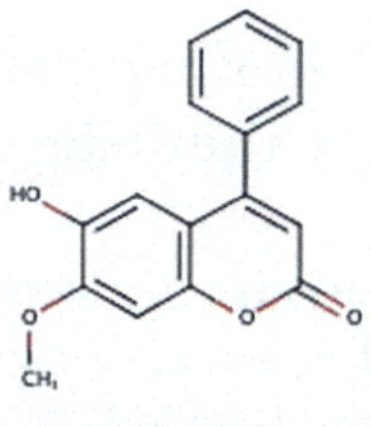

dalbergin

Fig. 5.6 Chemical compounds from *Anadenanthera colubrina* (Vell.) Brenan var. cebil (Griseb.) Altschul

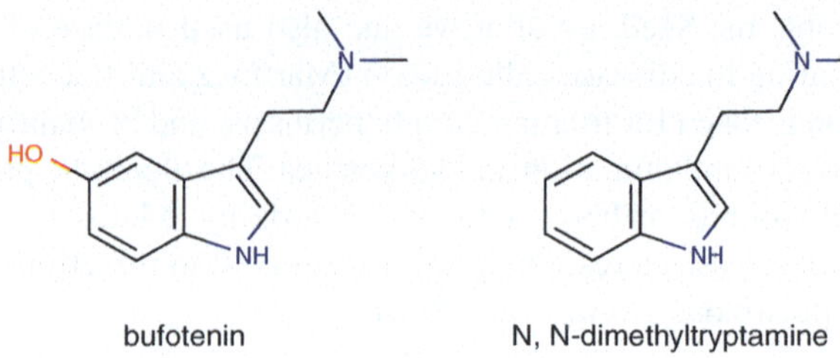

Fig. 5.7 Indole alkaloids from *Anadenanthera colubrina* (Vell.) Brenan var. cebil (Griseb.) Altschul

Staphylococcus aureus 358 (isolated from surgical wounds), and *S. aureus* ATCC25923 (Figueredo et al. 2013). Also, it was effective against *Pseudomona aeruginosa* (MIC 0.5 mg ml^{-1}) (Freitas Lima et al. 2014).

As antifungal it was effective against *Candida albicans* ATCC 76485 (MIC 0.31 mg ml^{-1}) and *C. albicans* ambulatorial lineages (MIC 1.0 mg ml^{-1}). Synergic effect was observed with fluconazole but not with nystatin (Nunes et al. 2015).

The aqueous (10% w/v) and ethanolic (12.5% w/v DMSO) extracts have *in vitro* anti-parasite action against larvae and nymphs from *Rhipicephalus microplus* (Boophilus) (Lopez da Silva Filho 2007). *R. microplus* is considered the most important tick parasite of livestock from the world. The aqueous extract was effective (85% mortality) against ticks at a concentration of 8.26 mg ml^{-1} during an exposure of 12 h. The ethanolic extract was effective (85% mortality) at 1.56, 6.25, and 12.5 mg ml^{-1} concentrations. *A. colubrina* is also efficient in agronomic crop pest control (Zanon de Moura et al. 2012). The leaf and bark aqueous extracts were active against *Meloidogyne incognita*, one of the main cotton pathogen nematodes, at 52.48 g l^{-1} and 70.72 g l^{-1} concentrations (Lima Fonseca et al. 2017). On the other hand, the extract affected the viability of the larval stage and influenced the fecundity of the lepidopteran *Spodoptera frugiperda* (Lepidoptera: Noctuidae) (Zanon de Moura et al. 2012).

The leaf ethanolic extract (5, 10, 20 mg ml^{-1}) showed allelopathic effect by interfering with the germination of *Brassica chinensis* and *Lactuca sativa* (Goncalves Silva et al. 2010).

5.6 *Ilex paraguariensis* A. St.-Hil var. paraguariensis (Aquifoliaceae)

Synonymy *Ilex mate* A. St.-Hil., nom. superfl., *Ilex curitibensis* Miers, *Ilex domestica* Reissek var. glabra, *Ilex sorbilis* Reissek, *Ilex theaezans* Bonpl. ex Miers, *Ilex paraguariensis* A. St.-Hil. f. latifolia, *Ilex paraguariensis* A. St.-Hil. f. parvifolia, *Ilex paraguariensis* A. St.-Hil. var. dentata, *Ilex paraguariensis* A. St.-Hil. var. idonea, *Ilex curitibensis* Miers var. gardneriana, *Ilex paraguariensis* A. St.-Hil. f. domestica, *Ilex paraguariensis* A. St.-Hil. f. sorbilis, *Ilex paraguariensis* A. St.-Hil.

f. confusa, *Ilex paraguariensis* A. St.-Hil. f. dasyprionata, *Ilex paraguariensis* A. St.-Hil. var. ulei, *Ilex paraguariensis* A. St.-Hil. f. glabra, *Ilex paraguariensis* A. St.-Hil. var. usitata, *Ilex paraguariensis* A. St.-Hil. var. genuina.

Vernacular name Yerba mate, mate, té de los Jesuitas, yerba del Paraguay, erva-mate, congonha, chá dos Jesuitas, thé du Paraguay, maté, ka'a, caá-mirim.

General information In Argentina several species of *Ilex*, e.g., *I. argentina* Lillo (roble tucumano, palo de yerba, árbol de yerba), *I. brevicuspis* Reissek (caona, cauna, siete sangrías), and *I. dumosa* var. Guaranina Loes (yerba señorita, caábera, caáchiri, caáverá), grow. However, the most popular is *I. paraguariensis* A. St.-Hill which is popularly taken as a beverage in Argentina and in neighboring countries (Uruguay, South Brazil, and Paraguay). In Argentina, *I. paraguariensis* grows in the provinces of COS and MIS where the plantations and processing facilities are located. A study demonstrated that plants that grow in plantations have a higher content of phenolic acids that those that grow in natural forests (Burris et al. 2012a). Some adulterants are *I. dumosa*, *I. theezans*, *I. brevicuspis*, *I. conocarpa*, and *I. microdonta* (Alonso and Desmarchelier 2015). Total RNA from *I. paraguariensis* was sequenced; multiple sequence analysis allowed estimating the existence of 32,355 genes and 12,551 gene variants or isoforms. Members of more than 100 metabolic pathways were identified and categorized. Several transcription factors and genes, involved in heat and oxidative stress, pathogen response, disease resistance, and hormone response, among others, were identified. The authors also provided a collection of over 10,800 simple sequence repeats accessible to the scientific community (Debat et al. 2014).

Botanical characteristics *I. paraguariensis* is a perennial dioecious tree, 3–8 m height. The bark is brown with white spots. The branches are glabrous, with perennial alternate coriaceous green leaves, 4–10 length, a remarkable central nerve and serrated margin. Petioles are up to 15 mm length (Fig. 5.8). Flowers are white, unisexual, with four petals. The fruit is a drupe, purple reddish with 4–5 seeds.

Ethnobotanical information The species was highly appreciated by the *Guaraní* natives that called it *caá mate*. When the Spanish colonizers arrived, they adopted the beverage. It was cultivated in the missions founded in the seventeenth century by the company of Jesus. Today it is the most popular beverage in Argentina, Uruguay, Paraguay, and south of Brazil. The leaves and small pieces of petiole are used to prepare an infusion, the *mate*, which is consumed mostly in a recipient, also called *mate*, using a metallic straw, the *bombilla*. It could also be taken cold (*tereré*) or as a tea in a cup (*mate cocido*). The popular use attributes to the infusion properties against influenza and scurf, as tonic, refrigerant, diuretic, digestive, stimulant, choleretic, eupeptic, hypolypemic, and hepatoprotective. The infusion made with the aerial parts is used as galactagogue, cicatrizing, and in cases of cephalea or fever (Toursarkissian 1980; Gupta 1995; Cansian et al. 2008).

Fig. 5.8 *Ilex paraguariensis* A. St.-Hil var. paraguariensis (Aquifoliaceae). *San Ignacio Miní* mission, Misiones, Argentina (**a** and **b**), and in Buenos Aires Botanical Garden (**c**). MA Alvarez personal collection

Chemical data The leaves contain saponins, alkaloids, and phenolics (Fig. 5.9). Among the phenolic compounds are caffeic acid, chlorogenic acid, 3,4-dicaffeoylquinic acid, 3,5-dicaffeoylquinic acid, 4,5-dicaffeoylquinic acid, and 5-caffeoylquinic acid. The flavonoids are quercetin, rutin, and kaempferol. The purine alkaloids are caffeine (1,3,7-trimethylxanthine), theobromine (3,7-dimethylxanthine), and theophylline (1,3-dimethylxanthine). The saponins identified were matesaponin, matesaponin 2 (ursolic acid 3-O-[β,D-glucopyranosyl(1 → 3)[α,L-rhamnopyranosyl(1 → 2)]]α,L-arabinopyranosyl] (28 → 1)-β,D-glucopyranosyl ester), matesaponin 3 (ursolic acid 3O[β,D-glucopyranosyl(1 → 3)α,L-arabinopyranosyl](28 → 1)[β,D-glucopyranosyl(1 → 6) β,D-glucopyranosyl] ester), matesaponin 4 (ursolic acid 3,O-[β,D-glucopyranosyl (1 → 3)[α,L-rhamnopyranosyl (1 → 2)]]α,L-arabinopyranosyl] (28 → 1)[β, D-glucopyranosyl(1 → 6) β,D-glucopyranosyl] ester), matesaponin 5, guaicin B, and nudicaucin (Martínez et al. 1997; Saldaña et al. 1999; Saldaña et al. 2002; Taketa et al. 2004; Barboza et al. 2009; Burris et al. 2012a). Also, several vitamins such as ascorbic acid, nicotinic acid, and beta carotenes, and minerals as iron, magnesium, manganese, calcium, copper, potassium, zinc, and sodium were found (Moreau 1973; Alikardis 1987; Clifford and Ramirez-Martinez 1990; Ashihara 1993; Gosmann et al. 1995; Kraemer et al. 1996; Ricco et al. 1991; Schenkel et al. 1996; Vera García et al. 1997; Paroul et al. 2002; Filip and Ferraro 2003; Filip et al. 2001; Cansian et al. 2008).

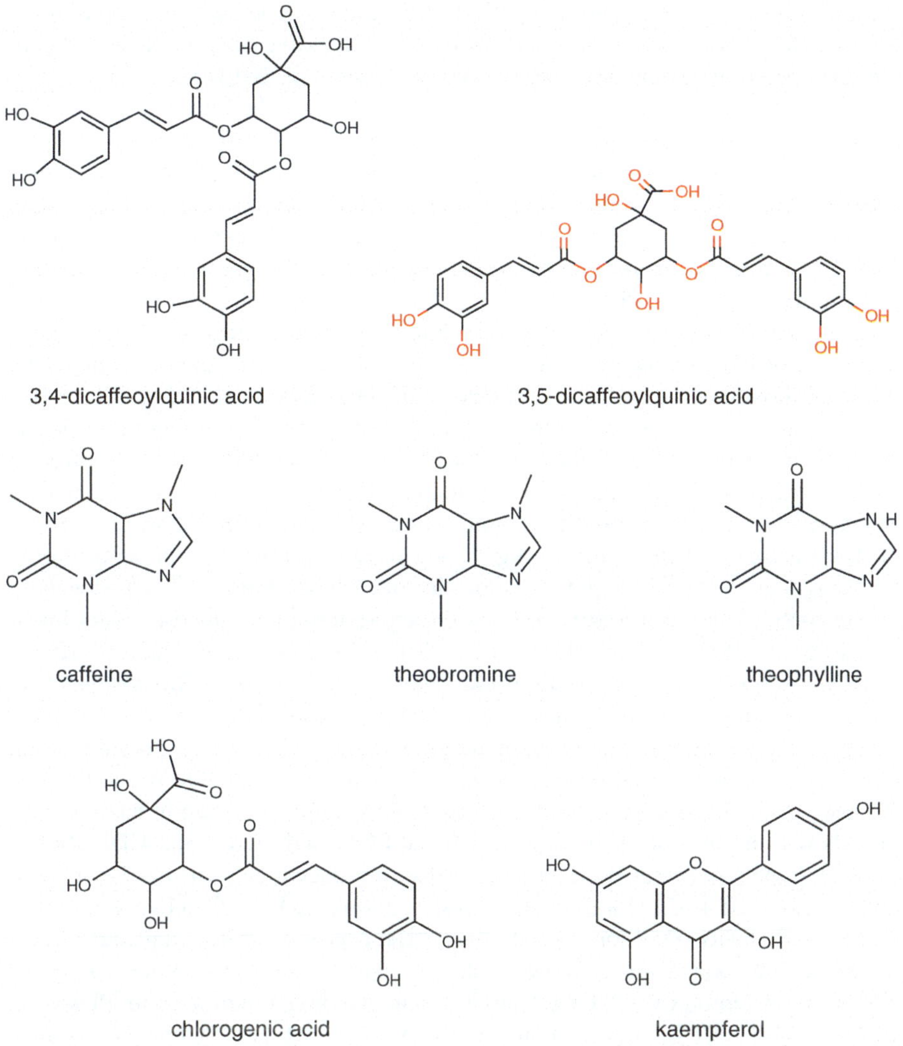

Fig. 5.9 Some chemical compounds found in *Ilex paraguariensis* A. St.-Hil var. paraguariensis. (Filip et al. 1998)

Biological activities The depurative, stimulant, and diuretic effect of *yerba mate* was attributed to its high purine content (Baltassat et al. 1984) and the antioxidant activity to its polyphenol content (Filip et al. 2000). Caffeine was reported as the responsible compound for the stimulant properties.

It was found that yerba mate produced a dose-dependent increase in alert or attentive wakefulness, a decrease in non-REM sleep, and shallow sleep episodes in adult cats (*Felis domesticus*) that orally received 0, 2.5, 5.0, 10.0, or 30.0%

I. paraguariensis solution over a period of 5 minutes. The authors proposed the use of different varieties or products for treating excessive sleepiness and for people that need to be awake for long periods (Falconi et al. 2013).

Antioxidant activity was tested by the inhibition of the luminal-induced chemiluminescence assay (Actis-Goretta et al. 2002), the inhibition of 2,2′-thiobarbituric-reactive substances production in liposomes (Filip et al. 2000), the peroxidase-like activity (Anesini et al. 2006), the ferric thiocyanate method (Bastos et al. 2006; Turner et al. 2011), and with 6-hydroxy-2,5,7,8- tetramethychroman-2-carboxylic acid. The extract obtained by spray drying also showed a strong free radical scavenger activity (Berté et al. 2011).

Leaf extracts were found to increase bile flow, have anticancer effect, trigger inhibition of LDL oxidation, and have a potential therapeutic and/or prophylactic use in cardiovascular disorders (Bracesco et al. 2003; Ruxton et al. 2007).

Assays performed with supercritical carbon dioxide extracts of *I. paraguariensis* on adult male Wistar rats with acute induced-peritonitis inhibited the neutrophil recruitment in the circulating blood (Pasquali et al. 2011). It also reduced acute lung inflammation in mice exposed to cigarette smoke (Lanzetti et al. 2008).

The treatment of rats with 1 g kg^{-1} body weight of instant *yerba mate* during 8 weeks attenuated gain weight and adiposity, decreased epididymal fat pad weight, and restored cholesterol, triglycerides, LDL cholesterol, and glucose serum levels (Arçari et al. 2009; Morgan-Martins et al. 2013). Besides, an inhibition of atherosclerosis progression in cholesterol-fed rabbits was informed (Mosimann et al. 2006). The methylxanthine fraction increased lipogenesis of the muscle and hepatic glycogen synthesis and lipolysis in adipose tissue, while the saponin fraction enhanced lipogenesis in adipose tissue and also fecal fat excretion in Wistar rats (De Resende et al. 2015). *I. paraguariensis* showed hypolipemic effect in healthy individuals with normo- or dyslipidemia, reduced LDL cholesterol, non-HDL cholesterol, and apolipoprotein B blood levels. Also, it reduced the LDL-C/HDL-C ratio and increased HDL-C (Morais et al.). Besides, it lowered LDL-C in hypercholesterolemic individuals (De Morais et al. 2009). The *I. paraguariensis* aqueous extract modulated the expression of genes related to obesity such as those that codify for TNF-α, IL-6, leptin, CCR2, CCL2, angiotensinogen, PAI-1, adiponectin, PPAR-γ2, PGC-1α, and UCP1 (Arçari et al. 2009). Also, it regulated adipogenesis in a β-catenin-dependent manner (Arcari et al. 2013a). Finally, it was reported an improvement of glucose tolerance in obese animals and a modulatory effect on insulin-related target genes in the liver of animals with a high-fat diet (Arcari et al. 2011, 2013b). Caffeine combined with the activities of the polyphenols and flavonoids was proposed of the responsible of the thermogenic activity observed (Ruxton et al. 2007).

Antimicrobial activity of the aqueous extracts from commercial yerba mate against pathogenic *E. coli* O157:H7 was reached at relatively low concentrations (5 to 10 mg ml^{-1}) supporting the use of *yerba mate* as a bio-based antimicrobial for foods and beverages. On the contrary, neither the crude ethanolic extract nor the saponin fraction, both at a 100 mg ml^{-1} concentration, showed antimicrobial activity (Burris et al. 2012b).

The anticancer activity of the infusions showed an inhibition of viability and cell growth of human colon adenocarcinoma cells SW480 and their metastatic derived SW620 cells (Pérez et al. 2014). The saponin inhibited colon cancer cells (HT-29) proliferation activating a specific intracellular apoptosis pathway and increased the expression of the pro-apoptotic protein Bax, decreased the expression of anti-apoptotic protein Bcl-2, and subsequently activated caspase- 3 (Puangpraphant et al. 2011). Also, an inhibitor effect on CaCo-2 adenocarcinoma cells growth was reported (González de Mejía et al. 2010).

The higher bioavailability of compounds in industrially processed samples (harvesting, zapecado, drying, and aging) than in green leaves was attributed to the conditions of temperature, humidity, cell disruption, and mechanical impact during the industrial process (Turner et al. 2011).

Beyond its traditional use, *yerba mate* has the potential to be used in the cosmetic and food industries. The increasing interest for *yerba mate* and its sub-products in the United States, Asia, and Europa, where it is exported, led to the development of a spray-drying technology that allow their conservation for long periods. The reconstructed extract had an IC50 of 2.52 mg ml^{-1}, high catalase-like activity, and a high concentration of caffeic acid (1.54 mg g^{-1}), 5-caffeoylquinic acid (91.40 mg g-[1]), rutin (5.38 mg g^{-1}), and total phenolics (178.32 mg g^{-1}) showing that the technology could be useful since it maintained the antioxidant activity and polyphenolic content (Berté et al. 2011).

Recently, the use of *yerba mate* as negative oral contrast in magnetic resonance cholangiopancreatography was explored using phantom devices (100 g l^{-1} infusion and 3:4, 1:2, and 1:4 dilutions). It was found that it produced an innocuous contrast for the patient, was effective for the elimination of the signal of the gastrointestinal tract, and improved the visualization of the anatomical structures of the biliary tract (Martin et al. 2006).

Toxicity The reports about toxicity of *yerba mate* are contradictory. Some studies reported an increased risk of bladder cancer and renal cell carcinoma associated with *mate* drinking in humans (Vassallo et al. 1985; De Stefani et al. 1998, 2007) and mutagenic and clastogenic activities in cell culture (Fonseca et al. 2000). The aqueous extract was genotoxic for *E. coli*, mutagenic for *Salmonella typhimurium*, and *in vitro* cyto- and genotoxic for lymphocytes (Wnuk et al. 2009). On the other hand, other reports established that *I. paraguariensis* is not genotoxic to liver, kidney, and bladder cells. Animals that received different doses of the aqueous extract showed unaltered levels of DNA damage. Moreover, animals that received regular ingestion of *mate* increased DNA resistance to H_2O_2-induced DNA strand breaks and improved DNA repair (Miranda et al. 2008). Also, it was reported that neither the infusions nor concoctions of *I. paraguariensis* showed cytotoxicity in the *Artemia salina* test (Bidau et al. 2004).

It was referred that at high doses, *yerba mate* causes insomnia, nervous excitation, gastritis, nausea, and colitis.

***In vitro* culture** *In vitro* cultures (Ross et al. 2017) and micropropagation of *I. paraguariensis* were achieved in 1/4 MS culture medium with the addition of 3% sucrose and BAP (0.44 µM). Different chemical and physical culture conditions affected the anatomy, histology (architecture, leaf vein pattern, leaf thickness, stomatal index, etc.), and physiological behavior of plants, being the temporary immersion system the most conservative treatment (Luna et al. 2017).

Legal status *Ilex paraguariensis* was included in the 6VIth edition of the FNA and in the Argentina's Alimentary Codex. *I. paraguariensis* also figures in the *British Herbal Pharmacopoeia* (1983–1996), Martindale (30th ed.), German Monographies commission E, List II from the European Council, and Ayurveda Pharmacopeia and is approved as dietary supplement by the FDA. In the United States, *yerba mate* is listed as GRAS (generally recognized as safe) (Newall et al. 1996).

5.7 *Tabebuia impetiginosa* (Mart. ex DC.) Standl. (Bignoniaceae)

Synonymy *T. avellanedae* Lorentz ex Griseb., *Tabebuia ipe* (Mart.), *Handroanthus avellanedae* (Lorente ex Griseb) Kuntze, *H. impetiginosus* (Mart. Ex DC.) Mattos, *H. heptaphylla* (Vell.).

Vernacular name Lapacho, lapacho rosado, lapacho dorado, palo de arco, taheebo.

General information *T. impetiginosa* grows in America from Mexico to the North and center of Argentina. Particularly, in Argentina it is found in the provinces of CAT, JUJ, SAL, and TUC. The wood is used as timber and to obtain a yellow dye (Bandoni et al. 1972; Tyler 1987). The native and exotic species of the genus are in general very required as ornamental, due to the striking brightness of its flowers. It is highly appreciated as timber for its density (higher than water), insect resistance, and durability (Mechler 2009). It is also called the "new mahogany," being the second most expensive timber and the most logged species in Brazil (Bonfim Silva-Junior et al. 2017). Hence, the species merit greater protection than provided by current generic forestry laws in Brazil and other Amazonian countries (Schulze et al. 2008). A well-curated but unfinished genome assembly was made providing a useful tool for genetic and functional analysis (Bonfim Silva-Junior et al. 2017).

Botanical characteristics *T. impetiginosa* is a caduceus tree, 5 to 50 m high, with a diameter of up to 1 m. The crown is stratified; leaves are opposite, petiolate, 10–14 cm length, 5–7 foliolate, with reticulate venation; the lower leaflet is the smallest, the terminal the largest. Leaflets are narrow, lanceolate, or elliptical, with an entire or slightly crenate margin. The gray brownish or yellowish external bark is suberified. The internal bark is pink whitish, 0.5–1.0 cm width. The flowers are short panicles, 15 cm length; calix is greenish, campanulate to tubular, 2.0–2.5 cm length, in general 5-lobed, looking like a trumpet. The corolla is pink or red, tubular by the base, and expanded in the upper section, 7–10 cm length; the outside surface is either glabrous or pubescent. Blooming take place in spring (August–September). The fruits are dark-brown dehiscent capsules, 15–38 cm long, 0.9–1.5 cm wide, smooth, with two lateral sutures, covered by numerous scales, with a persistent calix. The fruits contain numerous winged white seeds, 2–3 cm length. Plants are hermaphrodite (Gilman and Watson 1994; Grose and Olmstead 2007; Mechler 2009; Alonso and Desmarchelier 2015).

Ethnobotanical information In the pre-Columbian America, it was used by different native communities to treat diseases such as arthrosis, fever, and digestive and circulatory diseases and as a magical component for treating *mal de ojo*, etc. Later on, infusions from the aerial parts were used as antibacterial and the decoctions of the stem bark as astringent, vulnerary, and antiseptic. Also, in popular medicine it is taken to treat external diseases as ulcers, eczema, psoriasis, hemorrhoids, candidiasis, and skin tumors. Besides, it is used as antianemic, hypotensive, revitalizing, hypoglycemic, abortive, and anti-infective urinary laxative in diseases of respiratory tract (whooping cough) and as an antitumor (leukemia). Finally, it has applications as antinociceptive, anti-dermatogen, antiviral, antitrypanosome, antifungal, antibacterial, and antitumor (Ratera and Ratera 1980; Toursarkissian 1980; Barboza et al. 2009; Alonso and Desmarchelier 2015).

Chemical data The main active principle from the species of this family is lapachol, a quinone compound (Fig. 5.10). Other compounds found in *T. impetiginosa* are flavonoids, tannins, and other quinones. The dried heartwood contains, besides lapachol, the quinoids 1-hydroxyanthraquinone and 2-methyl anthraquinone. The dried bark contains the iridoid monoterpene ajugol, 6–0-(4-hydroxybenzoyl). Also, it contains anisaldehyde and the quinoid naphtho (2,3b) furan-4,9-dione, 2-acetyl (Fig. 5.11). The stem bark contains lapachol, α-and β-lapachone, the lignans cyclo-olivile, and aryltetraline-type lignans, and furanonaphtoquinones. The mayor volatile compounds isolated from dried inner bark are 4-methoxybenzaldehyde, 4-methoxyphenol, 5-allyl-1,2,3-trimethoxybenzene (elemicin), 1-methoxy-4-(1E)-

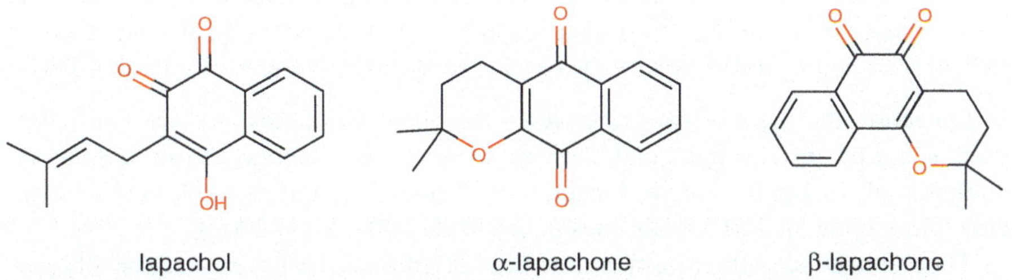

Fig. 5.10 Chemical structure of lapachol and other quinones found in *Tabebuia impetiginosa* (Mart. ex DC.) Standl

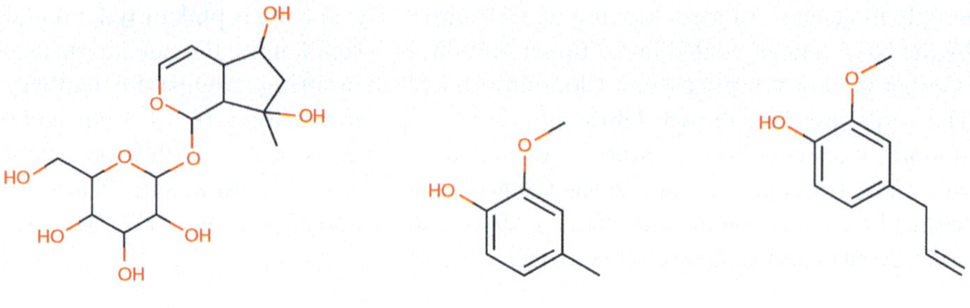

Fig. 5.11 Chemicals found on the dried inner bark of *Tabebuia impetiginosa* (Mart. ex DC.) Standl

1-propenylbenzene (trans-anethole), and 4-methoxybenzyl alcohol, 2-methoxy-4-methylphenol and eugenol (Park et al. 2003; Barboza et al. 2009).

Biological activities Dried plant material (inner bark) is commercialized to prepare infusions, pills, and syrups. *Impetiginosa* means that it was used to treat impetigo, an infection of the skin caused by *Staphylococcus* or *Streptococcus*. Also its activity was referred as antiulcer, leishmanicidal, anti-carcinogenic, anti-edema, anti-inflammatory, antimalarial, antiseptic, antiviral, bactericide, fungicide, pesticide schistosomicide, antinociceptive, and hypocholesterolemic in hyperlipidemic mice (Barboza et al. 2009). Some studies have shown that the preparations made with bark and leaf and the dried bark have antimicrobial activity, while leaves have antioxidant activity. *In vitro* antimicrobial activity assays were made testing ethanolic extracts from leaves, stem bark, and branches against *Staphylococcus aureus* ATCC 25923, *Salmonella enterica* ATCC 1307, *Acinetobacter baumannii* ATCC 17978, *Enterococcus faecalis* ATCC 29212, *Pseudomonas aeruginosa* ATCC 27853, and *Escherichia coli* ATCC 25922. The crude ethanolic extract of leaves has moderate antimicrobial activity against *S. aureus* (MIC $\geq$312.5 μg ml^{-1}) and a very weak inhibitory activity against *S. epidermidis* ($\geq$ 625 μg ml-*1*). For *Acinetobacter calcoaceticus* and *Enterococcus faecalis*, MIC was $\geq$1 μg ml-*1* (Moreira Vasconcelos et al. 2014). *In vitro* studies with the ethanolic and methanolic extracts and the infusion were made to test their free radical scavenging activity (DPPH), reducing power and inhibition of lipid peroxidation in brain homogenates. In all cases the free radical scavenging ability was negative (Suo et al. 2013; Vasconcelos et al. 2014).

Lapachol, the main active principle of the plant, was associated with activities such as antiulcer, leishmanicidal, anticarcinogenic, anti-edema, anti-inflammatory, antimalarial, antiseptic, antiviral, bactericide, fungicide, and pesticide schistosomicide (Hussain et al. 2007; Rodrigues de Almeida 2009; Silva 2009).

The methanolic extract from commercial dried inner bark was successively partitioned into hexane, chloroform, ethyl acetate, butanol, and water-soluble fractions, and then they were tested against *Helicobacter pylori* ATCC 43504. The hexane, ethyl acetate, and chloroform fractions had a strong antibacterial activity. The components present in the chloroform fraction 2-(hydroxymethyl) anthraquinone, anthraquinone-2-carboxylic acid, and lapachol were separately studied being 2-hydroxymethyl anthraquinone at a dose of 0.1 mg per disc, the one with the strongest activity against *H. pylori* (Park et al. 2006).

The antioxidant activity of the volatile compounds isolated from the dried inner bark showed a dose-dependent inhibitory activity at a concentration of 1 mg ml^{-1}. Among the components isolated from the extract, 2-methoxy-4-methylphenol and eugenol had the strongest antioxidant activity. The extract appeared to be more active in the hexanal/hexanoic acid assay than in the conjugated diene assay (Park et al. 2003).

Toxicity The ethanol extract of leaves and stem bark did not show cytotoxicity; however, the extract of twigs were cytotoxic (LC50 546.7 μM). The cytotoxicity was determined in four human tumor cell lines (MCF-7, NCI-H460, HeLa, and HepG2) and also in non-tumor cells (porcine liver primary cells, PLP2) (Pires et al. 2015). On the other hand, the genotoxic potential of the flower extracts (100, 300, and 500 mg kg^{-1} of body weight) on the blood and liver cells of male Wistar rats was evaluated after 24 h of administration. A significant increase on DNA damage was determined at the two highest doses mostly on liver cells; however, in both cases the response was dose-dependent (Lemos et al. 2012). When genotoxicity was tested in *D. melanogaster* by the somatic mutation and recombination test (SMART), it was shown that both lapachol and β-lapachone induced apoptosis by generating oxygen-reactive species affecting cell cycle checkpoints. However, the results showed that *T. impetiginosa* bark and stem powder dissolved in pure water were toxic but not genotoxic by themselves, having a potential carcinogenicity activity. More experiments are being performed in order to analyze interactions with chemotherapeutic drugs (de Sousa et al. 2009).

***In vitro* cultures** *In vitro* propagation was achieved in MS culture medium with Gamborg vitamins (MSG) (Murashige and Skoog 1962; Gamborg et al. 1968) and Woody Plant Medium (WPM) (Lloyd and McCown 1980). The optimal conditions were WPM supplemented with 20 μM BA and 1 μM IBA (MR: 4.37) and 40 d for rooting in 1/2 MSG using 50 μM IBA induction (83% rooting) (Larraburu et al. 2012).

5.8 *Allophylus edulis* (A. St.-Hil., A. Juss. & Cambess.) Hieron. ex Niederl. (Sapindaceae)

Synonymy *Allophylus edulis* (A. St.-Hil., A. Juss. & Cambess.) Hieron. ex Niederl. var. gracilis, *Allophylus pauciflorus* Radlk. var. rojasii, *Allophylus edulis* (A. St.-Hil., A. Juss. & Cambess.) Hieron. ex Niederl. var. rosae, *Schmidelia edulis* A. St.-Hil., A. Juss. & Cambess., *Urvillea seriana* Griseb.

Vernacular name Chal chal, coloradillo, conhú, kokú, picazú-rembiú (wild pigeon food), pitanga, guacú or huacú, chichita, bacú, cocú, codocoypu, coquy, cuquito, jocú, pucancho, puca-puca, sena, huaquito, vacú, vaquito (Ferrucci 2004).

General information The species is endemic to British Guiana, Brazil, Paraguay, Bolivia, Uruguay, and Argentina. In Argentina it is found in JUJ, SAL, TUC, CAT, FOR, CHA, MIS, COS, SFE, ERI, and BAI to the Martín García Island. It grows in humid and shady places, in *serrano* pastureland, and in the *Chaqueño* forest in the transition with the Yungas in *Valle de Lerma* (SAL). This plant is commonly used as a source of tannins and essential oils (Resico, 2007). Sometimes the wood is used as fuel (Tortorelli, op. cit.: 536). *Aloja de chalchal* is a fermented sweet drink made with the fruits (Ferrucci 2004). Fruits are appreciated by birds, inambues, agutíes, tangaráes, pigeons, thrushes, and *Artibeus lituratus* (Bodrati and Haene 2004). It is also considered an ornamental species. It was tested the development of the species in petroleum-contaminated soils; since roots and eophyll structure showed little variation, the authors concluded that *A. edulis* has the potential to be used for phytoremediation (Nogueira et al. 2011).

Botanical characteristics It is a perennial tree or shrub, 4–15 m high, 10–50 cm trunk diameter, and a thin bark. Bark is covered by scales. Branches are chestnut-gray, with a smooth bark and linear lenticels. Leaves are alternate, composite, trifoliate, with sessile or subsessile oblanceolate or elliptic foliates, mucronate, toothed-serrated, with an acute apex, and without stipules. Petiole is 1–4 cm length. Inflorescences are spiciform or sub-spiciform, puberulous, or pubescent, with a 1.0–4.5 cm peduncle and pubescent bracts. The male flower is 8 mm × 8 mm, with four free concave uneven green sepals, ovate the two external and subcircular the two internal; with four free white petals, with ovate 1 mm limb and a 0.77 mm nail and with long hairs in the internal surface particularly in the margins; and with seven stamen that surround the superior ovary. The feminine flowers are 2–4 mm long, with 1.25–2.00 mm glabre sepals; 1.25–2.5 ciliated petals, with a bifid scale, a gold-yellow or orange glabrous nectary disc; and 1.5–3.0 mm stamens in the staminate flower, smaller in the pistillate flower, pubescent filaments, and ovary with disperse glandular hairs. The fruit is subglobose or ovoid, 8 mm diameter, reddish when ripe. The seeds are subglobose, 4 mm diameter, brown with a small aril. Blooming is in September–October. Fructification is in November–January (Ferrucci 2004; Malizia et al. 2018).

Ethnomedical information Traditionally, the leaves were used as anti-inflammatory and for gastrointestinal disorders (Körbes 1995; Fatori Trevizan et al. 2016). Also, the leaves and stems were used in popular phytotherapy as a digestant and for hepatitis treatment (Amat and Vajia 1991). In the province of MIS, leaves are added to *tereré*, a beverage made with *I. paraguariensis*. Also, this is used in cases of digestive disorders. It was reported the use of leaves and stems by the Polish migrants in Misiones in cases of slow digestion, stomach and liver refreshing (humoral medicine), stomach cooling (humoral medicine), liver pains, hepatitis, liquid retention in organism (diuretic), and *empacho*; ocasionally, the fruits are taken as snack (Kujawska and Pieroni 2015). The decoction and the infusion made with the leaves are used as analgesic in cases of cardiac pain, cancer, and stomach-ache (Rondina et al. 2006).

Chemical data It has phenolic compounds, including bergenin, quercetin 3-O-(2”-O-galloyl)-glucoside, vitexin-2”-O-α-rhamnoside, isovitexin, vicenin, vitexin and quercetin-3-O-glucuronide dihydroisocoumarin, L-quebrachitol, 6,7-epoxycaryophyllene, spathunelol, sitosterone, lupeol, 2-oxo-13-hidroxy-neo-cleroda-3,14-diene, and sitosterol. Viridiflorol is the main component of the EO (30.88%) (Fig. 5.12). Other compounds, such as α-thujene, caryophyllene,

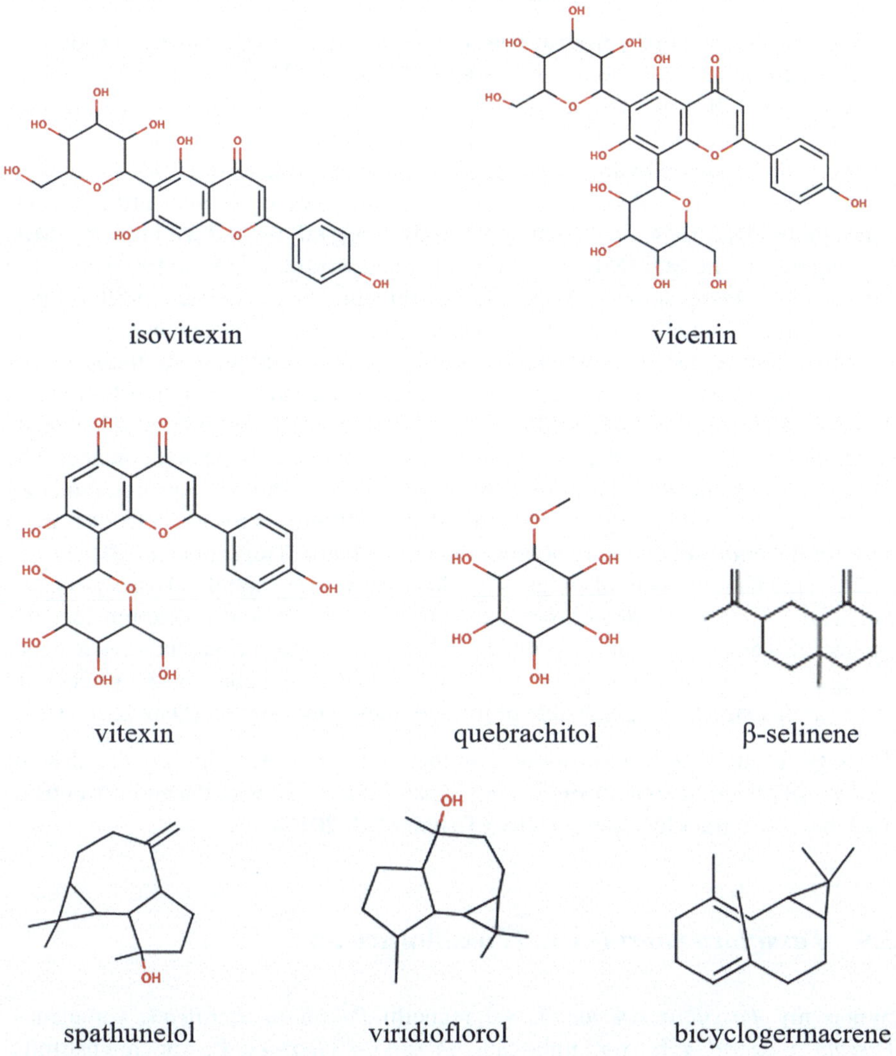

Fig. 5.12 Chemical compounds found in *Allophylus edulis* (A. St.-Hil., A. Juss. & Cambess.) Hieron. ex Niederl

β-selinene, bicyclogermacrene, δ-cadinene, β-atlantol, α-cadinol, and andledol, were present in concentrations ranging from 3.02% to 7.87%. Leaves also contain cyanogenic compounds and alkaloids (Bandoni et al. 1972; Ariswa et al. 1989; Rastogi and Mehrotra 1990, 1993; Yajía et al. 1999; Díaz et al. 2008; Ragonese et Milano, op. cit.: 1984; Fatori Trevizan et al. 2016).

Biological activities *A. edulis* extract have pharmacological activities such as anti-hepatotoxic, negative ionotropic effect, inhibition of angiotensin converting enzyme (ACE) and β-glucuronidase, cytotoxicity to KB cells, antioxidant activity, and antimicrobial effect against *S. aureus*, as well as low toxicity (Ariswa et al. 1989; Hoffmann-Bohm et al. 1992; Matsunaga et al. 1997; Tirloni et al. 2015a, b).

The EO obtained by hydrodistillation of fresh leaves and the isolated viridiflorol were *in vitro* tested as antioxidant using DPPH and ABTS reagents and as antimicrobial against *Mycobacterium tuberculosis* H37Rv (ATCC 27294). The samples exhibited moderate effect in both assays, comparable to those of the commercial antioxidants butylated hydroxytoluene (BHT) and ascorbic acid. As for the antimicrobial activity, MIC values were measured with respect to two standard antitubercular (anti-TB) drugs, isoniazid (MIC 0.05 mg ml^{-1}), and rifampicin (MIC 0.01 mg ml^{-1}). The EOs (MIC 157.5 mg ml^{-1}) and viridiflorol (MIC 190.0 mg ml^{-1}) exhibited promising activity. As for the anti-inflammatory activity, viridiflorol produced a significant inhibition of paw edema caused by carrageenan, and the oral administration of the EOs (30 and 100 mg kg^{-1}) inhibited leukocyte migration by 62 ± 5% and 35 ± 8%, respectively, 4 h after the carrageenan injection. Additionally, the oral administration of viridiflorol (3 and 30 mg kg^{-1}) also inhibited leukocyte emigration by 71 ± 5% and 57 ± 3%, respectively, in a dose-dependent manner. The authors have found that the pure viridiflorol and EOs showed anti-inflammatory activity in comparable doses, concluding that viridiflorol is responsible, at least in part, for the anti-inflammatory action of the EOs (Fatori Trevizan et al. 2016).

The ethanolic extracts of twigs were deterrent against aphids (*Myzus persicae*, *Epilachna paenulata*, *Rhopalosiphum padi*) and have a lower deterrent activity against chewers. Fractionation resulted in a lower activity against aphids and a higher one against chewers. A synergism against aphids and an antagonism against chewers in the crude extract are responsible of the activities demonstrated (Díaz et al. 2014).

Toxicity Preliminary assays showed an increase in liver weight only at a dose of 5 g kg^{-1} of ethanolic extract suggesting hepatotoxicity. Hence, the authors considered that the extract has low toxicity (Tirloni et al. 2015a, b).

5.9 *Passiflora caerulea* L. (Passifloraceae)

Synonymy *Passiflora caerulea* L. var. regnellii, *Passiflora caerulea* L. var. glauca, *Passiflora caerulea* L. var. imbricata, *Passiflora caerulea* L. var. angustifolia, *Passiflora caerulea* L. var. glaucophylla.

Vernacular name Granadilla, mburucuyá (fly hatchery), maracuyá (Portuguese), flor de la pasión (blue passion flower), pasionaria.

General information The name of the species comes from the first observations made by the Spanish conquerors that arrived to America. They saw the symbol of the crucifixion of Christ in the flower, hence the name of *pasionaria* (passion flower). In the polychrome filaments, they saw the crown of thorns and the five wounds of Christ; the three styles were the nails and the stamens the hammers that sank these nails. The sepals and petals, in number of ten, would represent the Apostles. The two that are not represented are Judas and Peter, one for his betrayal and the other, for having denied Christ. *P. caerulea* is a bulbous vine that grows in BAI, CAT, CHA, COR, COS, DFE, ERI, FOR, JUJ, LPA, LRI, MEN, MIS, SAL, SDE, SFE, SJU, SLU, and TUC. There were reported 18 species growing in Argentina. Among them, the species with medical properties are *P. caerulea*, *P. capsularis*, *P. cicinnata*, *P. edulis*, *P. elegans*, *P. foetida*, *P. misera* Kunth, *P. mooreana*, *P. palmatisecta*, *P. suberosa*, *P. tenuifolia*, and *P. umbilicata* (Barboza et al. 2009).

Botanical characteristics It is a woody, climbing plant with a glabrous stem, with tendrils to climb. The glabrous leaves are alternate, sectate palmate with five lanceolate lobes, dark green on the adaxial side, and pale bluish green on the abaxial one, and an entire margin. The petiole has nectaries and small and caducous stipules. The large flowers are solitary, hermaphrodites, with a calix with 4–5 free sepals and 4–5 free petals, with peduncles. The crown has filaments (*radii*), blue on the apex, white on the center, and purple on the base, five free or fused stamens, and a nectary (Fig. 5.13). The fruit is an ovoid fleshy berry, 6 cm-length, orange on the outside and an edible red endocarp. The obcordate or cuneate seeds have an intense perfume

Fig. 5.13 *Passiflora caerulea* L. flower. MA Alvarez personal collection

(Gupta 1995; Lahitte et al. 1999; Deginani 2001; Şesan et al. 2016). The striking corolla functions as attractant to pollinators by visual and olfactory stimuli (Varela et al. 2016).

Ethnobotanical information The dried aerial parts are used to prepare an infusion (usually 1%), with attributed properties as sedative, hypotensive, cardiotonic, antispasmodic, antiscorbutic, diuretic, anti-icteric, and emmenagogue, and to fight alcoholism. The decoction of the aerial parts is used to treat urinary and respiratory tract infections, diarrhea, catarrh, and pneumonia. Leaves, flowers, fruits, and stems are used as sedative, anxiolytic, cordial, anthelmintic, calmative, eupeptic, diuretic, emmenagogue, and contraceptive. The leaf is used as hypotensive, bradycardic, anticephalgia, and sedative. Fruits are taken as diuretic, antiscorbutic, and anti-jaundice. Roots are considered as antispasmodic, antiphlogistic, toxic, narcotic, emmenagogue, sedative, contraceptive, and anthelmintic. The infusion or the syrup made from the root is taken against pneumonia. Finally seeds are considered hypnotic. Sometimes the infusion is associated with *Melissa officinalis*, *Tilia* spp., *Piper methysticum*, or *Valeriana officinalis* for treating insomnia and with *Crataegus oxyacantha* to treat tachycardia and palpitations (Martínez Crovetto 1981; Barboza et al. 2009; Petenatti et al. 2014; Alonso and Desmarchelier 2015).

Chemical data The dried aerial parts contain caffeic acid, coumarin, umbelliferone, harman (passiflorine,1-methyl-9H-pyrido[3,4-b] indole), chrysin (5,7-dihidroxiflavone), tetraphyllin-4-sulphate, and epitetraphyllin- B4-sulphate. Leaves contain gynocardin and the flavone schaftoside. Vitexin and isoorientin were reported in leaves and calli. Greenery stems have steroids, saponins, peroxidases, and heterosides. Fruits contain glucose, fatty acids, the flavone chrysin, cyanogenic glucoside sulphate tetraphyllin 8–4-sulphate, and epitetraphyllin B-4-sulphate (Fig. 5.14). The alcoholic extracts contain more bioactive compounds than petroleum ether extracts; saponins were only determined in ethanolic extracts. Maximum flavonoid content was 12.82 mg Q g^{-1} DW (Farag et al. 2016; Şesan et al. 2016). The intense aroma from flowers were attributed to sesquiterpenes secreted by the periantum, α-copaene, 2(E),6(E)-farnesol, D-germacrene, (−)-α-cubebene, and to the ester methyl-linolenate (Martínez Crovetto 1981; Seigler et al. 1982; Speroni et al. 1996; Pereira et al. 2004; Áquila et al. 2005; Barboza et al. 2009; Varela et al. 2016; Hadas et al. 2017).

Biological activities The ethanolic extract from leaves had higher anticonvulsant, analgesic, anti-inflammatory, antipyretic, and antioxidant activities than the aqueous extract (Şesan et al. 2016; El-Askary et al. 2017; Hadas et al. 2017). After fractionated extraction from the ethanolic extract, the ethyl acetate fraction was the most active. The isolated compounds were lucenin, 2,4-hydroxycinnamic acid, and chrysin 6-C-β-D-glucoside. The biological activities were attributed to C-glycosyl flavones and hydroxycinnamic acid derivatives (El-Askary et al. 2017). The antibacterial activity was endorsed to flavonoids (Anesini and Perez 1993; Dhawan et al. 2004; Pérez Ibáñez et al. 2017).

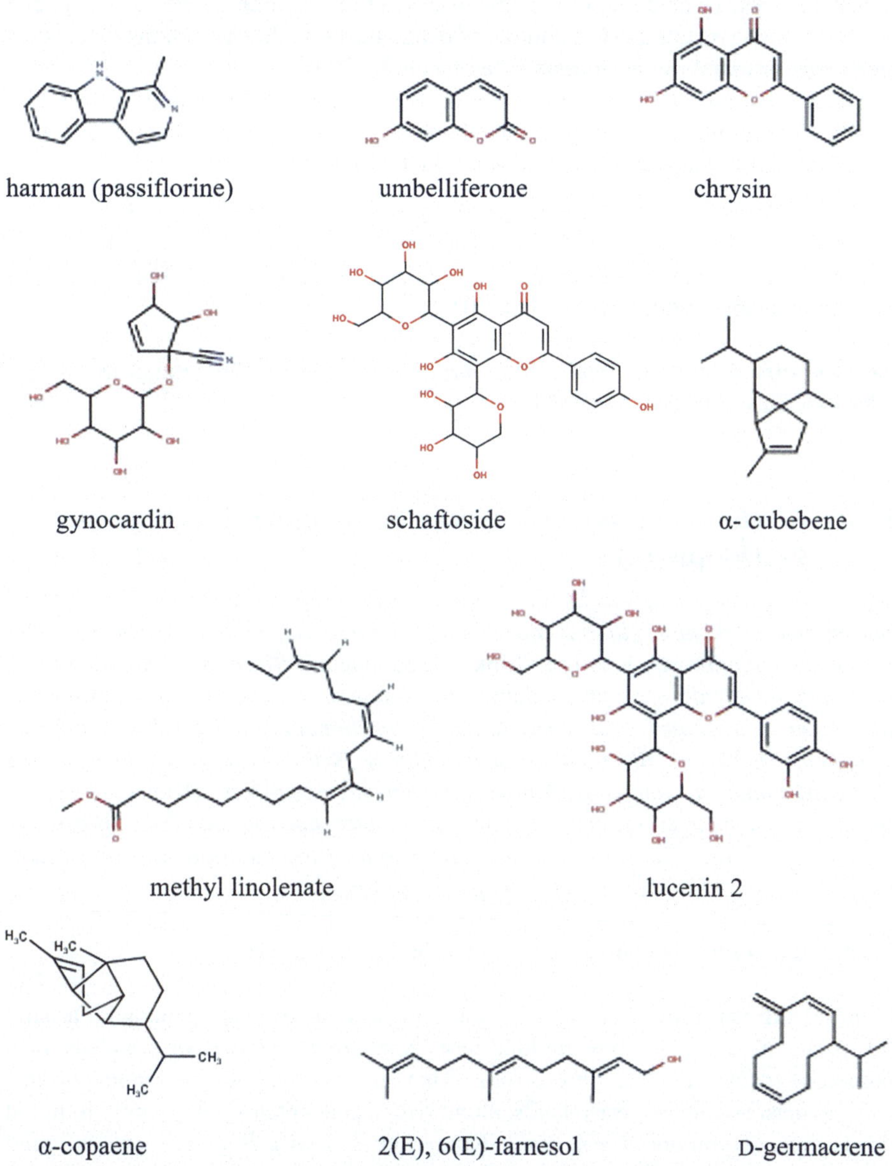

Fig. 5.14 Some of the chemical compounds determined in *Passiflora caerulea* L.

Chrysin (1 mg kg^{-1}) induced significant anxiolytic behavior in mice, without myorelaxant effect (dose range 0.6–30 mg kg-^{1}) suggesting that it was an anxiolytic devoid of sedative or muscle-relaxing collateral effect, unlike diazepam (Wolfman et al. 1994; Dhawan et al. 2004; Duke et al. 2009; Feliu-Hemmelmann et al. 2013). It was reported that chrysin prevented the expression of tonic-clonic seizures induced by pentylenetetrazol by acting as a ligand for central as well as peripheral benzodiazepine receptors. The content of benzodiazepine-like compounds in *P. caerulea* was assumed by the authors (Medina et al. 1990).

Other reported activities were as phytoestrogens with potential clinical and therapeutic utilizations against physiological and biochemical effects of aging (Dhawan et al. 2002) and as anthelmintic and acaricidal (Duke et al. 2009).

***In vitro* culture** *In vitro* cultures of *P. caerulea* were established. The *in vitro* response of different explants and biotypes on MSG culture medium with 1 mg l^{-1} of benzyl adenine (BA) as plant growth regulator produced calli or regenerated the plant (Nakayama 1966; Severin et al. 2011).

Legal status *P. caerulea* and *P. incarnata* were approved for human use by the ANMAT (Res. 2673/99, annex III).

5.10 *Blechnum occidentale* L. var. occidentale (Blechnaceae, Pteridophyta)

Synonymy *Blechnum glandulosum* Kaulf. ex Link, *Blechnum suburbicum* Vell., *Blechnum glandulosum* Kaulf. ex Link var. elongatum, *Blechnum caudatum* Cav., *Blechnum occidentale* L. var. caudatum, *Blechnum occidentale* L. var. pubirhachis, *Blechnum occidentale* L. var. lacerata, *Blechnum cognatum* C. Presl, *Blechnum cartilagineum* Schkuhr, *Blechnum appendiculatum* Willd., *Blechnum meridionale* C. Presl, *Lomaria campylotis* Kunze, *Mesothema campylotis* (Kunze) C. Presl, *Blechnum campylotis* (Kunze) J. Sm., *Blechnum mucronatum* Fée, *Blechnum lomarioides* Sodiro, *Blechnum scaberulum* Sodiro, *Blechnum glandulosum* Kaulf. ex Link var. pallida, *Blechnum flaccisquama* A. Rojas.

Vernacular name Doradilla, hammock fern, blechnum fern.

General information *Blechnum* L. is a worldwide distributed genus with around 50 species in America. They are land ferns, sometimes epiphyte or scandant, with numerous scales and often fibrous roots. They grow in rainy jungles, mountain forests, ravines, savannahs, wastelands, along rivers, and streams being rarely found in dry regions (Tryon and Tryon 1982; Kramer et al. 1990). *B. occidentale* is found from the south of North America to South America; in Argentina there are 20–21 taxa mainly in the provinces of JUJ, MIS, SAL, and TUC (Ponce 1996; Durán 1997).

Botanical characteristics *B. occidentale* is a perennial, terrestrial, or rare rupicolas (epipetric) species with an upright decumbent stiff with bicolor scales. The fronds are monomorphic, subdimorphic, or dimorphic with a petiole with triangular scales, mainly in the base. The leaf blade is pinnate at the base becoming pinnatisect towards the apex, truncated at the base, glabrous or with hyaline trichomes. The rachis is glabrous or with septate trichomes. Pinnae are in 13–28 pairs, 1.2–9.5 cm length × 0.6–1.6 cm wide with a free venation; veins are undivided or 1–4 bifurcated; pinnae have an entire to serrate margin, not articulated to the rachis. Sori are placed on a long or rarely short linear vascular commissure, parallel and close to the main vein of each segment, sometimes extending beyond the commissure. Sporangia have a three-rowed stalk with ellipsoidal to spheroidal monolete spores (kidney-shaped grains with a single elongate scar) (Dittrich et al. 2007; Pessôa Santiago et al. 2014; Dittrich et al. 2015).

Ethnobotanical information In Brazil this fern has been used to treat inflammatory and pulmonary diseases, urinary infections, and liver diseases (Barros and Andrade, 1997). In Argentina the entire plant was used to treat lung and urinary disorders. A not specified part was taken as an infusion and decoction to dissolve renal calculi and for treatment of nervous hysteria (Barboza et al. 2009).

Chemical data The dried leaf contains rhodoxanthin, lutein epoxide, lutein, adonixanthin, zeaxanthin, violaxanthin, and antheraxanthin (Fig. 5.15) (Barboza et al. 2009).

Biological activities The methanolic extract from air-dried and powdered blades was tested on male Wistar rats or Swiss Webster mice for its activity as anti-inflammatory (paw edema model in mice) and antinociceptive (writhing test). The administration resulted in an inhibition of the inflammatory effect caused by carrageenan and in an antinociceptive effect in acetic acid-induced writhing and formalin tests. The authors suggested that those effects are related to the arachidonic acid cascade and/or modulation of pro-inflammatory molecule production (Nonato et al. 2009). The entire plant had hypothermic, diuretic, and antibacterial activity (Barboza et al. 2009).

Toxicity The acute toxicity of the methanolic extract was tested giving to mice a 100 or 1000 mg kg^{-1} intraperitoneal dose. After 14 days any death was produced. Besides, any motor performance alteration was not detected (Nonato et al. 2009).

5.11 *Maytenus ilicifolia* Mart. ex Reissek (Celesteraceae)

Synonymy *Celastrus spinifolium* Larrañaga, *Monteverdia ilicifolia* (Mart. ex Reissek) Biral, *Maytenus aquifolium* Mart., *Maytenus hassleri* Briq., *Maytenus ilicifolia* Mart. ex Reissek f. angustior Briq., *Maytenus pilcomayensis* Briq.

Vernacular name Pus pus, congorosa, concorosa, quebrachillo, molle morotí, cangorosa, capororoca, molle espinoso, cancerosa, sombra de toro.

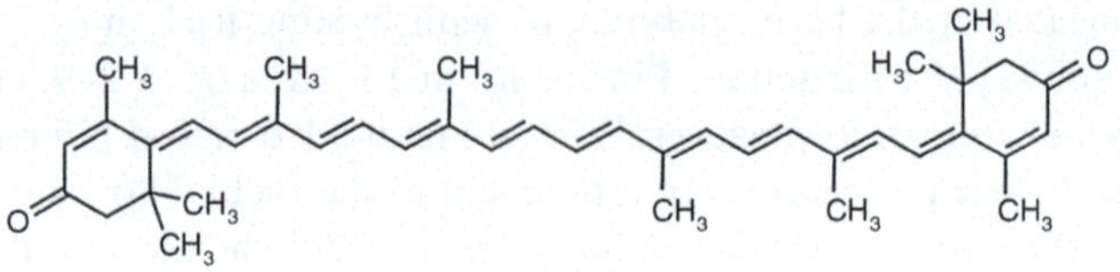

rhodoxanthin

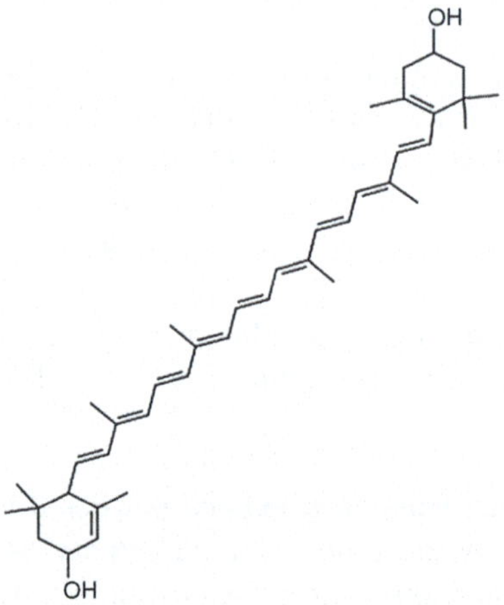

lutein

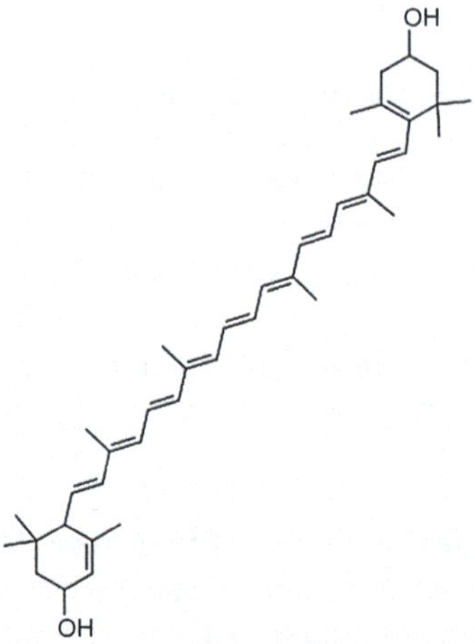

zeaxanthin

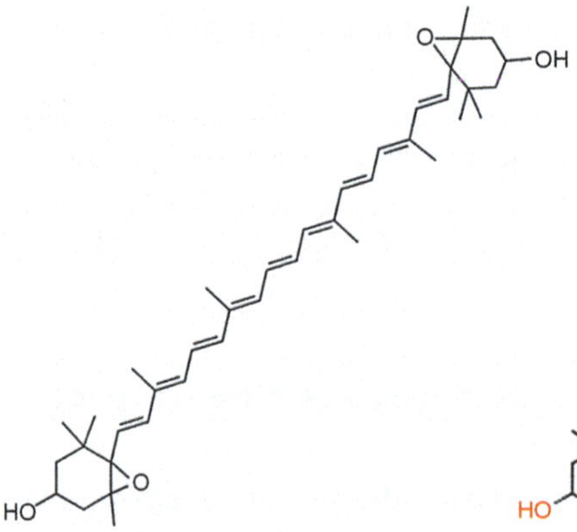

violaxanthin

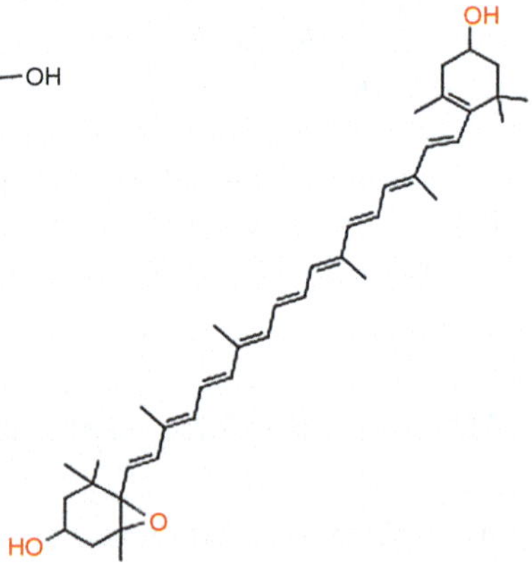

antheraxanthin

Fig. 5.15 Chemicals found in *Blechnum occidentale* L. var. occidentale

General information The genus *Maytenus* include about 200 American species of which 11 are in Argentina and 77 in Brazil (Hurrell and Bazanno, 2003). *M. ilicifolia* is native from South Brazil, Paraguay, Bolivia, Uruguay, and Argentina where it is mainly found in the provinces of BAI, CHA, COS, ERI, FOR, MIS, SAL, and SFE. It grows in subtropical temperate climates, with well-drained clay soils and a high content of organic matter, up to 1200–2000 m.a.s.l (Hurrell and Bazanno 2003).

Botanical characteristics *M. ilicifolia* is a dioecious perennial shrub, up to 5 m height (up to 10 m height in the Mata Atlántica). The root is long and pivoting (Brussa Santander and Grela González 2007). Leaves are simple, entire, coriaceous, with alternate phyllotaxy, a short petiole and a lanceolate laminae (2–7 cm length and 1.5–3.0 cm width). The leaf apex is acute with toothed thorny margins (2–7 pairs), the leaf base is cuneate or rounded, the midrib is bright green more prominent in the abaxial side which is paler (Fig. 5.16). The inflorescence is solitary or in axillar yellow fascicles with reddish bracts. Flowers are small, actinomorphic, with a small pedicel, five reddish sepals united by the base, five free yellow ovate or entire or dented petals, five stamen, five emarginated discs, and two loculate ovary with four included or emerging ovules. The fruit is a reddish ovoid or ellipsoid 1 cm length capsule with a thin aryl containing 1–4 reddish seeds. Blooming is during spring and fructification in summer; fertilization is entomophile (Cabrera 1965; Hurrell and Bazzano 2003; Duarte and Debur 2005).

Fig. 5.16 *Maytenus ilicifolia* Mart. ex Reissek, Celesteraceae. Buenos Aires, Botanical Garden. MA A Personal collection

Ethnobotanical information The infusion made with the entire plant is used in folk medicine for fertility regulation and as antiasthmatic. The *Guaraníes* have taken it as contraceptive. The aerial parts are used as vulnerary, contraceptive, hypotensive, cardiotonic, antidiarrheal, decongestant, and odontalgic, for lumbago, kidney disorders, gastric ulcers and gastritis, and articular pains. The decoction made with the flowers is taken as anti-inflammatory and those from leaves as antiasthmatic and antiseptic. The infusion of the leaves is considered as emmenagogue, sialagogue, astringent, antispasmodic, and contraceptive. The roots are taken as diuretic, contraceptive, and emmenagogue. In Argentina and South Brazil, the decoction of leaves and stems is added to the beverage *mate* in cases of bloody ulcers, hypertension, and articular pains, as depurative; as sialagogue, to treat asthma; and as antitumoral (Bandoni 1976; Toursarkissian 1980; Martínez-Crovetto 1987; Arenas 1997; Hurrell and Bazzano 2003; Alonso 2004; Scheffer et al. 2005; Barboza et al. 2009; Alonso and Desmarchelier 2015).

Chemical data The leaves have terpenes, phenols, phytosterols, and alkaloids (Figs. 5.17 and 5.18). The dried leaves contain the terpenoids α-amyrin, cangorosin A and B; the triterpenoids maytefolins A, B, C, uvaol-3-caffeate, erythrodiol, betulin, betulin-3-caffeate, moradiol, erythrodiol-3-caffeate, 20α-hydroxymaytenin, 22β-hydroxymaytenin, maytenin, celastrol, and pristimerin; the tannins afzelechin, epiafzelechin, catechin, epicatechin, gallocatechin and epigallocatechin, 4O-methyl-(epi)-catechin, and 4'O-methyl (epi) catechin; the flavones quercetin and kaempferol; and the maytansinoid alkaloids maitansine, maytanprine, and maytanbutin. Also, it was found friedelin, friedelan-3-ol, α-tocopherol, simiarenol, lupeol, lupenone, β-sitosterol, stigmasterol, campesterol, ergosterol, brassicasterol, squalene, hexadecanoic acid, T1 and T2 tocopherols, cangorin F, G, H, I, J, phytol, vitamin E, dodecanoic acid, and geranyl acetate, and Fe, K, Mg, S, Na, and Ca. The dried root bark contains the sesquiterpenes cangorin A, B, C, D, E, F, G, H, I, and J and the triterpenes friedoolean-24-al-3-en-3-ol-2-on-29-oic acid (cangoronine), friedoolean-1-en-29-ol-3-one (ilicifoline), maytenoic acid, D:B friedoolean-5-en 3β, 29-diol, D:A friedoolean-29-ol-3-one, pristimerin, salasperimic acid, isopristimerin, isotingenone, and 6-oxotingenol. The roots contain the triterpenoids milicifolines AD (Marini-Bettòlo 1981; Ahmed et al. 1981; Itokawa et al. 1991; Silva and Récio 1992; Shirota et al. 1994; Cordeiro et al. 1999; Niero et al. 2001; Buffa Filho et al. 2002; Alonso 2004; Ohsaki et al. 2004; Soares et al. 2004; Pereira et al. 2005; Gutiérrez et al. 2007; Barboza et al. 2009; Leite et al. 2010; Santos et al. 2012).

Biological activities The dried leaves extracts have barbiturate potentiation, antiulcer, antimicrobial, antitumoral, antioxidant, antinociceptive, anti-inflammatory, and relaxant activities. The root bark has activity as antioxidant, myeloperoxidase inhibitor, and as cytotoxic against V79, KB, L1210, VJ300, KU 1920, and P 388 cells (Pereira et al. 1992; Jorge et al. 2004; Barboza et al. 2009; Leite et al. 2010; Alonso and Desmarchelier 2015). The antiulcerogenic and antigastritis effects were attributed to the terpene friedelin, to the tri- and tetraglycosid flavonoid derivatives, and to the combined action of different phytocomplexes (Souza-Formigoni et al. 1991; Cunha 2003).

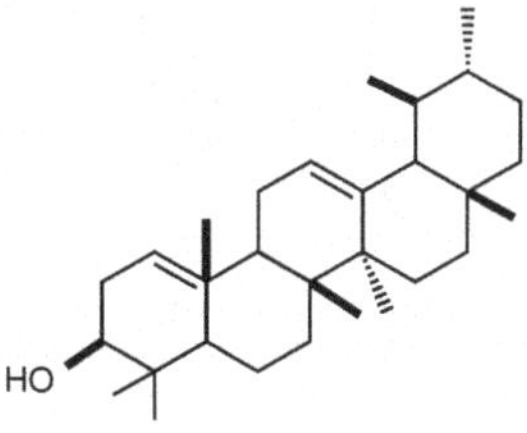

α-amyrin

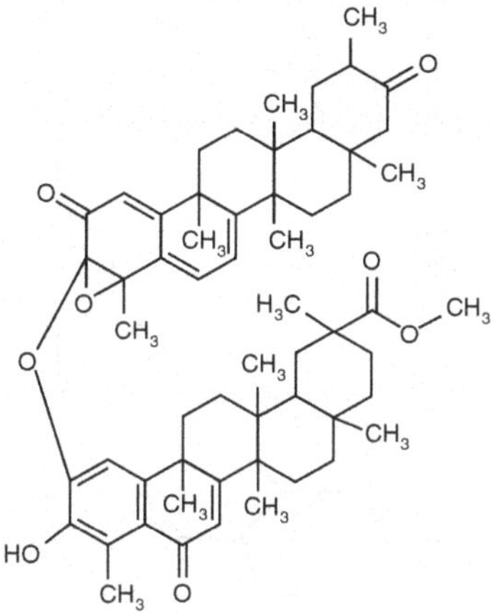

cangorosin B

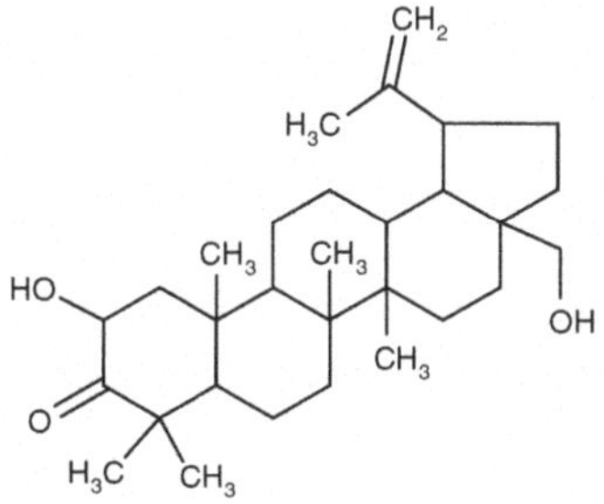

maytefolin A

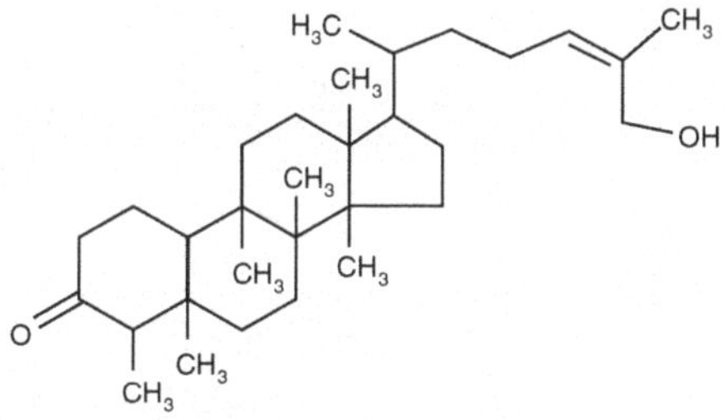

maytefolin C

Fig. 5.17 Some terpenoids from *Maytenus ilicifolia* Mart. ex Reissek leaves

The leaf infusion inhibited gastric secretion induced by histamine and acidity and reduced dyspepsia symptoms (Carlini et al. 1988; Baggio et al. 2007). Also, the oral administration of spray-drying powders maintained antiulcer activity in Wistar male rats (Martins et al. 2003). The lyophilized aqueous extract from leaves inhibited the acid secretion in isolated frog gastric mucosa (Ferreira et al. 2004). The arabinogalactan isolated from *M. ilicifolia* leaves showed a dose-related potent gastroprotective activity against gastric lesions induced by ethanol after oral administration (10–100 mg kg^{-1} body weight) to female Wistar rats and female Swiss mice (Baggio et al. 2012).

The ethanolic extract from dried leaves (intraperitoneally, 200 mg kg^{-1} day^{-1} for 20 days and oral administration, 300 mg kg^{-1} day^{-1} for 30 days) did not arrest spermatogenesis in Swiss mice, although minor ultrastructural alterations were observed (Montanari et al. 1998). Antitumoral and antioxidant effects were attributed to quinonemethide triterpenes, 22β-hydroxymaytenin, and maytenin (Buffa Filho et al. 2004). Patients with advanced stage of neoplasia were treated with 150 μg kg−1

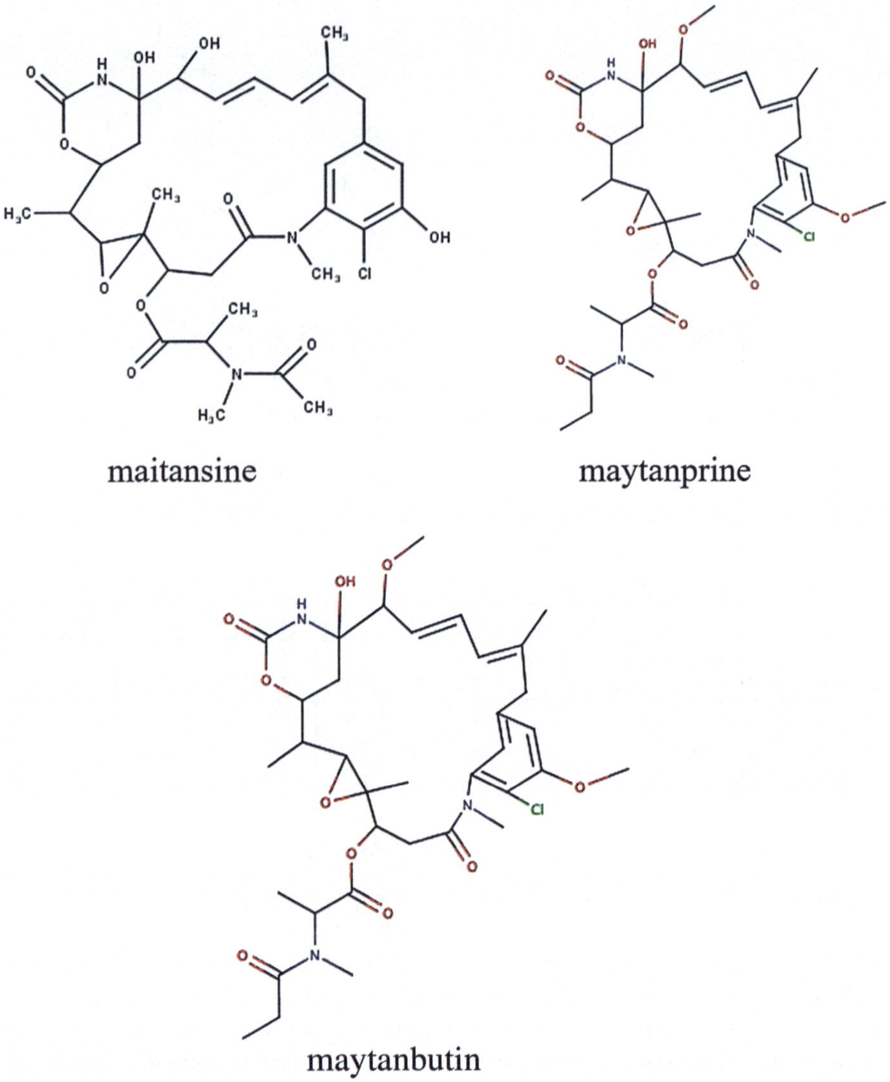

Fig. 5.18 Maytansinoid alkaloids from *Maytenus ilicifolia* Mart. ex Reissek

day−1 and 450 μg kg−1 day−1 of triterpenes extracted from *Maytenus* sp. enriched in maiteina, by intravenous route. Patients with carcinoma epidermoid of the pillars of the amygdala or tonsil, the base of the tongue, and the larynx were experimented an injury reduction of around 40–60% as well as the disappearance of bleeding. Subjective improvements were verified with respect to pain, asthenia, and anorexia. It was not observed a positive response for carcinomas of uterus and ovary, osteosarcoma, and chondrosarcoma (Santos-Oliveira et al. 2009).

Toxicity There are contradictory reports about the *M. ilicifolia* interference with embryo-fetal development. On one side, it was reported that *M. ilicifolia* interferes with uterine receptivity to the embryo for its estrogenic activity in pregnant mice that produced an uterotropic effect (Montanari and Bevilacqua 2002). On the other side, it was reported that *M. ilicifolia* hydroacetonic extract (70:30) resulted nontoxic to Wistar pregnant rats apparently not interfering with pregnancy. Furthermore, fertility of female rats was not altered (Oliveira et al. 1991; Cunha et al. 2014).

***In vitro* cultures** *M. ilicifolia* was micropropagated from leaves in MS medium (Murashige and Skoog 1962) supplemented with BAP 0.4 mg l^{-1} and NAA 0.2 mg l^{-1}. Calli were obtained from leaf explants from micropropagated plants placed on MS medium supplemented with Kin 1.0 mg l^{-1} and 2,4-D 1.0 mg l^{-1} (Buffa Filho et al. 2004).

Legal status: The species in not included in FNA. However, it was included in the Primary Health Care Program from the province of MIS in 2003. In Brazil it is included in the Brazilian Pharmacopeia as phytotherapic (Alonso and Desmarchelier 2015).

References

Actis-Goretta L, Mackenzie GG, Oteiza PI, Fraga CG (2002) Comparative study on the antioxidant capacity of wines and other plant-derived beverages. Ann N Y Acad Sci 957(1):279–283

Agra MF, Locatelli E, Rocha EA et al (1996) Plantas Medicinais dos Cariris Velhos, Paraíba, Brasil, Parte II: Subclasses Magnoliidae, Caryophyliidae, Dilleniida e Rosidae. Ver Bras Farm 77(1):97–102

Ahmed MS, Fong HHS, Soejarto DD, Dobberstein RH, Waller DP, Moreno-Azorero R (1981) High-performance liquid chromatographic separation and quantitation of maytansinoids in *Maytenus ilicifolia*. J Chromatogr A 213(2):340–344

Alikardis F (1987) Natural constituents of *Ilex* species. J Ethnopharmacol 20:121–144

Alonso J (2004) Tratado de Fitofármacos y Nutracéuticos. Corpus, Buenos Aires

Alonso J, Desmarchelier CJ (2015) Plantas medicinales autóctonas de la Argentina. Bases científicas para su aplicación en atención primaria de la salud. 1a ed. – Ciudad Autónoma de Buenos Aires: Corpus Libros Médicos y Científicos

Amat AG, Vajia ME (1991) Plantas medicinales y etnofarmacología en la provincia de Misiones (Argentina). Acta Farm Bonaer 10:153–159

Amorín JL (1974) Plantas de la Flora Argentina relacionadas con alucinógenos americanos. Publicaciones de la Academia Argentina de Farmacia y Bioquímica 1:33–63

Amorin JL, Rosow R (1988) Guia taxonómica con plantas de interés farmacéutico. Dominguezia 10(1):35–40

Anesini C, Perez C (1993) Screening of plants used in argentine folk medicine for antimicrobial activity. J Ethnopharmacol 39:119–128

Anesini C, Ferraro G, Filip R (2006) Peroxidase-like activity of *Ilex paraguariensis*. Food Chem 97:459–464

Anonimous (2017) La Misión. Los 30 pueblos guaraníes. Su estructura cultural y política. San Ignacio-Misiones-Argentina, pp 56

Áquila S, Confort P, Spegazzini E, Volonté M, Debenedetti S. (2005) Determinación de flavonoides de *Passiflora caerulea* por TLC y HPLC para su control de calidad. 1° Congreso de Fitoterápicos del Mercosur. Montevideo, Uruguay. 28 nov-2 dic 2005

Aragão DMO, Guarize L, Lanini J, da Costa JC, García RMG, Scio E (2010) Hypoglycemic effects of *Cecropia pachystachya* in normal and alloxan-induced diabetic rats. J Ethnopharmacol 128:629–633

Aragão DM, Lima IV, da Silva JM, Bellozi PM, da Costa J de C, Cardoso GM, de Souza-Fagundes EM, Scio E (2013) Anti-inflammatory, antinociceptive and cytotoxic effects of the methanol extract of *Cecropia pachystachya* Trecul. Phytother Res 27:926–930

Arçari DP, Bartchewsky W, dos SantosTW OKA, Funck A, Pedrazzoli J, de Souza MF, Saad MJ, Bastos DH, Gambero A, Carvalho Pde O, Ribeiro ML (2009) Antiobesity Effects of yerba maté Extract (*Ilex paraguariensis*) in High-fat Diet–induced Obese Mice. Obesity 17:2127–2133. https://doi.org/10.1038/oby.2009.158

Arcari DP, Bartchewsky W Jr, dos Santos TW, Oliveira KA, De Oliveira CC, Gotardo EM, Pedrazzoli J Jr, Gambero A, Ferraz LF, Carvalho Pde O et al (2011) Anti-inflammatory effects of yerba mate extract (*Ilex paraguariensis*) ameliorate insulin resistance in mice with high fat diet-induced obesity. Mol Cell Endocrinol 335:110–115

Arcari DP, Santos JC, Gambero A, Ribeiro ML (2013a) The *in vitro* and *in vivo* effects of yerba mate (*Ilex paraguariensis*) extract on adipogenesis. Food Chem 141:809–815

Arcari DP, Santos JC, Gambero A, Ferraz LF, Ribeiro ML (2013b) Modulatory effects of yerba mate (*Ilex paraguariensis*) on the PI3K-AKT signaling pathway. Mol Nutr Food Res 57:1882–1885

Arenas P (1997) Las fuentes actuales y del pasado para la etnobotánica del gran Chaco. Monograf Jard Bot Córdoba 5:17–25

Argentina National Pharmacopeia (ANP) Codex Medicamentarius Argentino VIth edition (1978) 111–112

Ariswa M, Morinaga Y, Nishi Y, UenoH SS, Hayashi T, Shimizu M, Yoshizaki M, Morita N, Berganza LH (1989) Chemical and pharmaceutical studies on medicinal plants:constituents of angiotensin converting enzyme inhibitory fraction from "Cocu," *Allophylus edulis* RADLK. Shoyakugaku Zasshi 43:78–80

Ashihara H (1993) Purine metabolism and the biosynthesis of caffeine in maté leaves. Phytochemistry 33(6):1427–1430

Baggio CH, Freitas CS, Otofuji GM, Cipriani TR, de Souza LM, Sassaki GL, Iacomini M, Andrade Marques MC, Mesia-Vela S (2007) Flavonoid-rich fraction of *Maytenus ilicifolia* Mart. ex. Reiss protects the gastric mucosa of rodents through inhibition of both H+, K+-ATPase activity and formation of nitric oxide. J Ethnopharmacol 113(3):433–330

Baggio CH, Freitas CS, Twardowschy A, dos Santos AC, Mayer B, Potrich FB, Cipriani TR, Sassaki GL, Iacomini M, Marques MCA, Mesia-Vela S (2012) *In vivo/in vitro* Studies of the Effects of the Type II Arabinogalactan Isolated from *Maytenus ilicifolia* Mart. ex Reissek on the Gastrointestinal Tract of Rat. Z Naturforsch C 67:405–410

Baltassat F, Darbour N, Ferry S (1984) Study on the purine levels in caffeine drugs. Plantes Medicinales et Phytotherapie 18:195–203

Bandoni AL, Mendiondo ME, Rondina RV, Coussio JD (1972) Survey of Argentine medicinal plants. I Folklore and phytochemical screening Lloydia 35:69–80

Bandoni A (1976) Survey of Argentine medicinal plants. Folklore and Phytochemical Screening II. Econ Bot 30:161–166

Barboza GE, Cantero JJ, Núñez C, Pacciaroni A, Ariza Espinar L (2009) Medicinal Plants: A general review and a phytochemical and ethnopharmacological screening of the native Argentine Flora. Kurtziana 34(12):7365

Barros ICL, Andrade LHC (1997) Pteridófitas medicinais (samambaias, avencas e plantas afins). Universitária da Universidade Federal de Pernambuco, Recife

Bastos DHM, Ishimotoa EY, Marques MOM, Ferri F, Torres EAFS (2006) Essential oil and antioxidant activity of green mate and mate tea (*Ilex paraguariensis*) infusions. J Food Compos Anal 19:538–543

Berté AKAS, Beux MR, Spada PKWDS, Salvador M, Hoffman-Ribani R (2011) Chemical composition and antioxidant activity of yerba-mate (*Ilex paraguariensis* A.St.-Hil., Aquifoliaceae) extract as obtained by spray drying. J Agric Food Chem 59:5523–5527

Bidau CJ, Amat AG, Yajía M, Martí DA, Riglos AG, Silvestroni A (2004) Evaluation of the Genotoxicity of aqueous extracts of *Ilex paraguariensis* St. Hil. (Aquifoliaceae) using the Allium Test. Cytologia 69(2):109–117

Bodrati A, Haene E (2004) Chal-chal o koku. Plantas que atraen aves, Chal-chal o Koku. Nat Conserv 26:16–17

Bonfim Silva-Junior O, Grattapaglia D, Novaes E, Collevatti RG (2017) Genome assembly of the Pink Ipeˆ (*Handroanthus impetiginosus*, Bignoniaceae), a highly valued, ecologically keystone Neotropical timber forest tree. Gigascience 7:1–16

Bracesco N, Dell M, Rocha A, Behtash S, Menini T, Gugliucci A, Nunes E (2003) Antioxidant activity of a botanical extract preparation of *Ilex paraguariensis*: prevention of DNA double-strand breaks in Saccharomyces cerevisiae and human low density lipoprotein oxidation. J Altern Complement Med 9(3):379–387

Brango-Vanegas J, Costa GM, Ortmann CF, Schenkel EP, Reginatto FH, Ramos FA, Arévalo-Ferro C, Castellanos L (2014) Glycosyl flavonoids from *Cecropia pachystachya* Trécul are quor-um sensing inhibitors. Phytomedicine 21:670–675

Brussa Santander CA, Grela González IA (2007) Flora arbórea del Uruguay: con énfasis en las especies de Rivera y Tacuarembó. COFUSA, Rivera (Uruguay)

Buffa Filho W, Corsino J, da Silva BV, Furlan M, Pereira AMS, Castro França S (2002) Quantitative determination of cytotoxic -oleanane derivatives from five morphological types of Celastraceae by reverse-phase high-performance liquid chromatography. Phytochem Anal 13(2):75–78

Buffa Filho W, Silva Bolzani V, Furlan M, Vaz Pereira SI, Soares Pereira AM, Castro França S (2004) *In vitro* propagation of *Maytenus ilicifolia* (Celastraceae) as potential source for antitumoral and antioxidant quinomethide triterpenes production. A rapid quantitative method for their analysis by reverse-phase high-performance liquid chromatography. Arkivoc 2004(vi):137–146

Burris KP, Harte MH, Davidson PM, Neal Stewart C Jr, Zivanovic S (2012a) Composition and bioactive properties of yerba mate (*Ilex paraguariensis* A. St.-Hil.): a review. Chilean J Agric Res 72(2):268–274

Burris KP, Davidson PM, Stewart CN Jr, Zivanovic S, Harte FM (2012b) Aqueous extracts of yerba mate (*Ilex paraguariensis*) as a natural antimicrobial against Escherichia coli O157:H7 in a microbiological medium and pH 6.0 apple juice. J Food Prot 75(4):753–757

Cabrera A (1965) Flora de la Provincia de Buenos Aires. Colección INTA. Tomo IV, Buenos Aires, pp 136–137

Carlini EA, Macaubas CIP, Oliveira MGM, Barbosa VP (1988) Toxicologia préclínica da espinheira-santa (*Maytenus ilicifolia*). In: Carlini, ELA, coord. Estudo da ação antiúlcera gástrica de plantas Brasileiras (*Maytenus ilicifolia* "espinheira-santa" e outras). CEME/ AFIP, Brasília, p 49–66

Cansian RL, Mossi AJ, Mosele SH, Toniazzo G, Treichel H, Paroul N, Oliveira JV, Oliveira D, Mazutti MA, Echeverrigaray S (2008) Genetic conservation and medicinal properties of mate *Ilex paraguariensis* St Hil. Pharmacognosy Rev 2:326–338

Clifford MN, Ramirez-Martinez JR (1990) Chlorogenic acids and purine alkaloids contents of Maté (*Ilex paraguariensis*) leaf and beverage. Food Chem 35(1):13–21

Consolini AE, Migliorisi GN (2005) Cardiovascular effects of the South American medicinal plant *Cecropia pachystachya* (ambay) on rats. J Ethnopharmacol 96:417–422

Consolini AE, Ragone MI, Migliori GN, Conforti P, Volonté MG (2006) Cardiotonic and sedative effects of *Cecropia pachystachya* Mart. (ambay) on isolated rat hearts and conscious mice. J Ethnopharmacol 106:90–96

Cordeiro PJM, Vilegas JHY, Lanças FM (1999) HRGC-MS Analysis of Terpenoids from *Maytenus ilicifolia* and *Maytenus aquifolium* ("Espinheira Santa"). J Braz Chem Soc 10(6):523–526

Costa GM, Schenkel EP, Ortmann CF, Reginatto FH (2011) An HPLC-DAD method to quan-tification of main phenolic compounds from leaves of *Cecropia* species. Braz Chem Soc 26:1096–1102

Cunha SC (2003) Ação farmacológica da espinheira-santa: usos e precauções. Theses, Universidade Federal de Lavras, Brasil

Cunha AL, Auharek SA, Oliveira RJ, Siqueira JM, Vieira MC, Leite VS, Portugal LC (2014) Effects of *Maytenus ilicifolia* on reproduction and embryo-fetal development in Wistar rats. Genetics and Molecular Research 13(2):3711–3720

Debat HJ, Grabiele M, Aguilera PM, Bubillo RE, Otegui MB, Ducasse DA, Zapata PD, Marti DA (2014) Exploring the genes of yerba mate (*Ilex paraguariensis* A. St.-Hil.) by NGS and *De Novo* transcriptome assembly. PLoS One 9(10):e109835. https://doi.org/10.1371/journal.pone.0109835

Deginani NB (2001) Las especies argentinas del género *Passiflora* (Passifloraceae). Darwin 39(1–2):43–129

De Morais EC, Stefanuto A, Klein GA, Boaventura BC, de Andrade F, Wazlawik E, di Pietro PF, Maraschin M, da Silva EL (2009) Consumption of yerba mate (*Ilex paraguariensis*) improves serum lipid parameters in healthy dyslipidemic subjects and provides an additional LDL-cholesterol reduction in individuals on statin therapy. J Agric Food Chem 57:8316–8324

De Resende PE, Kaiser S, Pittol V, Hoefel AL, D' Agostini Silva R, Vieira Marques C, Kucharski LC, González Ortega G (2015) Influence of crude extract and bioactive fractions of *Ilex paraguariensis* A. St. Hil. (yerba mate) on the Wistar rat lipid metabolism. J Funct Foods 15:440–451

De Stefani E, Boffetta P, Deneo-Pellegrini H, Correa P, Ronco A, Brennan P, Ferro G, Acosta G, Mendilaharsu M (2007) Nonalcoholic beverages and risk of bladder cancer in Uruguay. BMC Cancer 29:7–57

De Stefani E, Fierro L, Mendilaharsu M, Ronco A, Larrinaga M, Balbi J, Alonso S, Deneo-Pellegrini H (1998) Meat intake, 'mate' drinking and renal cell cancer in Uruguay: a case-control study. Br J Cancer 78:1239–1243

Dhawan K, Kumar S, Sharma A (2002) Beneficial effects of chrysin and benzoflavone on virility in 2-year-old male rats. J Med Food 5:43–47

Dhawan K, Dhawan S, Sharma A (2004) *Passiflora*: a review update. J Ethnopharmacol 94:1–23

Díaz M, González A, Castro-Gamboa I, Gonzalez D, Rossini C (2008) First record of L-quebrachitol in *Allophylus edulis* (Sapindaceae). Carbohydr Res 343:2699–2700

Díaz M, Castillo L, Diaz CE, Alvarez RG, González-Coloma A, Rossini C (2014) Differential deterrent activity of natural products isolated from All*ophylus edulis* (Sapindaceae). Adv Biol Chem 4:168–179. https://doi.org/10.4236/abc.2014.42021

Dittrich VAO, Heringer G, Salino A (2007) Blechnaceae. In: Cavalcanti, T.B.; Ramos, A.E. (org.). Vol. 6. Flora do Distrito Federal, Brasil. Embrapa Recursos Genéticos e Biotecnologia, Brasília pp 91–108

Dittrich VAO, Salino A, Monteiro R (2015) The *Blechnum occidentale* (Blechnaceae, Polypodiopsida) species group in Southern and southeastern Brazil. Phytotaxa 231(3):201–229

Di Bitetti MS, Placci G, Dietz LA (2005) Situación ambiental en la ecorregión del bosque atlántico del Alto Paraná (selva paranaense) In: La situación ambiental Argentina 2005

Domínguez JA, Soto M (1925) Monografías farmacológicas argentinas: el ambay. Trabajos del Instituto de Botánica y Farmacología, Editorial Peuser, Buenos Aires 43:1–47

Duarte MR, Debur MC (2005) Stem and leaf morphoanatomy of *Maytenus ilicifolia.* Fitoterapia 76:41–49

Duke JA, Bogenschutz-Godwin MJ, Ottesen AR (2009) Dukeís handbook of medicinal plants of Latin America. CRC Press, New York Taylor & Francis Group, Boca Raton, p 498

Durán ML (1997) Estudios morfológicos, taxonómicos y biosistemáticos en el género *Blechnum.* Blechnaceae – Pteridophyta. Tesis doctoral. Universidad Nacional de Córdoba

El-Askary HI, Haggag MY, Abou-Hussein DR, Hussein SM, Sleem AA (2017) Bioactivity guided study of *Passiflora caerulea* L. leaf extracts. Iran J Pharm Res 15:46–57

Falconi A, Guitiérrez M, Benedetto L, Abin-Carriquiry JA, Bracesco N, Torterolo P (2013) Waking-promoting action of yerba mate (*Ilex paraguariensis*). Sleep Sci 6(1):0–00

Fatori Trevizan LN, do Nascimento KF, Santos JA, Leite Kassuya CA, Lima Cardoso CA, Vieira MC, Martinez Figueira Moreira F, Croda J, Nazari Formagio AS (2016) Anti-inflammatory, antioxidante and anti- *Mycobacterium tuberculosis* activity of viridiflorol: the major constituent of *Allophylus edulis* (A.St.-Hil., A.J uss. & Cambess.) Radlk. J Ethnopharmacol 192:510–515

Farag MA, Otify A, Porzel A, Michel CG, Elsayed A, Wessjohann LA (2016) Comparative metabolite profiling and fingerprinting of genus *Passiflora* leaves using a multiplex approach of UPLC-MS and NMR analyzed by chemometric tools. Anal Bioanal Chem 408(12):3125–3143

Feliu-Hemmelmann K, Monsalve F, Rivera C (2013) *Melissa officinalis* and *Passiflora caerulea* infusion as physiological stress decreaser. Int J Clin Exp Med 6:444–451

Ferreira PM, Nonato de Oliveira C, Braga de Oliveira A, Lopez MJ, Alzamora F, Ribeiro Vieira MA (2004) A lyophilized aqueous extract of *Maytenus ilicifolia* leaves inhibits histamine-mediated acid secretion in isolated frog gastric mucosa. Planta 219:319–324

Ferrucci MS (2004) Sapindaceae Juss. Aportes botánicos de Salta- Ser. Flora 7(4):1–45

Figueredo FG, Ferreira EO, Lucena BFF, Torres CMG, Dl L, ECP L, JMFL S, FAV S, Medeiros CR, GMM O, Colares AV, JGM C, PRL F, EFF M (2013) Modulation of the antibiotic activity by extracts from *Amburana cearensis* A.C. Smith and *Anadenanthera macrocarpa* (Benth.) Brenan. BioMed Res Int 2013:640682. https://doi.org/10.1155/2013/640682

Filip R, Ferraro G (2003) Researching on new species of Mate: *Ilex brevicuspis*. Phytochemical and pharmacology study. Eur J Nutr 42:50–54

Filip R, López P, Coussio J, Ferraro G (1998) Mate substitutes or adulterants: study of Xanthine content. Phytother Res 12:129–131

Filip R, López P, Giberti G, Coussio J, Ferraro G (2001) Phenolic compounds in seven South American *Ilex* species. Fitoterapia 72:774–778

Filip R, Lotito SB, Ferraro G, Fraga CG (2000) Antioxidant activity of *Ilex paraguariensis* and related species. Nutr Res 20:1437–1446

Fonseca C, Otto S, Paumgartten F, Leitão A (2000) Non-toxic, mutagenic, and clastogenic activities of mate-chimarrão (*Ilex paraguariensis*). J Environ Pathol Toxicol Oncol 19:333–346

de Freitas Lima R, Ponchet Alves E, Rosalen PL, Tasca Gois Ruiz AL, Teixeira Duarte MC, Fernandes Furletti Góes V, Dantas de Medeiros AC, Vieira Pereira J, Pina Godoy G, Melo de Brito Costa EM (2014) Antimicrobial and Antiproliferative potential of Anadenanthera colubrina (Vell.) Brenan. Evid Based Complement Alternat Med 2014:1–7

Gamborg OL, Miller RA, Ojima K (1968) Nutrient requirements of suspension cultures of soybean root cells. Exp Cell Res 50(1):151–158

Gazal M, Flach Ortmann C, Amelia Martins F, Streck EL, Quevedo J, Machado de Campos A, Stefanello FM, Kaster MP, Ghisleni G, Reginatto FH, Lencina CL (2014) Antidepressant-like effects of aqueous extract from *Cecropia pachystachya* leaves in a mouse model of chronic unpredictable stress. Brain Res Bull 108:10–17

Gilman EF, Watson DG (1994) *Tabebuia impetiginosa* Purple Tabebuia. Forest Service US Department of Agriculture. Fact Sheet ST-617, pp 1–3

Giraudo AR, Povedano H, BelgranoMJ KE, Pardiñas U, Miquelarena A, Ligier DH, Baldo D, Castelino M (2003) Biodiversity status of the Interior Atlantic Forest of Argentina. In: Galindo Leal C, De Gusmao CI (eds) The Atlantic Forest of South America: biodiversity status, threats and outlook (State of the Hotspots, 1). Center for Applied Biodiversity Science at Conservation International, Island Press, Washington, DC, pp 160–180

Goncalves Silva RM, Saraiva TS, Boreli Silva R, Goncalves LA, Pereira Silva L (2010) Allelopathy potential of ethanolic extract of *Anadenanthera macrocarpa* and Astronium graveolens. Biosci J Uberlândia 26(4):632–637

González de Mejía E, Song YS, Heck CI, Ramírez-Mares MV (2010) Yerba mate tea (*Ilex paraguariensis*): Phenolics, antioxidant capacity and *in vitro* inhibition of colon cancer cell proliferation. J Funct Foods 2:23–34

Gosmann G, Guillaume D, Toshirrico Cardoso Taketa A, Schenkel EP (1995) Triterpenoid Saponins from *Ilex paraguariensis*. J Nat Prod 58(3):438–441

Grose SO, Olmstead RG (2007) Taxonomic revisions in the polyphyletic genus *Tabebuia* s. l. (Bignoniaceae). Syst Bot 32(3):660–670

Gupta MP (1995) 270 Plantas Iberoamericanas. Editorial Presencia, Colombia, pp 407–408

Gutiérrez F, Estévez-Braun A, Ravelo AG, Astudillo L, Zárate R (2007) Terpenoids from the medicinal plant. J Nat Prod 70(6):1049–1052

Hadas E, Ozarowski M, Derda M, Thiem B, Cholewinski M, Skrzypczak L, Gryszczynska A, Piasecka A (2017) The use of extracts from *Passiflora* spp. In helping the treatment of Acanthamoebiasis. Acta Pol Pharm Drug Res 74(3):921–928
Hikawczuk V, Saad J, Guardia T, Juarez A, Giordano O (1998) Anti-inflammatory activity of compounds isolated from *Cecropia pachystachya*. Anal Asoc Quím Argentina 86:167–170
Hoffmann-Bohm K, Lotter H, Seligmann O, Wagner H (1992) Antihepatotoxic -Glycosylflavones from the leaves of Allophyllus edulis var. edulis and gracilis. Planta Med 58(06):544–548
Hurrell J, Bazzano D (2003). Arbustos 1: nativos y exóticos. Buenos Aires, L.O.L.A
Hussain H, KrohnK AVU, Miana GA, Greend IR (2007) Lapachol: an overview. ARKIVOC 2(1):145–171
Itokawa H, Shirota O, Ikuta H, Morita H, Takeya K, Itaka Y (1991) Triterpenes from *Maytenus ilicifolia*. Phytochemistry 30:3713–3716
Ivorra MD, Paya M, Villar A (1989) A review of natural products and plants as potent antidiabetic drugs. J Ethnopharmacol 27:243–275
Jorge RM, Leite JPV, Oliveira AB, Tagliati CA (2004) Evaluation of antinociceptive, anti-inflammatory and antiulcerogenic activities of *Maytenus ilicifolia*. J Ethnopharmacol 94:93–100
Justiniano MJ, Fredericksen TS (1998) Curupaú *Anadenanthera colubrina* (Vell. Conc.) Benth. Mimosoideae In: Ecología y silvicultura de especies menos conocidas
Körbes, VC (1995) Plantas medicinais, Francisco Beltrão: Associação de Estudos, Orientação e Assistência Rural (ASSESOAR) 51ed. p 188
Kramer KU, Chambers TC, Hennipman E (1990). Blechnaceae. In: Kramer, K.U. & Green, P.S. (eds.). Pteridophytes and gymnosperms. In: Kubitzki, K. (ed). The families and genera of vascular plants. vol 1. Springer-Verlag, Berlin. pp 60–68
Kraemer KH, Taketa ATC, Schenkel EP, Gosmann G, Guillaume D (1996) Matesaponin 5, a highly polar saponin from *Ilex paraguariensis*. Phytochemistry 42(4):1119–1122
Kujawska M, Pieroni A (2015) Plants used as food and medicine by polish migrants in Misiones, Argentina. Ecol Food Nutr 54:255–279
Lahitte HB, Hurrell JA, Valla JJ, Jankowski LS, Bazzano D, Hernández AJ (1999) Árboles urbanos. Biota Rioplatense IV. Inventario de la biota de la región del Delta del Paraná, Isla Martín García y Ribera Platense. Literature of Latin America (L.O.L.A.). Buenos Aires, Argentina. 320 p
Lanzetti M, Bezerra FS, Romana-Souza B, Brando-Lima AC, Koatz VLG, Porto LC, Valenca SS (2008) Mate tea reduced acute lung inflammation in mice exposed to cigarette smoke. Nutrition 24:375–381
Larraburu EE, Apóstolo NM, Llorente BE (2012) In vitro propagation of pink lapacho: response surface methodology and factorial analysis for optimization of medium components. Int J For Res 2012:1–9
Leite JPV, Braga FC, Romussi G, Persoli RM (2010) Constituents from *Maytenus ilicifolia* leaves and bioguided fractionation for gastroprotective activity. J Braz Chem Soc 21:248–254
Lemos AO, Sanches JCM, Silva IEF, Silva MLA, Vinhólis AHC, Feliz MAP, Santos RA, Cecchi AO (2012) Genotoxic effects of *Tabebuia impetiginosa* (Mart. Ex DC.) Standl. (Lamiales, Bignoniaceae) extract in Wistar rats. Genet Mol Biol 35(2):498–502
Lima Fonseca W, de Almeida FA, Tiburtino Leite ML, Ap PR, Petter FA, de Oliveira AM, Martins Carvalho R, de Alcântara Neto F, Fernandes Pereira F, Mendonça Junior AF (2017) Bioactivity of aqueous extracts of *Anadenanthera macrocarpa* to *Meloidogyne incognita* in cotton crop. AJCS 11(02):156–161
Lloyd GB, McCown BH (1980) Commercially-feasible micropropagation of mountain laurel, *Kalmia latifolia*, by use of shoot tip culture. Int Plant Prop Society Proc J 30:421–426
Lopes da Silva Filho M (2007) Avaliação *in vitro* da ação antiparasitária do extrato aquoso e etanólico do angico preto (*Anadenanthera macrocarpa*) (Benth.) Brenan sobre o carrapato *Rhipicephalus* (Boophilus) *microplus* (Canestrini, 1887) / Dissertação (Mestrado) – Universidade Federal do Piauí

Lopez da Silva Filho (2007) Avaliação in vitro da ação antiparasitária do extracto aquoso e etanólico do angico preto (Anadenanthera macrocarpa) (Benth.) Brenan sobre o carrapato Rhipicephalus (Boophilus) microplus (Canestrini, 1887) Thesis Mestre em Ciência Animal, Área de Concentração: Clínica Médico – Cirúrgica de Animais de Interesse EconômicoEstado do Piauí – Brasil Agosto – 2007

Lorenzi H, Matos FJA (2002a) Plantas Medicinais no Brasil: nativas e exóticas. Instituto Plantarum, São Paulo, pp 189–190

Lorenzi H, Matos FJA (2002b) Plantas Medicinais no Brasil: nativas e exóticas. Instituto Plantarum, São Paulo, pp 189–190

Luna CV, Gonález AM, Mroginski LA, Sansberro PA (2017) Anatomical and histological features of *Ilex paraguariensis* leaves under different *in vitro* shoot culture systems. Plant Cell Tiss Organ Cult 129(3):457–467. https://doi.org/10.1007/s11240-017-1191-x

Malizia LR, Cuykens E, Blundo C, Berno A, Pacheco S, Brown A (2018). http://www.ambiente-forestalnoa.org.ar/userfiles/especies/pdf/Allophylusedulis.pdf

Maquiaveli CC, da Silva ER, Francescato HDC, Costa RS, Silva CGA, Casarini DE, Fonchi AF, Coimbra TM (2014) Brazilian embauba (*Cecropia pachystachya*) extract reduces renal lesionin 5/6 nephrectomized rats. J Renin Angiotensin Aldosterone Syst. 15 (4): 430-439

Marini-Bettòlo GB, Nicoletti M, Patamia M, Galeffi C, Messana I (1981) Plant screening by chemical and chromatographic procedures under field conditions. J Chromatogr A 213(1):113–127

Martínez MAP, Pelloto JP, Basualdo N (1997) Distribution of flavonoid aglycones in *Ilex* species (Aquifoliaceae). Biochem Syst Ecol 25:619–622

Martínez-Crovetto RM (1987) Plantas reguladoras de la fecundidad utilizadas en la medicina popular del nordeste Argentino. Am Ind 47:279–293

Martínez Crovetto R (1981) Plantas utilizadas en medicina en el N.O. de Corrientes. Fund. Miguel Lillo, Tucumán, 139 pp

Martínez MR, Pochettino ML (1999) El valor del conocimiento etnobotánico local: aporte a la curricula educativa en el área de biología en las escuelas de Molinos. Valles Calchaquíes, provincia de Salta. Cuadernos del Instituto Nacional de Antropología y Pensamiento Latinoamericano 18:257–270

Martin EE, Battezzati P, Bruno M, Sarroca D, Usero S, Bruno C (2006) Yerba mate (*Ilex paraguariensis*) como contraste oral negativo para colangiopancreatoresonancia. Estudio preliminar. Revista Argentina de Radiología 70(4):323–330

Martins AG, Guterres SS, González Ortega G (2003) Anti-ulcer activity of spray-dried powders prepared from leaf extracts of *Maytenus ilicifolia* Martius ex Reiss. Acta Farm Bonaer 22(1):39–44

Matos FJA (2ed) (1997) O Formulário fitoterápico do professor Dias da Rocha:informações sobre o emprego na medicina caseira, de plantas do Nordeste,especialmente do Ceará. Fortaleza: EUFC. method for broth dilution antifungal susceptibility testing of yeasts. Aproved standard – third edition M-27A3. Clinical and Laboratory Standards Institute, Wayne, PA

Matsunaga K, Sasaki S, Ohizumi Y (1997) Excitatory and inhibitory effects of paraguayan medicinal plants *Equisetum giganteum*, *Acanthospermum australe*, *Allophylus edulis* and *Cordia salicifolia* on contraction of rabbit aorta and Guinea-pig left atrium. Nat Med 51:478–481

Mechler E (2009) *Tabebuia impetiginosa*, *Tabebuia avellanedae* (Bignoniaceae). In: Medicinal plants in tropical countries II. Traditional Use-Experience-Facts. Christian Intercultural Health Ministries, Wettenberg, pp 1–9. http://www.mmh-mms.com/downloads/mp19tmpabebuiaimpetiginosa.pdf

Medina JH, Paladini AC, Wolfman C, Levide Stein M, Calvo D, Diaz LE, Pena C (1990) Chrysin (5,7-di-OH-flavone) a naturally occurring ligand for benzodiazepine receptors, with anticonvulsant properties. Biochem Pharmacol 40:2227–2231

Mello Cruz E, da Silva ER, Maquiaveli CC, Sousa Sobral Alves E, Jr Lucon JF, Gonçalves dos Reis MB, Mendes de Toledo CE, Guaré Cruz F, Vannier-Santos MA (2013) Leishmanicidal activity of *Cecropia pachystachya* flavonoids: arginase inhibition and altered mitochondrial DNA arrangement. Phytochemistry 89:71–77

Mendonça ED, da Silva J, dos Santos MS, Carvalho P, Kick Mausolff Papke D, Flach Ortmann C, Picada JN, Reginatto FH, de Barros Falcao Ferraz A (2016) Genotoxic, mutagenic and anti-genotoxic effects of *Cecropia pachystachya* Trécul aqueous extract using *in vivo* and *in vitro* assays. J Ethnopharmacol 193:214–220

Miranda DD, Arcari DP, Pedrazzoli J Jr, Carvalho Pde O, Cerutti SM, DHM B, Ribeiro ML (2008) Protective effects of mate tea (Ilex paraguariensis) on H2O2-induced DNA damage and DNA repair in mice. Mutagenesis 23(4):261–265

Montanari T, de Carvalho JE, Dolder H (1998) Effect of *Maytenus ilicifolia* Mart. ex. Reiss on Spermatogenesis. Contraception 57:335–339

Montanari T, Bevilacqua E (2002) Effect of *Maytenus ilicifolia* Mart. on pregnant mice. Contraception 65:171–175

Moreau F (1973) Alcaloides y plantas alcaloideas. Oikos-Tau, Barcelona

Morgan-Martins MI, Silva VS, Licks F, Colares JR, Schemitt EG, Forgiarini LF, Rosa DP, Marroni NP (2013) *Ilex paraguariensis* St. Hil. (yerba mate) improves the metabolic profile of and protects against oxidative stress in exercised and non-exercised rats. Int J Pharma Bio Sci 4(4):B1303–B1311

Mors WB, Rizzini CT, Pereira NA (2000) Medicinal plants of Brazil. Reference Publications, Michigan

Mosimann ALP, Wolhelm-Filho D, Silva EL (2006) Aqueous extract of *Ilex paraguariensis* (mate) attenuates the progression of atherosclerosis in cholesterol-fed rabbits. Biofactors 26:59–70

Murashige T, Skoog F (1962) A revised medium for rapid growth and bioassay with tobacco tissue cultures. Physiol Plant 15(3):473–497

Nakayama F (1966) Cultivo *in vitro* de tejidos de *Passiflora caerulea*. Revista de la Facultad de Agronomía de la Universidad Nacional de La Plata 42:63–74

Newall CA, Anderson LA, Phillipson JD (1996) Herbal medicines. The Pharmaceutical Press, London, pp 189–190

Niero R, Moser R, Busato ACB, Yunes RA, Reis A, Cechinel Filho V (2001) A comparative chemical study of *Maytenus ilicifolia* Mart. Reiss and Maytenus robusta Reiss (Celastraceae). Z. Naturforsch 56c:158–161

Nogueira L, Charvet Inckot R, de Oliveria SG, de Souza LA, Bona C (2011) Phytotoxicity of petroleum-contaminated soil and bio-remediated soil on *Allophylus edulis*. Rodriguésia 62(3):459–466

Nonato FR, Almeida Barros TA, Lucchese AM, Cordeiro Oliveira CE, Ribeiro dos Santos R, Pereira Soares MB, Villareeal CF (2009) Antiinflammatory and antinociceptive activities of *Blechnum occidentale* L. extract. J Ethnopharmacol 125:102–107

Nunes LE, Porto Viana AP, Vieira da Rocha RV, Sena Cunha VD, Ramalho Catao RM, Melo de Brito Costa EM (2015) *In vitro* evaluation of antifungal activity and interactive effect of *Anadenanthera colubrina* (Benth). Afr J Microbiol Res 9(36):2006–2012

Ohsaki A, Imai Y, Naruse M, Ayabe S, Komiyama K, Takashima J (2004) Four new triterpenoids from *Maytenus ilicifolia*. J Nat Prod 67(3):469–471

Oliveira MG, Monteiro MG, Macaubas C, Barbosa VP, Carlini EA (1991) Pharmacologic and toxicologic effects of two *Maytenus* species in laboratory animals. J Ethnopharmacol 34:29–41

Oyarzabal M, Clavijo J, Oakley L, Biganzoli F, Tognetti P, Barberis I, Maturo HM, Aragón R, Campanello PI, Prado D, Oesterheld M, León RJC (2018) Unidades de vegetación de la Argentina. Ecol Austral 28(1):040–063

Palmeira JD, Ferreira SB, SOUZA JH, Almeida JM, Figueiredo MCP, Pequeno AS, Arruda TA, Antunes RMP, Catão RMR (2010) Avaliação da atividade antimicrobiana in vitro e determinação da concentração inibitória mínina (CIM) de extratos hidroalcóolico de angico sobre cepas de Staphylococcus aureus. Rev Bras Anal Clin 42:33–38

Pasquali TR, Macedo SMD, Roman SS, Dal Prá V, Cansian RL, Mossi AJ, Oliveira VJ, Mazutti MA (2011) Acute toxicity and anti inflammatory effects of supercritical extracts of *Ilex paraguariensis*. Afr J Pharm Pharmacol 5(8):1162–1169

Park B-S, Lee K-G, Shibamoto T, Lee S-E, Takeoka GR (2003) Antioxidant activity and characterization of volatile constituents of Taheebo (*Tabebuia impetiginosa* Martius ex DC). J Agric Food Chem 51:295–300

Park B-S, Lee K-K, Lee S-E, Piao X-L, Takeoka GR, Wong RY, Ahn YJ, Kim JH (2006) Antibacterial activity of *Tabebuia impetiginosa* Martius ex DC (Taheebo) against *Helicobacter pylori*. J Ethnopharmacol 105:255–262

Paroul N, Buogo de Jesús G, Atti Serafini L, Moyna P, Cancian R (2002) Essential oils from two different populations of *Ilex paraguariensis*. Trabajo presentado en el 1° Congreso Latinoamericano de Fitoquímica. Buenos Aires

Pérez Gollán JÁ, Gordillo I (1993) Alucinógenos y sociedades indígenas del noroeste argentino. Anales de Antropología 30:299–350

Pereira AMS, Rodrigues DC, Cerdeira RM, de M França SC (1992) Isolamento de metabólitos de *Maytenus* associados à ação anti-úlcera gástrica. XII Simpósio de Plantas Medicinais Brasileiras. Anais, p 55 UPFR, Curitiba, Brazil

Pereira C, Yariwake J, Lancas F, Wauters J, Tits M, Angenot L (2004) A HPLC densitometric determination of flavonoids from *Passiflora alata*, *P. edulis*, *P. incarnata* and *P. coerulea* and comparison with HPLC method. Phytochem Anal 15:241–248

Pereira MAS, Januário AH, Queiroz MEE, Biondo R, França SC (2005) Evaluation of *Maytenus aquifolia* Mart. and *Maytenus ilicifolia* Mart. chemotypes for tannins, total phenols and triterpenes. Rev Bras Pl Med, Botucatu 8(1):13 17

Pérez JM, Maldonado ME, Rojano BA, Alzate F, Sáez J, Cardona W (2014) Comparative antioxidant, antiproliferative and apoptotic effects of *Ilex laurina* and *Ilex paraguariensis* on Colon Cancer cells. Trop J Pharm Res 13(8):1279–1286

Perez-Guerrero C, Herrera MD, Ortiz R, Alvarez de Sotomayor M, Fernandez MA (2001) A pharmacological study of *Cecropia obtusifolia* Bertol aqueous extract. J Ethnopharmacol 76:279–284

Pérez Ibáñez NM, Tillett S, Taddei A (2017) C-Glucosylflavone of three species of the genus *Passiflora*. Antibacterial, antioxidant activity and quantification. Revista de la Facultad de Farmacia, Universidad Central de Venezuela 79(1–2):68–76

Pessôa Santiago AC, Carneiro Leão Barros I, de Oliveira Dittrich VA (2014) Ferns and lycophytes of Pernambuco state, Brazil: Blechnaceae. Rodriguésia 65(4):861–869

Petenatti ME, Gette MA, Popovich MC, Garro MF, Camí G, Aragón L, Del Vitto LA, Petenatti EM (2014) Identificación de C-glicosilflavonas por HPLC y FTIR en extractos acuoso y metanólico de *Passiflora caerulea* y *P. foetida* (Passifloraceae). Dominguezia 30(1):35–40

Piacente S, Balderrama L, De Tommasi N, Morales L, Vargas L, Pizza C (1999) Anadanthoside: a favanol-3-O-β-D-xylopyranoside from *Anadenanthera macrocarpa*. Phytochemistry 51:709–711

Pires TCSP, Dias MI, Calhelha RC, Carvalho AM, Queiroz MJRP, Barros L, CFR F (2015) Bioactive Properties of *Tabebuia impetiginosa*-Based Phytopreparations and Phytoformulations: A Comparison between Extracts and Dietary Supplements. Molecules 20:22863–22871. https://doi.org/10.3390/molecules201219885

Placci LG, Giorgis P (1994) Estructura y diversidad de la selva del Parque Nacional Iguazú, Argentina, VII Jornadas Técnicas sobre Ecosistemas Forestales Nativos: Uso, Manejo y Conservación, El Dorado, pp 123–138

Ponce MM (1996) Pteridophyta. In: Zuloaga, F.O. & Morrone, O. (eds.) Catálogo de las plantas vasculares de la República Argentina. I: Pteridophyta, Gymnospermae y Angiospermae (Monocotyledoneae). Monographs in Systematic Botany from the Missouri Botanical Garden 60: i–xviii, 1–323

Puangpraphant S, Berhow MA, de Mejjía G (2011) Mate (*Ilex paraguariensis* St. Hilaire) saponins induce caspase-3-dependent apoptosis in human colon cancer cells *in vitro*. Food Chem 125:1171–1178

Ragonese A, Milano V (1984) Vegetales y Sustancias Tóxicas de la Flora Argentina. In: Enciclopedia Argentina dela Agricultura y Jardinería, 2nd edn. Edit. Acme, Buenos Aires, pp 256–257

Rastogi RP, Mehrotra BN (1990) Compendium of Indian Medicinal plants CDRI Lucknow and, CSIR, New Delhi, 1990; 1:25

Ratera EL, Ratera MO (1980) Plantas de la flora argentina empleadas en medicina popular. Ed. Hemisferio sur S. A., Buenos Aires

Reginatto F, Kauffman C, Schripsema J, Guillaume D, Gosmann G, Schenkel EP (2001) Steroidal and triterpenoidal glucosides from *Passiflora alata*. J Braz Chem Soc 12:32–36
Resico C (2007) Estado actual de la informacion sobre productions forestales no madereros. Food and Agriculture Organization of the United Nations. http://www.fao.org//OCREP//oo6//AD3935//AD393s13.htm (accessed 22 nov 2007).\par
Ricco RA, Wagner ML, Gurni AA (1991) Estudio comparativo de flavonoides en seis especies austrosudamericanas del género Ilex. Acta Farm Bonaer 10:29–35
Rodrigues de Almeida E (2009) Preclinical and clinical studies of Lapachol and Beta-Lapachone. Open Nat Prod J 2:42–47
Rondina RVD, Al B, Coussio JD (2006) Especies medicinales argentinas con potencial actividad analgésica. Dominguezia 24(1):47–69
Ross S, Arriaga ME, Pechi E (2017) Establecimiento *in vitro* de Yerba mate (*Ilex paraguariensis* A. St.-Hil.) nativa de Uruguay. Agrociencia Uruguay 21(1):15–23
Ruxton CHS, Kirkwood L, McMillan B, St John D, Evans CEL (2007) Effectiveness of a herbal weight loss supplement (Zotrim™) for weight management. Br Food J 109(6):416–428
Saldaña MDA, Mohamed RS, Baer MG, Mazzafera P (1999) Extraction of purine alkaloids from maté (*Ilex paraguariensis*) using supercritical CO2. J Agric Food Chem 47:3804–3808
Saldaña MDA, Zetzl C, Mohamed RS, Brunner G (2002) Extraction of methylxanthines from guaraná seeds, maté leaves, and cocoa beans using supercritical carbon dioxide and ethanol. J Agric Food Chem 50:4820–4826
Santos-Oliveira R, Coulaud-Cunha S, Colaçao W (2009) Revisão da *Maytenus ilicifolia* Mart. ex Reissek, Celastraceae. Contribuição ao estudo das propriedades farmacológicas. Braz J Pharm 19(2B):650–659
Santos VAFFM, Regasini LO, Nogueira CR, Passerini GD, Martínez I, Bolzani VS, Graminha M, Cicarelli RMB, Furlan M (2012) Antiprotozoal Sesquiterpene pyridine alkaloids from *Maytenus ilicifolia*. J Nat Prod 75(5):991–995
Scheffer MC, Corrêa Júnior C, Radomski M (2005) Domesticaçao de plantas medicinais: a experiencia da espinheira-santa (*Maytenus ilicifolia* Mart ex Reiss). Agrociencia IX(1, 2):53–59
Schenkel E, Montanha J, Gosman G (1996) Triterpene saponins from mate, *Ilex paraguariensis*. Adv Exp Med Biol 405:47–56
Schinella G, Aquila S, Dade M, Giner R, Recio Me C, Spegazzini E, de Buschiazzo P, Tournier H, Ríos JL (2008) Anti-inflammatory and apoptotic activities of pomolic acid isolated from *Cecropia pachystachya*. Planta Med 74:215–220
Schultes RE, Hofman A (1993) Les plantes des dieux. Ed du Lézard, France
Schulze M, Grogan J, Uhl C, Lentinia M, Vidal E (2008) Evaluating ipe^ (Tabebuia, Bignoniaceae) logging in Amazonia: sustainable management or catalyst for forest degradation? Biol Conserv 141:2071–2085
Seigler DS, Spencer KC, Statier WS, Conn EE, Dunn JE (1982) Tetraphyllin B and epitetraphyllin B sulphates: novel cyanogenic glucosides from *Passiflora caerulea* and *P. alato-caerulea*. Phytochemistry 21:2277–2282
Şesan TE, Sarbu A, Smarandache D, Oancea F, Oancea A, Savin S, Toma A, Ştefan L, Negru G, Bira AF, Vlascenau G, Ghiurea M, Jecu L, Vasilescu G, Pomaohaci CM (2016) Botanical and phytochemical approach on *Passiflora* spp. – new nutraceutical crop in Romania. J Plant Develop 23:97–126
Severin C, Bueno M, Santin F, Giubileo G (2011) Respuesta *in vitro* de diferentes biotipos y explantos de *Passiflora caerulea* L. Rev Colomb Biotecnol 13(1):73–79
Shirota O, Morita H, Takeya K, Itokawa H (1994) Cytotoxic aromatic triterpenes from *Maytenus ilicifolia* and *Maytenus ilicifolia*. J Nat Prod 57:1675–1681
Silva C, Récio R (1992) Coleta e avaliação dos compostos fitoquimicos da espinheira santa (*Maytenus ilicifolia*). Comunicacion al XII Simpósio de Plantas Medicinais do Brasil. Curitiba, Brasil
Silva MAB (2009) Levantamento etnobotânico e triagem hipolipidêmica de plantas medicinais do Cerrado matogrossense: avaliação da atividade hipolipidêmica da *Simaba ferruginea* St. Hil. e Vochysia ru-fa Mart [dissertação]. Cuiabá: Faculdade de Ciências Médicas, Uni-versidade Federal de mato Grosso

Soares LAL, Oliveira AL, González Ortega G, Petrovick PR (2004) Development and validation of a LC-method for determination of catechin and epicatechin in aqueous extractives from leaves of Maytenus ilicifolia. J Pharm Biomed Anal 36(4):787–790

Soraru SB, Bandoni AL (1978) Plantas de la medicina popular argentina. Editorial Albatros, Buenos Aires, pp 76–78

de Sousa NC, de Rezender AA, da Silva RMG, Guterres ZR, Gral U, Kerr WE, Spano MA (2009) Modulatory effects of *Tabebuia impetiginosa* (Lamiales, Bignoniaceae) on doxorubicin-induced somatic mutation and recombination in *Drosophila melanogaster*. Genet Mol Biol 32(2):382–388

Souza-Formigoni MLO, Oliveira MGM, Goldnadel Monteiro M, Gomes da Silveira-Filho N, Braz S, Carlini EA (1991) Antiulcerogenic effects of two *Maytenus* species in laboratory animals. J Ethnopharmacol 34:21–27

Speroni E, Billi R, Perellino NC, Minghetti A (1996) Role of chrysin in the sedative effects of *P. incarnata* L. Phytother Res 10:S98–S100

Suo M, Ohta T, Takano F, Jin S (2013) Bioactive phenypropanoid glycosides from *Tabebuia avellanedae*. Molecules 18:7336–7345

Taketa ATC, Breitmaier E, Schenkel EP (2004) Triterpenes and triterpenoidal glycosides from the fruits of *Ilex paraguariensis* (Maté). J Braz Chem Soc 15:205–211

Tirloni CAS, Macorini LFB, dos Santos UP, da Rocha PS, Barros SV, de Melo AMMF, Vieira MC, Souza KP, dos Santos EL (2015a) Evaluation of the antioxidant activity, antimicrobial effect and acute toxicity from leaves of *Allophylus edulis* A. St.-Hil., A. Juss. Cambess & Hieron. ex Niederl. Afr J Pharm Pharmacol 9(11):353–362

Tirloni CAS, Traesel GK, Costa HMG, Victol NS, Takegava VF, Donatini RS, Souza KP, Santos EL (2015b) Chemical profile and evaluation of acute toxicity of *Allophylus edulis* A. St.-Hil., A. Juss. Cambess & Hieron. ex Niederl. 23rd Congress of the International Union for Biochemistry and Molecular Biology 44th Annual Meeting of the Brazilian Society for Biochemistry and Molecular Biology Foz do Iguaçu, PR, Brazil, August 24th to 28th, 2015

Tryon RM, Tryon A (1982) Ferns and allied plants, with special reference to tropical America. Springer Verlag, New York

Toursarkissian M (1980) Plantas Medicinales de la Argentina. Ed. Hemisferio Sur, Buenos Aires, Argentina

Turner S, Cogoi L, Isolobella S, Filip R, Anesini C (2011) Evaluation of the antioxidant activity and polyphenols content of *Ilex paraguariensis* (mate) during industrialization. Adv J Food Sci Technol 3(1):23–30

Tyler V (1987) The new honest herbal. Stickley G Co (Ed), Philadelphia (PA)

Uchoa VT, de Paula RC, Krettl LG, Santana AEG, Krettli AU (2010) Antimalarial activity of compounds and mixed fractions of *Cecropia pachystachya*. Drug Dev Res 71:82–91

Varela G, Cocucci A, Sérsic NA (2016) Función de la corona en *Passiflora caerulea* (Passifloraceae) como atrayente de sus vectores de polinización. Bol Soc Argent Bot 51(1):99–110

Vasconcelos CM, Chaves Vasconcelos TL, Póvoas Fabiani Tenório X, Evangelista Pires dos Santos RD, da Costa Maynart WH, Gomes de Almeida T, Fragoso da Silva Oliveira J, Dalboni França AD, Sales Santos Veríssimo RC, Lins Thais H, de Araújo-Júnior JX, de Assis Bastos ML (2014) Antimicrobial, antioxidant and cytotoxic activity of extracts of *Tabebuia impetiginosa* (Mart. ex DC.) Standl. J Chem Pharm Res 6(7):2673–2681

Vassallo A, Correa P, De Stefani E, Cendan M, Zavala D, Chen V, Carzoglio J, Deneo-Pellegrini H (1985) Esophageal cancer in Uruguay: a case-control study. J Natl Cancer Inst 75:1005–1009

Velázquez E, Tournier HA, Mordujovich de Buschiazzo P, Saavedra G, Schinella GR (2003) Antioxidant activity of Paraguayan plant extracts. Fitoterapia 74:91–97

Vera García R, Basualdo I, Peralta I, de Herebia M, Caballero S (1997) Minerals content of Paraguayan yerba mate (*Ilex paraguariensis*). Arch Latinoam Nutr 47:77–80

Viglione A, Vallejo N (2011) CEBIL: an American hallucinogenic plant in Botany books by different authors. 40th international congress for the history of pharmacy. September 2011, Berlin, Germany

Wnuk M, Lewinska A, Oklejewicz B, Bugno M, Slota E, Bartosz G (2009) Evaluation of the cyto- and genotoxic activity of yerba mate (I lex paraguariensis) in human lymphocytes in vitro. Mutat Res Genet Toxicol Environ Mutagen 679(1–2):18–23

Wolfman C, Viola H, Paladini A, Dajas F, Medina JH (1994) Possible anxiolytic effects of chrysin, a central benzodiazepine receptor ligand isolated from *Passiflora caerulea*. Pharmacol Biochem Behav 47:1–4

Woodrow A (1985) Los Jesuitas. Historia de un dramático conflicto Sudamericana-Planeta, pp 186

Yajía ME, Martí DA, Bidau CJ, Amat AG, Riglos AG, Silvestroni A (1999) Genotoxicity evaluation of *Allophylus edulis* (Camb.) Radlk. (Sapindaceae) aqueous extract. Acta Hortic 501:31–36

Zanon de Moura M, Pádua LE, Ramalho e Silva PR, Silva AA, Maggioni K (2012) Extrato de folhas de *Anadenanthera macrocarpa* sobre a biologia de *Spodoptera frugiperda* criada em dieta artificial. Comunicata Scientiae 3(4):249–254

Zuloaga F, Morrone O, Belgrano M (2000) Características biogeográficas de la provincia de Misiones, Instituto de Botánica Darwinion, Informe para Fundación Vida Silvestre Argentina

Chapter 6
Puna: A Surrealistic Landscape in the Argentinean Highlands

6.1 Introduction: The Puna

The Puna is a geographical region located in the northwest of Argentina (the provinces of JUJ, SAL, CAT, TUC, LRI, and SDE). It is an extensive highland, or high mountain plateau, typical of the central area of the cordillera de los Andes (Fig. 6.1). The limits of the Argentine Puna range from the Cordillera de San Buenaventura (North of the San Francisco mountain pass, in the province of CAT) to the South, to the extreme North of the province of JUJ, and from the bordering cordillera with Chile, by the West, to the mountains of the Eastern Cordillera of the provinces of CAT, SAL, and JUJ, by the East.

The lagoons in there form closed basins without drainage that were the origin of extensive salt pans such as de *Salinas grandes* in the provinces of JUJ and SAL (Fig. 6.2). The hills, of great height and rounded shape (volcanic morphology), exceed 5000–6000 meters in height with volcanoes such as *Pissis* (6882 m), *Llullaillaco* (6739 m), *Ojos del Salado* (6864 m), *Bonete Chico* (6759 m), *Tres Cruces* (6749 m), and Walter Penck (6658 m).

At altitudes higher than 4300–4500 m.a.s.l. predominates the geological and bioclimatic conditions of the high Andes eco-region. The climate is cold and dry, with a large daily temperature range. The annual average temperature is less than 8 °C; in summer the temperature can reach 30 °C, in winter less than −20 °C. The annual average rain is 100–200 mm.

The fauna is well characterized; the domestic species are the llama (*Lama glama*): the wild species are puma (*Puma concolor*), the guanaco (*Lama guanicoe*), and numerous endemics species such as the vicuña (*Vicugna vicugna*), huemul del Norte or taruca (*Hippocamelus antisensis*), Andean cat (*Leopardus jacobitus*), and real skunk (*Conepatus chinga rex*). Among the rodents are chinchillas (*Chinchilla chinchilla*). The great diversity of birds is associated to high-altitude wetlands (salt flats and lagoons); there are flamingos such as the parina James (*Phoenicoparrus jamesi*) and big parina (*Ph. andinus*), ducks such as the guayata (*Chloephaga melanoptera*)

M. A. Alvarez, *Pharmacological Properties of Native Plants from Argentina*,
https://doi.org/10.1007/978-3-030-20198-2_6

Fig. 6.1 Puna, section in the province of Jujuy, Argentina. Personal collection P G Alvarez

and puna duck (*Spatula puna*), gallareta cornuda (*Fulica cristata*), tero serrano (*Vanellus resplendens*), Andean becasina (*Gallinago andina*), and the plover from the Puna (*Charadrius alticola*). Among the terrestrial birds are the suri (*Rhea pennata*, ñandú petiso), partridges such as quiula puneña (*Tinamotis pentlandii*), and small species such as *Geositta punensis* and *Muscisaxicola juniensis*.

Flora is substantially affected by the climatic conditions. The lack of rain defines a vegetal carpet of bushes and herbs, and in some areas soils without vegetation. The dominant flora is the shrub steppe with dispersed bushes. Local edaphic microclimates determine the existence of small sectors with different physiognomies.

Fig. 6.2 Salinas grandes (**a**) and llama (*Lama lama*) in the province of Jujuy (**b**), Argentina. Personal collection PG Alvarez

Soils are stony or saline, scarcely covered by vegetation, such as *Maihueniopsis glomerata*, very susceptible to erosion. In the Northern and Eastern areas, where the annual rainfall exceeds 400 mm, predominate the *tolas* (*Baccharis incarum*). Towards the South and West, in the high Andean steppe dominate the *chijua* (*Baccharis boliviensis*), *tolilla* (*Fabiana densa)*, *añaguilla* (*Adesmia tucumanensis*), *rica-rica* (*Acantholippia hastulata*), *suriyanta* (*Nardophyllum armatum*), *airampu* (*Opuntia soehrendsii*), and ca*nguía* or *cailla* (*Tetraglochin cistatum*) among other species. Open groves of *queñoa* (*Polylepis tomentella*) grow on slopes and ravines between 3800 and 4300 m height.

The *cordillera de los Andes* has two parallel cords, Western and Eastern, separated by ravines and valleys. The Western cord shows higher heights (Chañi 6200 m). The Eastern cord is of less height (Cerro Negro de Zucho around 5000 m). The ravines, elongated and narrow valleys, are the communication routes of this area.

The main economic activities in the Puna are livestock (herds of sheep and llamas raised for wool and meat production), subsistence agriculture, and mining (gold, copper, lithium). The Puna reunites the optimal environmental conditions to develop solar energy with minimal impact on the ecosystem.

6.2 *Acacia caven* (Molina) Molina var. caven (Fabaceae)

Synonymy *Acacia adenopa* Hook & Arn., *Acacia cavenia* Colla, hom. illeg., *Acacia cavenia* (Molina) Hook & Arn, *Acacia farnesiana* (L.) Wild., *Acacia farnesiana* (L.) Willd. var. heterocarpa, F. cavenia, *Mimosa cavenia* Molina, *Mimosa caven* Molina, *Vachellia farnesiana* (L.) Wight & Arn. f. brachypoda, *Vachellia farnesiana* (L.) Wight & Arn. f. cavenia.

Vernacular name Espinillo, espinillo negro, aromo criollo, tusca, aromo, aromo criollo, churqui, aromita, aromito, cavén, espino.

General information *A. caven* grows in Argentina, south of Brazil, Bolivia, Chile, Paraguay, and Uruguay. In Argentina, it could be found in the provinces of BAI, CAT, CHA, COR, COS, ERI, FOR, JUJ, LPA, LRI, MEN, MIS, SAL, SDE, SFE, SJU, SLU, and TUC. It develops in flat areas from forests and in mountain range areas (*serranías*), in the hillocks with high drainage, and also in previously flooded areas, being removed when the strong humidity returns. Among other applications, fruits, seeds, and pods are used to make high protein content flours, roasted seeds to make a coffee-like beverage, and flowers to make a tea. It is also used for firewood, to make utility poles, and handicrafts; it is considered a melliferous species and also of use in perfumery (Ortiz 1966; Aronson and Ovalle 1989; Carrere 1990; Benedetti 2012). Additionally, it is used to feed cattle.

Botanical characteristics *A. caven* is a deciduous tree, 2–6 m height, with a short trunk and a wide tree top (Fig. 6.3). It has extended lenticellate and glabrescent or pubescent branches, dark brown bark, with deep longitudinal cracks. Stipules are whitish, with thin and acute 5.0–30.0 mm length light gray pubescent thorns in the base, dark and glabrescent in the apex. Leaves are bipinnately compound with 3–15 pinnate pairs (1.0–2.5 length), numerous opposed sessile 1–4 mm length leaflets with incospicuous veins and entire margins, glabrous or with scarce hairs on the margins, usually with one petiolar gland. The inflorescence is a densely flowered globose head with orange-yellowish 1 cm diameter flowers. Flowers are hermaphrodite, actinomorphic, gamopetalous, pentamers, sessile, with pubescent peduncles. The sparkling blackish or green-yellowish fruit is an indehiscent legume, with 3–5 mm seeds, 25–30 per fruit, surrounded by a spongy substance. Fruit size and shape is highly variable. Blooming is before foliation in spring, and fructification is in summer (Spegazzini 1924; Cialdella 1994; Cialdella 1996; Lahitte et al. 1999; Marzocca 1997; Ebinger et al. 2000; Pometti et al. 2007).

Ethnobotanical information The root is considered purgative: the bark, detergent, cicatrizing, and antidiarrheal. The dried leaf and stems are used as antiseptic, to treat hypertension, diabetes, skin mycosis, and irritations, and as cicatrizing. The decoction of the leaf is taken as antidiarrheal and diuretic. Flowers are sedative. The combination of seeds and leaf is astringent, antiseptic, anti-asthmatic, and

Fig. 6.3 *Acacia caven* (Molina) Molina var. caven (Fabaceae) tree (**a**), fruits and leaves (**b**), *A. caven* parasited by *Ligaria cuneifolia* (**c**). MA Alvarez personal collection

antitussive. Some specimens were found with cyanogenic activity (Aronson and Nash 1989; Ebinger et al. 2000; Martínez 2007; Barboza et al. 2009).

Chemical data The dried leaf has the flavonols quercetin and quercitrin. Fresh flowers contain p-anisaldehyde, (E,E)-farnesyl acetate, eugenol, benzyl alcohol, methyl salicylate, and cuminyl alcohol (4-propan-2-ylphenyl methanol) (Fig. 6.4) (Lamarque et al. 1998; Barboza et al. 2009).

Biological activities The ethanolic extract from leaves and stems has cytotoxic, antimicrobial, and antifungal activity (Quiroga et al. 2004). The infusion of leaves showed an arterial contractile effect in a dose-dependent manner in pre-anesthetized rats, which was prevented by the α-adrenergic antagonist phenoxybenzamine (Adrados et al. 1997; Alonso and Desmarchelier 2015). The dried seeds are β-glucuronidase inhibitors (Barboza et al. 2009). The cortex, rich in tannins, is used as astringent and to cure blows and wounds (Ortiz 1966). The seeds are considered digestive and stimulants (Ortiz 1966). A macerate made with fresh flowers and alcohol is used as sedative and in cardialgias. The infusion made with the fruits is antihemorroidal (Carrere 1990; Benedetti 2012).

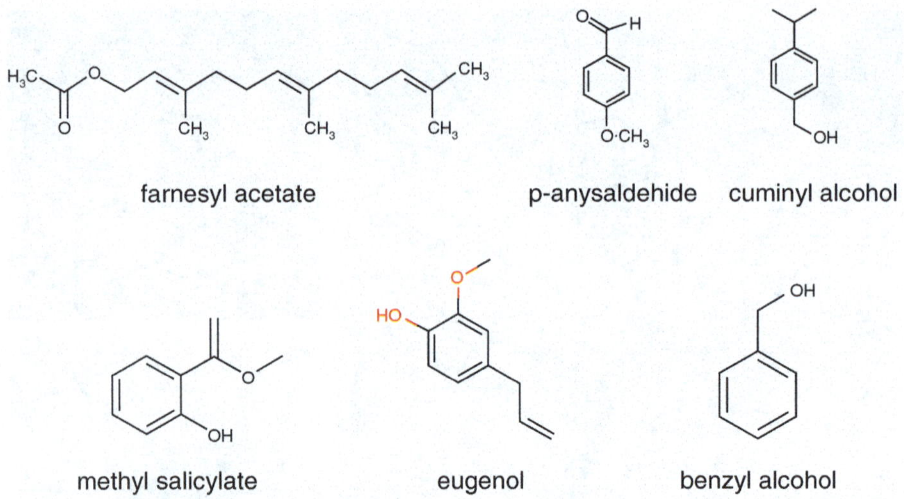

Fig. 6.4 Chemicals from *Acacia caven* (Molina) Molina var. caven

In vitro culture Micropropagation was attained from axenic seedlings in Broadleaved Tree Medium with IBA (0.49 and 1.2 μM) and BAP (0.44, 1.10, and 2.22 μM). Calli differentiated and developed longer roots with the addition of 9 and 10% sucrose without plant growth regulators (Abedini et al. 2000). Somatic embryogenesis was also achieved (Sharry et al. 2010).

6.3 *Chenopodium ambrosioides* L. (Chenopodiaceae)

Synonymy *Dysphania ambrosioides* (L.) Mosyakin & Clemants, *Ambrina ambrosioides* (L.) Spach., *Ambrina parvula* Phil., *Ambrina. spathulata* Moq., *Atriplex ambrosioides* (L.) Crantz, *Blitum ambrosioides* (L.) Crantz, *Chenopodium anthelminticum* L., *C. integrifolium* Boros., C. spathulatum Sieber, *Chenopodium suffruticosum* Willd.

Vernacular name Paico, paico macho, hierba hedionda, té de los jesuitas, pichim tirao, paico hembra, yerba de Santa María, ambrosía, apasote, epazote, té de México.

General information *C. ambrosioides* is a native species from temperate America, although today it is found in almost all the temperate regions of the world (Lorenzi and Matos 2002). The plant is considered by the World Health Organization as one of the most used among traditional medicines in the world (Lorenzi and Matos 2002). It grows in the edges of cultivation lands and borders of gardens and parks. It adapts well to clay, sandy, xerophytic, and sub-xerophytic soils and could be found at altitudes between 0 and 2760 m.a.s.l. (Alban 1984). It has a wide distribution in the country, growing in the provinces of BAI, CHA, OR, COS, DFE, ERI, FOR, JUJ, LPA, LRI, MEN, MIS, RNE, SAL, SDE, SFE, SJU, SLU, and TUC (Barboza et al. 2009).

Botanical characteristics *C. ambrosioides* is an annual, perennial aromatic herb, 40–100 cm height, pubescent. Stem is branched with erect or prostrate branches on their lower section (Fig. 6.5). Leaves are ovate-lanceolate, serrated-toothed, pinnately nerved, 3–10 cm length, 1–5 cm width, with petiole. The inflorescence is a spike, with numerous green flowers in a pyramidal panicle that bloom in summer. The calix has 5 membranoceous or herbaceous sepals free up to the half, 5 stamens, 2–5 long stigmas. The fruit is a utricle; the shining black seeds are elongated (Gupta 1995; Giusti 1997; Tolaba 2006).

Ethnomedical information Decoctions and infusions of inflorescences, leaves, and roots of *C. ambrosioides* have been used for centuries as traditional anthelmintic and dietary condiments against intestinal worms (Hernández 1571; Ratera and Ratera 1980; Martínez Crovetto 1981a, b; Kliks 1985; Carballo et al. 2005). Also, those preparations are taken as emmenagogue, digestive, diuretic, antispasmodic, hypotensive, and diaphoretic (Bandoni et al. 2009; Barboza et al. 2009; González et al. 2009; Da Silva et al. 2014; Navone et al. 2014; Neiva et al. 2014; Alonso and Desmarchelier 2015; Flores and Pellegrini 2015; Moya and Escudero 2015). The infusion made with the aerial parts is considered antitussive, antiarthritic, antipyretic, and hemostatic. Also, it is used for stomachache, for asthma and tumors, and as appetitive (Gómez Castellanos 2008; Barboza et al. 2009). In Northern Brazil the leaves are blended with milk and taken for treating flu (Morais et al. 2005). Leaves are used in cases of muscle pain and bone fractures (Santayana et al. 2005;

Fig. 6.5 *Chenopodium ambrosioides* L. leaves and inflorescences. MA Alvarez personal collection

Garcia et al. 2010). Ascaridol and commercial preparations of the essential oil are prescribed for the treatment of intestinal worms (Patel 2017).

Chemical data The aerial parts have (-)(1R',4S')-1,4-dihydroxy-p-menth-2-ene, (-)(1R', 2S', 3S', 4S')-1,2,3,4-tetrahydroxy-p-menthane. The EOs from the aerial parts have aritasone, limonene, ascaridol, cis-isascaridol, α-terpinene, camphor, trans-ascaridol, p–cymene, pinocarvone, α-and β-pinene, geraniol, α- and γ-terpineol, thymol, β-caryophyllene, γ-gurjunene, myrcene, phellandrene, safrole, n-docosane, n-hentriacontane, n-heptacosane, n-octacosane, p-cymol, spinasterol, terpinyl-acetate, terpinyl-salicylate, and triacontyl-alcohol. Also, stigmasterol, β-sitosterol, octadecanoic acid, scopoletin and 1-piperoylpiperidine T kaempferol, isorhamnetin, quercetin, kaempferol 3-rhamnoside-4′-xyloside, and kaempferol 3-rhamnoside-7- xyloside (Jain et al. 1990; Shah and Khan 2017). The entire plant has saponins, trimethylamine, citric acid, salicylic acid, tartaric acid, succinic acid, butyric acid, ferulic acid, malic acid, methyl salicylate, urease, and vainillic acid (Fig. 6.6). The fruits have quercetin, kaempferol rhamnoside, anethole, and santonin. The roots, contain heterosides, and betaine (Kiuchi et al. 2002; MacDonald et al. 2004; Barboza et al. 2009; Bakker et al. 2011; Neiva et al. 2011; Barros et al. 2013; Moya and Escudero 2015; Patel 2017; Shah 2014; Shameem et al. 2019).

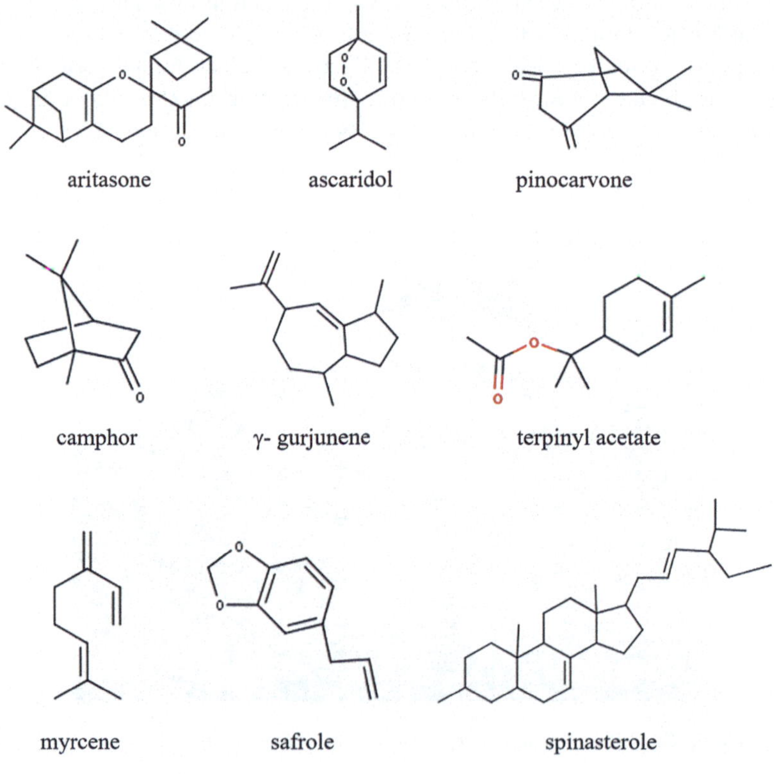

Fig. 6.6 Chemicals found in *Chenopodium ambrosioides* L. essential oil

Biological activities The EO of the aerial parts has antioxidant (Kumar et al. 2007; Barros et al. 2013), antiparasitic (Avila-Blanco et al. 2014; Moya and Escudero 2015; Patel 2017), antimalarial (Kiuchi et al. 2002), anti-leishmania (França et al. 1996; Monzote et al. 2007, 2011, 2014, 2018), anti-*Trypanosoma cruzi* (Rojas et al. 2010), cardiotonic, hypotensive, antialatoxigenic (Kumar et al. 2007), and spasmogenic activity (da Silva et al. 2014). Also, it has fumigant (Chu et al. 2011; Zhu et al. 2012) and repellent activity against head lice.

The aqueous extracts of bark and fruits showed a high antioxidant potential determined by the ABTS assay, whereas petroleum ether bark extracts showed the maximal % DPPH (Ajaib et al. 2016). The infusion of aerial parts gave higher DPPH scavenging activity, β-carotene bleaching, and thiobarbituric acid reactive substances (TBARS) inhibition than the methanolic extract (Barros et al. 2013).

The aerial parts have carcinogenic, giardicidal (Neiva et al. 2011), antiulcerous, antimalarial (Pollack et al. 1990), anti-leishmania (Monzote et al. 2007), and antibacterial activity (MacDonald et al. 2004; Sá et al. 2016; Jesus et al. 2018). The crude hydroalcoholic extract had plasmodicidal potential since it inhibited the parasite growth *in vitro*. *In vivo* it decreased parasitemia in infected BALB/c mice (Nunes Cysne et al. 2016).

The leaves have antimicrobial, analgesic, and anti-inflammatory activity, and the combination of leaves with flower has an antispasmodic activity.

The antiparasitic activity included activity against *Ancylostoma duodenale*, *Trichuris tricura*, *Ascaris lumbricoides*, and *Giardia lambia* (Giove 1996; Neiva et al. 2014). The EO and the ethanol extract from leaves were useful for controlling canine *Ancylostoma* spp. (Moraes Monteiro et al. 2017), gastrointestinal parasites of fighting cocks (*Gallus domesticus*) (Álvarez et al. 2011; Cazorla and Morales 2013), young cattle (Clavijo-López et al. 2016), and the cattle tick *Rhipicephalus* (Boophilus) *microplus* (Cunha dos Santos et al. 2013).

Intraperitoneal and oral treatment with the EO (30 mg kg^{-1}) had better anti-leishmania effect than the treatment with the reference drug, amphotericin B (1 mg kg^{-1}) (Monzote et al. 2007).

The bark macerated in petroleum ether was active against pathogens such as *S. aureus*, *P. aeruginosa*, *Brevibacillus agri*, *Trichophyton mentagrophytes*, and C. *albicans* with MICs ranging from 0.25 to 0.80 μg ml^{-1}. In the case of *M. tuberculosis* MIC was 0.5 μg ml^{-1} (Lall and Meyer 1999; Mabona et al. 2013; Jesus et al. 2018). Against *Staphylococcus aureus* the minimal zone of inhibition was 0.009 ± 0.02 mm at 0.7 μg ml^{-1}.

The n-hexane, ethyl acetate, dichloromethane, as well as n-butanol fractions showed moderate to significant activities especially against *B. subtilus*, *K. pneumoniae*, and *S. epidermidis* (Shah 2014). *C. ambrosioides* also has antifungal properties (Fenner et al. 2006; Jardim et al. 2010). The maximal antifungal zone of inhibition against *Aspergillus niger* (16 ± 1.5 mm) was attained with methanol extracts of fruits (Ajaib et al. 2016).

The *in vitro* activity from EOs of the aerial parts against various *Candida* spp. showed that all the microorganisms were sensitive in a concentration-dependent way, being *C. albicans* the most sensitive ATCC 2091 (MIC: 2 μg ml^{-1}). However, in induced vaginal candidiasis rat model (0.1, 1.0, 10% weight volume^{-1}), the response was not dose-dependent, and rats recovered after 12 days of treatment (Chekem et al. 2010).

In osteoarthritis experimentally produced to Wistar rats, the crude hydroalcoholic extract of leaves produced a reduction of synovial inflammation and pain with ascaridol probably acting as antagonist of the N-methyl-D-aspartate (NMDA) receptor (Calado et al. 2015). The ethanolic extract of aerial parts exhibited anti-inflammatory effects in different inflammation animal models with extended effects both to vascular and cellular events (inhibition of myeloperoxidase, decrease of adenosine-deaminase, reduction of nitric oxide levels, inhibition of TNF-α in pleural fluid after carrageenan injection) of the inflammatory process (Trivellato Grassi et al. 2013). A novel use of a gel made with lyophilized aqueous extract as a bone graft substitute for fracture osseointegration was reported. The treatment, which was tested in rabbits, stimulates bone alkaline phosphatase activity during early fracture healing in a time-dependent manner promoting early bone formation (Pinheiro Neto et al. 2017).

The methanolic extract from lyophilized plant material revealed higher antioxidant activity and antitumor effect against colon (HCT-15), cervical (HeLa), and hepatocellular carcinoma (HepG2) cell lines (Barros et al. 2013; Ruffa et al. 2002). The intraperitoneal administration of hydroalcoholic extract (5 mg kg^{-1}) from leaves produced Ehrlich tumor inhibition in Swiss mice (Nascimento et al. 2006). The hydrodistilled essential oils from leaves had antiproliferative activity against A549 and MCF-7 lines (Shameem et al. 2019) and cytotoxic activity (IC50 = 1.0 μg ml^{-1}) against Burkitt's lymphoma (RAJI) cells. The dichloromethane fraction (IC50 34.0 μg ml^{-1}) and ethanol extract (IC50 47.0 μg ml^{-1}) were active against myeloid leukemia (K562) cells (Degenhardt et al. 2016).

Ascaridol, which is present in essential oils from seeds and leaves, is considered to be responsible of the anthelmintic, antiplasmodial, insecticide, pain-relieving, sedative, and antifungal activities (Smillie and Pessoa 1924; Johnson and Croteau 1984; MacDonald et al. 2004; Cafferata et al. 2005; Potawale et al. 2008; Patel 2017). Ascaridol was also active on tumor cell lines such as human promyelocytic leukemia cells (HL60), human leukemia cell lines (CCRF-CEM), and breast cancer cell line (MDA-MB-231) (Efferth et al. 2002; Patel 2017). As for the anti-inflammatory activity, it was attributed to ascaridol and p-cymene (Degenhardt et al. 2016). There are numerous studies related to its application as biopesticide. The aqueous extract (2 and 20%, 24 h of treatment) was effective to reduce the population of the phytonematode *Pratylenchus brachyurus*; however, the concentration resulted phytotoxic to the parasited soybean plants (Mello et al. 2006). Alcoholic extracts (5%) of leaves reduced the number of *Tuta absoluta* (Lepidoptera: Trichogrammatideae) after 24 h of spraying tomato plants (Barbosa et al. 2011). The effect was attributed to flavonoids and terpenoids (Cruz et al. 2007). Also, a 1.0 and 2.0% (P/P) extract had insecticide effect on *Sitophilus zeamais* Mots. (Coleoptera: Curculionidae) (Silva et al. 2005). When the EO (0.1–0.5%) and aqueous extracts (5.0%) were incorporated into the diet of the cabbage pest *Copitarsia decolorata* Guenée 1852 (Lepidoptera: Noctuidae) larvae, the essential oils reduced larval weight (33%), increased the larval period (19–20%), and reduced fecundity (88%) and fertility (93%). Also, 0.5% essential oil reduced the mean survival time (53%). Aqueous extracts only reduced fecundity (70%) and fertility (75%) (Barbosa et al. 2011; Vázquez-Covarrubias et al. 2015). The EO (0.36 μl ml^{-1}) resulted in 100% repellency and mortality of *Callosobruchus chinensis* L. and *C. maculatus* F. on stored pigeon pea seeds (LD50 2.8 μl and 2.5 μl, respectively) (Pandey et al. 2014). The 5% ethanolic extracts eliminated 54% of *C. maculatus* (Fabricus) adult and reduced oviposition rate (72%) (Kemabonta and Okogbue 2002); the 10% aqueous extract reduced oviposition (98%) of *Plutella xylostella* (L) (Lepidoptera: Plutellidae) (Medeiros et al. 2005). The insecticide activity from EOs varied according to the *P. xylostella* developmental stage and the treatment method (Wei et al. 2015). It also eliminated the second instar larvae of *Anticarsia gemmatalis* (Hübner) (Barbieri and Fiuza 2004). Similarly, EOs from leaves influenced the behavior of *Diabrotica speciosa* (Coleoptera: Chrysomelidae) reducing the feeding activity causing its death (de Andrade Santiago et al. 2014). The EOs (16.75% in oil dispersion) sprayed on leaves of ornamental crops (*Viola* x *wittrockiana* var. Lubega F1 Mix and Hebe 'Purple Pixie') showed a slow rate but significant reduction of the aphids *Myzus persicae* and *Aphis gossypii* (Smith et al. 2018).

As for mosquito, it was effective against *Culex pipiens* larvae (EC50 = 0.750 ppm), *C. pipiens* (EC50 3.097 ppm) (Harraz et al. 2014), and *Anopheles gambiae* Giles (Diptera: Culicidae) larvae (a vector of malaria), with a 100% of mortality (at 200 and 300 ppm) and a 100% of repellent effect (200 ppm) against adult mosquitoes (Bigoga et al. 2013).

C. ambrosioides infusion and ethanolic extract from aerial parts prevented microbial development and lipid oxidation of raw food (for its flavonoids and citric acid content), particularly raw ground pork maintained at 4 °C. As that could increase shelf-life of raw ground meat it was proposed as a natural antioxidant for lipids and myoglobin (Villalobos-Delgado et al. 2017).

It was reported an allelopathic activity of the EO from leaves against the weed *Avena fatua* L. (Sangeeta et al. 2017). Moreover, crude aqueous extract from aerial parts (1 g 100 ml^{-1} roots and 1.5 g 100 ml^{-1} leaves) inhibited hypocotyl growth of *Amaranthus hypochondriacus*. Methanolic extracts from roots and leaves (0.552–0.509 μl per dish) inhibited hypocotyl growth (55%) and germination of *A. hypochondriacus*. Also, the EO (0.15 ml l^{-1}) was detrimental to plant growth (root dying, inhibition of root growth, smallest and greenest leaves) of *Phaseolus acutifolius* after 12 h of treatment. After 30 d of treatment with the lowest concentration, biomass was 30% less than the control while biomass loss was more than 50% at higher concentrations (Jiménez Osornio et al. 1996).

Toxicity The extracts from aerial parts showed activity against *Artemia salina* Lech (Sousa et al. 2012). The infusion was safer than the EO in healthy individuals; however, it was presumed that it could potentiate hepatic and renal disorders (MacDonald et al. 2004; da Silva et al. 2014). The subchronic treatment of Swiss mice with the hydroalcoholic extract (5 mg kg^{-1}) and of NIH Swiss mice with the infusion (32, 64 and 134 mg ml^{-1}) was not lethal and did not induce toxic alterations at therapeutic doses (Pereira et al. 2010; Moreno Mendoza et al. 2013). The EO at high doses was toxic (Montoya-Cabrera et al. 1996; da Silva et al. 2014). The *in vitro* exposure of human lymphocytes to *C. ambrosioides* extracts increased the frequency of chromosomal aberrations and reduced the mitotic index (Gadano et al. 2002, 2007). Signs of toxicity were evident only in the animals treated by intraperitoneal route (Monzote et al. 2007). Ascaridol, which constitutes more than 50% of the EO, produced hypothermia and decreased locomotor activity in mice and was lethal at high doses (Patel 2017). In Egyptian toad (*Bufo regularis*), the EO administrated for 3 months produced hepatocellular carcinomas (el-Mofty et al. 1992).

Carvacrol, caryophyllene oxide, and ascaridol inhibited the mitochondrial electron transport chain. Carvacrol and caryophyllene oxide affected mitochondrial electron transport by directly inhibiting the electron-transferring complex I, whereas the effects of ascaridol depended strongly on the availability of redox active Fe^{2+} (Monzote et al. 2009).

As for genotoxicity, when human lymphocyte cell cultures where treated with different decoction concentrations, the chromosomal aberrations and sister chromatid exchange increased, and the mitotic index decreased (Gadano et al. 2002).

Legal status The EO was considered officinal by the FNA sixth edition. According to the National Institute of Drugs (INAME-ANMAT), it is a toxic species (Flores and Pellegrini 2015). The EO of *C. ambrosioides* is included in the United States National Formulary and in the British Pharmacopoeia.

6.4 *Anemia tomentosa* (Savigny) Sw. var. anthriscifolia (Schrad.) Mickel (Anemiaceae)

Synonymy *Anemia fulva* auct. non (Cav.) Sw., *Anemia anthriscifolia* Schrad., *Anemia ferruginea* Kunth var. anthriscifolia, *Hemianemia anthriscifolia* (Schrad.) C. F. Reed.

Vernacular name Doradilla.

General information The family Anemiaceae comprises about 100 species of monophyletic ferns included in only one genus, *Anemia*. It is an aromatic fern with a wide distribution, from the Atlantic coast of Brazil to Bolivia and Argentina. In Argentina it could be found in the provinces of BAI, CAT, CHA, COR, COS, FOR, JUJ, MIS, SAL, SDE, SFE, and TUC.

Botanical characteristics *A. tomentosa* is a perennial species up to 70 cm long with creeping horizontal rhizomes, 3–7 mm diameter with 1–7 mm orange long hairs. It has frond dimorphism; the fertile frond is longer than the sterile one and has pilose stipes brown at the base. Fronds are deltoid to ovo-deltoid blades, bipinnate-pinnatifid, with 8–13 pairs of chartaceous o subcoriaceous pinna and multicellular trichomes, thin tomentose rachis, and free venation. Sporangia are distributed in two rows on the last segments, with trilete spores, ornamented exosporium, and thin perisporium (Mickel 1962; de la Sota and Mickel 1968; Martínez et al. 2003; Di Pasquo et al. 2016).

Ethnobotanical information It is used to improve blood circulation (Martínez Crovetto 1981a, b). The leaf is used as digestive, antitussive, antigripal, and expectorant (Martínez 2005; Barboza et al. 2009; Scarpa and Cassá 2015; Trillo 2016).

Chemical data Aerial parts contain EOs with spathulenol, caryophyllene oxide, α-bisabolol, α-bisaboloxide, 14-hydroxy-9-epi-(E) caryophyllene, neral, geranial, α-pinene, camphene, 6-methyl-5-hepten-2-one,1,8-cineole, pinocarveol, isoafricanol, and triquinane sesquiterpenes such as epi-presilphiperfolan-1-ol and silphiperfol-6-ene (Figs. 6.7 and 6.8) (Santos et al. 2003; Juliani et al. 2004; Santos et al. 2006; Barboza et al. 2009; Pinto et al. 2009; Joseph-Nathan et al. 2010).

Biological activities The antimicrobial activity against *Mycobacterium tuberculosis* H37Rv (ATCC-27294) of the EO (200 μg ml^{-1}) (MIC 100 μg ml^{-1}) was mainly attributed to triquinane sesquiterpenes (Pinto et al. 2009). The EO of the aerial parts also exhibited activity against *Bacillus cereus* (18 mm inhibition zone) (Demo et al. 2005). The repellent activity against *Aedes aegypti* at a 90% essential oil concentration was attributed to monoterpenoids and sesquiterpenes (Gillij et al. 2008).

In vitro culture Spore-derived gametophytes developed into sporophytes in MS (Murashige and Skoog 1962) culture media and ½ MS culture media without plant growth regulators. The *in vitro* sporophytes showed an inverted volatile profile regarding the wild-grown plants. In wild-grown plants, sesquiterpenes represented 97.5% of the volatiles produced, whereas in *in vitro* plants, the monoterpene profile prevailed (69.8–89.8%) over sesquiterpenes (9.4–28.7%). The main monoterpene identified in those cultures was α-pinene followed by *trans*-pinocarveol, pinocarvone, and myrtenyl acetate. The trinquinane sesquiterpenes found were silphiperfol-6-ene followed by α-guaiene, β-barbatene, and 9-epipresilphiperfolan-1-ol. The plant growth regulator IAA was not adequate to promote biomass production while jasmonic acid promoted gametophyte and sporophyte development.

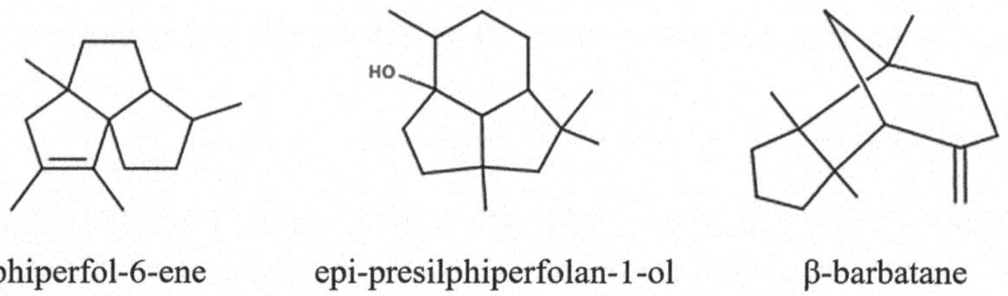

Fig. 6.7 Some triquinane sesquiterpenes found in *Anemia tomentosa* (Savigny) Sw. var. anthriscifolia (Schrad.) Mickel

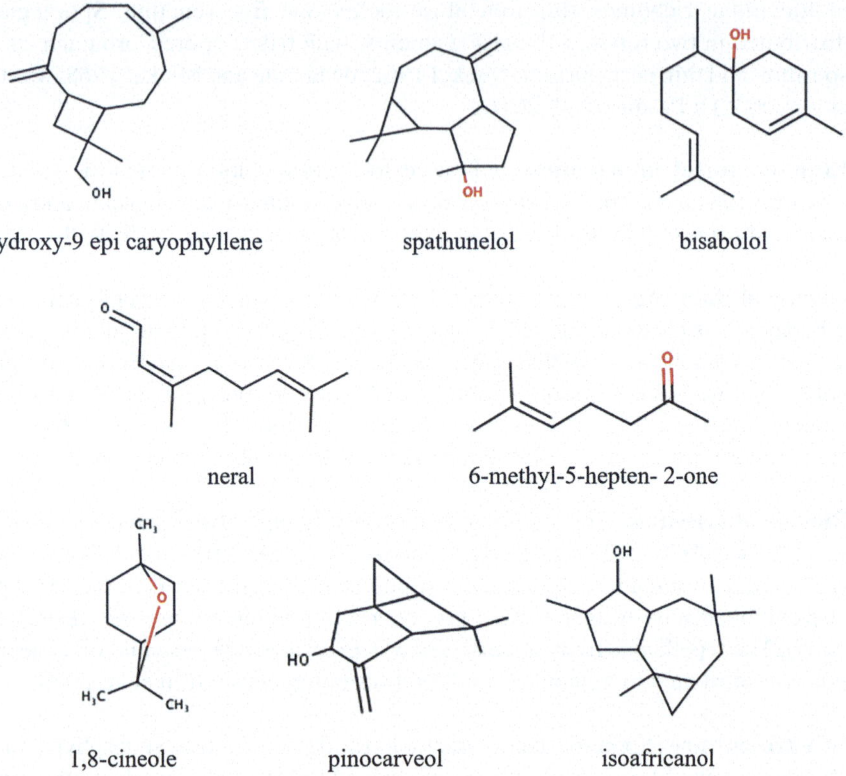

Fig. 6.8 Chemical compounds found in *Anemia tomentosa* (Savigny) Sw. var. anthriscifolia (Schrad.) Mickel essential oils

None of the plant growth regulators were able to reproduce the monoterpene/sesquiterpene balance of wild-grown plants a phenomenon that it was found related to the different irradiance levels of *in vitro* cultures, to the necessity of differentiated plant tissues at a specific moment, or to the development phase of the plant (Pinto et al. 2013; Velasco-Castilho et al. 2018).

6.5 *Plantago australis* Lam. subsp. Australis (Plantaginaceae)

Vernacular name llantén, siete varas, lengua de vaca, llantén blanco.

Synonymy *Plantago macropus* Pilg., *Plantago denudata* Pilg., *Plantago hirtella* Kunth var. platensis, *Plantago candollei* Rapin, *Plantago macrostachys* Decne., *Plantago myosuros* Lam. var. latifolia, *Plantago accrescens* Pilg., *Plantago kurtzii* Pilg., *Plantago refracta* Pilg., *Plantago stuckertii* Pilg., *Plantago macrostachys*

Decne. var. angustifolia, *Plantago macrostachys* Decne. var. denudata, *Plantago macrostachys* Decne. var. platensis, *Plantago macrostachys* Decne. var. stuckertii, *Plantago macrostachys* Decne. var. catamarcensis, *Plantago macrostachys* Decne. var. typica, *Plantago macrostachys* Decne. var. brachypus, *Plantago stuckertii* Pilg. var. catamarcensis, *Plantago hirtella* Kunth var. glabrescens, *Plantago macrostachys* Decne. f. brachypus, *Plantago australis* Lam. ssp. macrostachys (Decne.) Rahn, *Plantago australis* Lam. ssp. angustifolia (Pilg.) Rahn, *Plantago bicallosa* Decne. var. angustifolia, *Plantago macrostachys* Decne. var. accrescens, *Plantago gigantea* Decne., *Plantago macrostachys* Decne. var. gigantea, *Plantago durvillei* Fisch. & C.A. Mey.

General information *Plantago australis* is a popular plant widely spread in Latin America. In Argentina it grows in the provinces of BAI, CAT, CHU, COR, COS, ERI, JUJ, LPA, LRI, MEN, NEU, RNE, SAL, SCR, SFE, SJU, SLU, and TUC.

Botanical characteristics *P. australis* is a perennial herb with an ephemeral main root that is replaced by numerous secondary roots. Leaves are simple, alternate or whorled, rarely opposite, elliptic-lanceolate, without stipules. The blade margin is entire or toothed. Primary veins are prominent to the hipophile; the petiole is long. Flowers are actinomorphe, small, anemophilous, generally in spikes, rarely solitary, in the axil of each bract. Perianth has four sepals, a gamopetal corolla, with four imbedding lobules. Androecium has four stamens, with versatile introrse anthers. Ovary is superior, bicarpelar, uni- or bilocular with uni-to-pluri-ovules; style is short with a long filiform stigma. Fruit is a pyxidium (pyxis). Seeds, 3–4 for capsule, are oblong and albuminous (Rahn 1996; Bayón et al. 2000; Moroni et al. 2018).

Ethnomedical information *P. australis* is used in folk medicine as pectoral, anti-catarrhal, emollient, analgesic, anti-inflammatory, for postpartum, bladder and kidney diseases, for cystitis, wounds, and vaginal discharge, to healing wounds and to treat contaminated (thick or bad) blood. The root is used in case of infections, hernia, and prolapse. Leaves are taken as astringent, vulnerary, abluent, to facilitate birth, and the seeds, as laxative. It is also referred its use to treat flu and common cold by Polish immigrants from the province of MIS (Toursarkissian 1980; Bayón et al. 2000; Barboza et al. 2009; Gheno-Heredia et al. 2011; Kujawska and Pieroni 2015).

Chemical data It has aucubin (iridoid monoterpene), verbascoside, isoverbascoside, salidroside, plantamajoside, baicalein, hispidulin, ursolic acid, and oleanolic acid (Ronsted et al. 2000; Barboza et al. 2009; Flores et al. 2016a, b). Verbascoside or acetoside is a phenolic glycoside compound with four chemical groups: caffeic acid, 4,5-hydroxyphenylethanol linked to β-(D)-glucopyranoside, rhamnose, and glucose (Korkina et al. 2007), that is the main active component from extracts (Fig. 6.9).

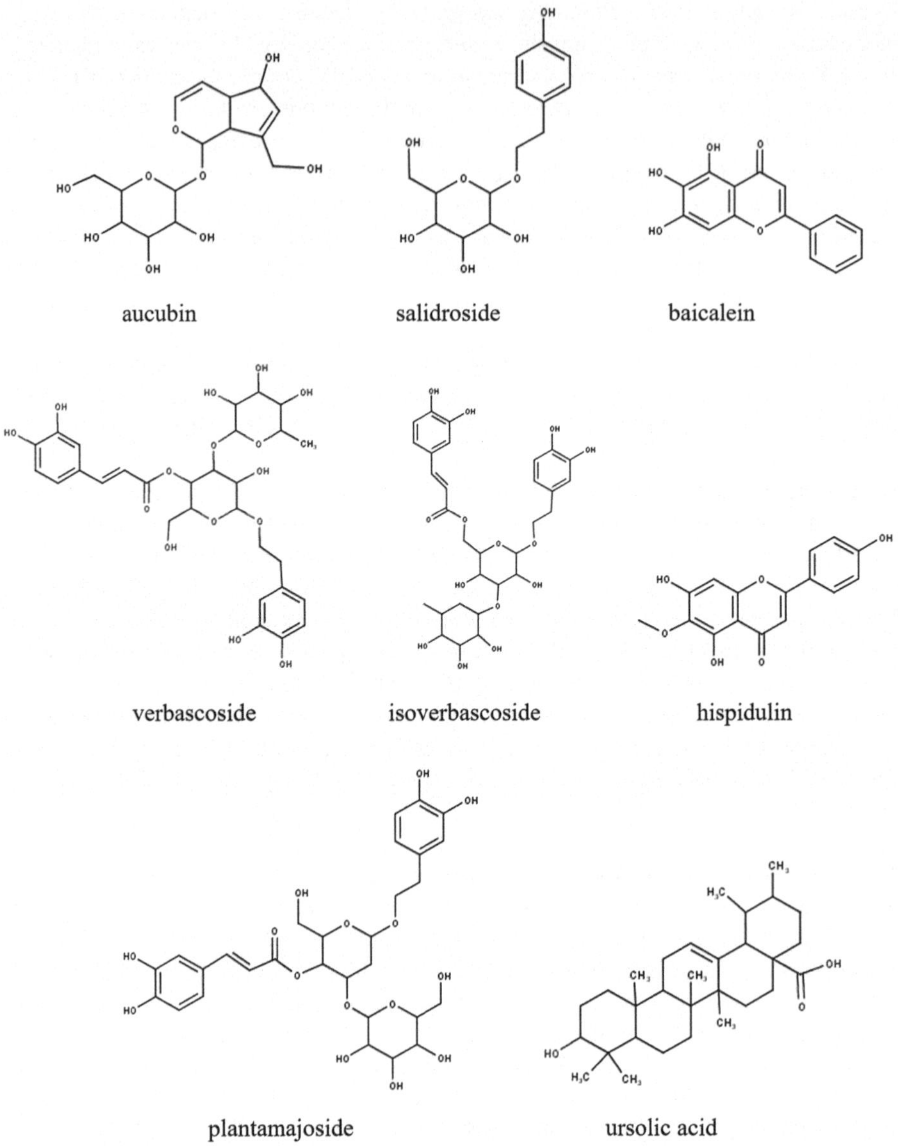

Fig. 6.9 Chemical compounds from *Plantago australis* Lam. subsp. Australis (Plantaginaceae)

Biological activities The antioxidant, anti-inflammatory, healing, and anti-nociceptive properties were attributed to verbascoside (Lee et al. 2007; Hausmann et al. 2007; Vertuani et al. 2011; Yoou et al. 2015; de Moura Sperotto et al. 2018). The dried fruit, dried root, leaf, and fruits have antiulcer and anti-inflammatory activity (Barboza et al. 2009). *In vitro* treatment of keratinocyte cells HaCaT cells (scratch test) with several concentrations of the hydroalcoholic extract (standardized on verbascoside) or verbascoside for 24 h promoted migration of keratinocytes at all concentrations tested. On the other hand, *in vivo* tests administrating different concentrations of the hydroalcoholic extract to male Wistar rats and with a LPS inflammation model in N9 microglial cells showed wound healing promotion. The extract decreased viability at 10 times higher concentrations (1000 μg kg^{-1}) and promotes a faster wound healing than the verbascoside. On the other hand, both the extract and verbascoside reduced the oxidative parameters (SOD, CAT in N9 cells LPS-activated) at the concentrations tested (de Moura Sperotto et al. 2018). Previously, the positive cicatrizing effect of the leaf extract on induced wounds in mouse (*Mus musculus*) was attributed to flavonoids and tannins (Asto-Guamán et al. 2017) or to prostaglandin synthesis inhibition (Palmeiro et al. 2002).

The analgesic and anti-inflammatory activity of the hydroalcoholic extract (70%) from leaves, roots, and fruits from *P. australis* growing in Southern Brazil were tested on carrageenan-induced rat hind paw edema (250, 500, and 1000 mg kg^{-1}) and 1.2% acetic acid (10 mg kg^{-1})- induced writhing. The oral administration inhibited both inflammation and writhes significantly (Palmeiro et al. 2003). The crude ethanolic extract (500, 1000 mg kg^{-1}) obtained from leaves was tested against ethanol-, indomethacin-, and cold restrain-induced stress ulcers showing a reduction of the lesion and ulcer indexes produced by ethanol and an increase of mucous at the highest concentration tested in cold restrain-induced stress ulcer. In this case the responsible mechanism proposed was an interference with histamine and leukotriene C4 deleterious effect on the gastric mucosa (Bürger et al. 2002).

The *in vivo* anti-inflammatory activity on oral lesions of a pharmaceutical formulation (solution or cream) containing an ethanolic extract (10%) of *P. australis* showed that the topical use reduced signs and symptoms of recurrent aphthous stomatitis, lichen planus, and actinic cheilitis (Flores et al. 2016a). Also, the effect on the growth of *Candida albicans* on resilient denture liners was verified (Pereira-Cenci et al. 2014).

Toxicity studies *In vitro* cytotoxicity of different concentrations of leaf crude extract was reported. The hydroethanolic extract on *in vitro* models showed toxicity in the *Salmonella*/microsome assay. No cytotoxicity nor mutagenicity was detected at the concentrations tested; however, when the toxicity was studied in V79 cells by the dimethyl thiazolyl diphenyl tetrazolium salt test (MTT) and neutral red uptake (NRU) assay, both showed cytotoxicity at the highest concentrations (1000 μg ml^{-1} for the hydroalcoholic extract and 50 mg ml^{-1} for the verbascoside). As for *in vivo* assays, acute oral toxicity and subchronic tests made on healthy female and male Wistar rats did not detect clinical signs of toxicity. Behavior, body weight, histopathological analysis, and biochemical parameters did not show significant differences respect to the controls (Henn et al. 2019). Similar results were obtained with *P. australis* aqueous extract; however, in this last case, a significant increase of aspartate aminotransferase, alanine aminotransferase, and creatinine serum levels was found suggesting a hepatic function alteration (Palmeiro et al. 2003). On the other hand, when the MTT was tested on a 3 T3 mouse fibroblast cell line with serial dilutions (10^{-1}, 10^{-2}, 10^{-3}, 10^{-4}, and 10^{-5}) of the 70% ethanol extract, the stronger cytotoxic effect was obtained as the concentration increased (Flores et al. 2016b).

As it was proposed the use of the extract in phytocosmetics, phototoxicity was also tested. The neutral red uptake (NRU) test in 3T3 did not show signs of verbascoside phototoxicity at the concentrations tested (1.0, 2.5, 5.0, 10, 25, 50, and 100 $\mu g\ ml^{-1}$) (Henn et al. 2019).

As for genotoxicity and mutagenicity, neither the hydroethanolic extract nor the verbascoside generated DNA-strand breaks or alterations in micronucleus, bud frequency, or polychromatic erythrocytes/normochromic erythrocytes relationship at the tested concentrations (Henn et al. 2019).

6.6 *Alternanthera pungens* Kunth (Amaranthaceae)

Synonymy *Achyranthes lorentzii* (Uline) Standl., *Achyranthes mucronata* Lam., nom. illeg., *Achyranthes radicans* Cav., nom. illeg., *Achyranthes repens* L., *Alternanthera achyrantha* (L.) R. Br. ex Sweet, nom. illeg., *Alternanthera achyrantha* (L.) R. Br. ex Sweet var. leiantha, *Alternanthera echinata* Sm., *Alternanthera lorentzii* Uline, *Alternanthera pungens* Kunth var. leiantha, *Alternanthera pungens* Kunth f. pauciflora, *Alternanthera repens* (L.) Link, comb. superfl., *Celosia echinata* Willd. ex Schult., *Desmochaeta sordida* Bunbury, *Guilleminea procumbens* Rojas Acosta, *Illecebrum achyranthum* L., nom. illeg., *Illecebrum pungens* (Kunth) Spreng., *Pityranthus crassifolius* Mart., *Pupalia sordida* (Bunbury) Moq., *Telanthera pungens* (Kunth) Moq., *Achyranthes leiantha* (Seub.) Standl., *Paronychia achyrantha* (L.) Desf., (Amorin and Rosow 1988).

Vernacular name Yerba de pollo, guambarusa, raíz colorada, lagunilla, hierba del pollo, llapá reliñ, achawál kachú (pasto del pollo), ierwé achawál (hierba del pollo) in the Araucanian-Pampa language; taasó, taa-só in toba language; bashé umpatpatelét (espinita pegadita a la tierra) in vilela language; caá-pé, quisca-yuyu, torito, sanguinaria, colchón de perro, yerba de carretero (Cuba), adorna jardín (Puerto Rico), caroca, carauca, sangradera (Venezuela), yerba del moro hembra (Perú), khakiweed, creeping chaffweed, spingflower alternanthera (English).

General information Native to South America, *A. pungens* is characterized by its pungent easily shedding tepals that become harder at maturity. It grows in tropical, subtropical, semiarid, and warmer temperate environments, and also in disturbed sites, bare areas, roadsides, parks, lawns, waste areas, watercourses, turf grasses, orchards, and occasionally in native pastures and grasslands. It was naturalized in Australia, northern Africa, Asia (China, Bhutan, Myanmar, Thailand, and Papua New Guinea), and Pacific islands (Hawaii and New Caledonia) (Iamonico and Sánchez del Pino 2016; Peña Rivera and Ferrer-Gallego 2016). In Argentina it usually grows as a dense mat of prickly vegetation in the provinces of BAI, CHA, COS, ERI, FOR, JUJ, SAL, SDE, SFE, SJU, and TUC (Carrizo and Isasmendi 1998; Barboza et al. 2009). It has nutritional value (Raju et al. 2004; Hundiwale Jogendra et al. 2012). Flowering is in summer and autumn (Arenas 1981). It propagates by seed and vegetatively.

Botanical characteristics *A. pungens* is a small, perennial, creeping, pubescent herbaceous plant with zigzag stems up to 60 cm long that springs from a tough and thickened central crown. Stem is hirsute (abundant little short and soft hairs in the nodes). Leaves are shortly petiolate, opposite, slightly asymmetric, egg-shaped or elliptic with a rounded tip, and an entire or slightly waved margin. Leaves are usually glabrous in the main vain, with few hairs on the veins of the underside, which are prominent in the hipophile. Greenish or greenish-yellow flowers are clustered as axillary spikes (8–12 mm long), lanceolate or ovate oblong bracts, with 5 tepals, the two external larger, with 5 stamens, a globular ovary with stigma. The fruit is a utricle, shiny orange or yellowish lenticular seeds usually remained hidden inside the old, prickly, flower parts (Filippa and Espinar 1993). The seeds are contained inside a burr that is transported attached to animals, clothing, and other objects (e.g., vehicle tires). They may also be dispersed by water, in contaminated agricultural products (e.g., fodder and pasture seed). The stem fragments can be spread by machinery, livestock, or cultivation (Owa et al. 2016).

Ethnomedical information *A. pungens* is used in case of measles, against flu, headache, and abdominal pain. A spoonful of the decoction, made with the inflorescences, is taken as vermifuge before breakfast once a day (Hundiwale Jogendra et al. 2012). Root and leaves have activity as diuretic, depurative, digestive, and hepatic. The infusion made with leaves and stems is active in cases of viral diseases, hepatitis, epidemic parotitis, hemorrhagic fever, and influenza. There are also references about its properties to treat gastric, liver, and intestinal problems (Domínguez 1928; Saggese 1959; Soraru and Bandoni 1978; Ratera and Ratera 1980; Toursarkissian 1980; Hieronymus 1882; Amorín 1988; Boelcke 1989; Roig 1992). In Argentina, infusions and decoctions made with roots, leaves, and flowering tips are commonly taken to cure *empacho* and indigestion. The preparations must be handled with care when administered to children because the diuretic activity could cause dehydration. It has depurative mildly laxative and diuretic properties, regularizing stomach and intestinal functions (Orfila 1971; Martínez Crovetto 1981a, b; Amorín 1988). It is also used in Ayurveda medicine (Kalpana et al. 2018).

Chemical Data The chemical compounds found in the entire plant are choline, oleanolic acid, and β-spinaestrol (de Ruiz et al. 1991; Rondina and Coussio 1969, 1981; Dogra et al. 1977) and in the flowers, azulene, α-curcumene, bornyl acetate, α-borneol, camphene, camphor, 1,8-cineol, and α-terpineol (Fig. 6.10). In the dried fruit, it was identified oleanolic acid and rutin (de Ruiz et al. 1991; Barboza et al. 2009). As for the carotenoids, it was found violaxanthin, lutein, zeaxanthin, and β-carotene (Raju et al. 2007).

Biological activities The dried aerial parts showed antitumor, antiviral (HIV and anti-Hantaan) activity, and reverse transcriptase inhibition (Barboza et al. 2009). It was also cited that the leaf extracts have immunological reactivity (Gayathri et al. 2001).

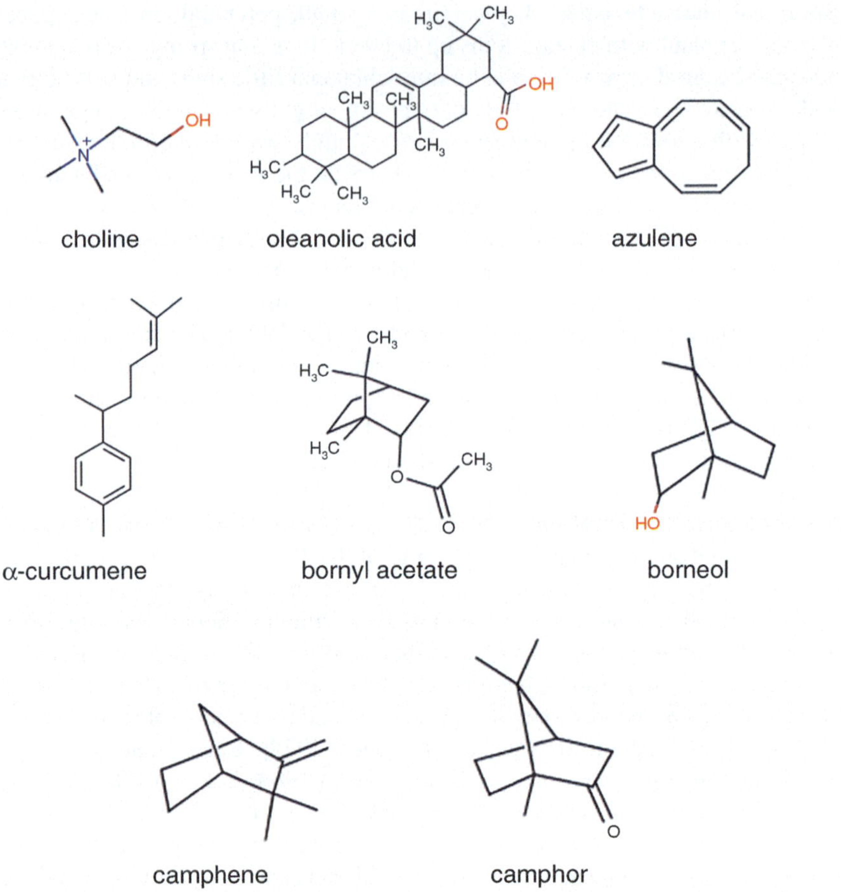

Fig. 6.10 Chemical compounds found in *Althernanthera pungens* Kunth (Amaranthaceae)

The antimicrobial activity from different crude extracts was analyzed against several bacteria and fungi. The crude plant extracts (200, 100, 50, and 25 mg ml^{-1}) from aerial parts showed antibacterial activity against *Klebsiella pneumoniae* MTCC NO 109 (ethanolic extract), *Mycobacterium smegmatis* MTCC NO 992 (acetone and ethanolic extract), *Bacillus subtilis* MTCC NO 2057 (aqueous extract), and *Chromobacterium violaceum* MTCC NO 2656 (acetone extract) with inhibition zones >10 mm. Antifungal activity was shown only against *Aspergillus fumigatus* MTCC NO 3002 (acetone and aqueous extracts) (Jakhar and Dahiya 2017).

The ethanolic extract exhibited highest antioxidant activity (DPPH assay) (IC50 100.79 mg ml^{-1}) than the acetone (IC50 203.56 mg ml^{-1}), aqueous (IC50 324.43 mg ml^{-1}), and petroleum ether extracts (IC50 931.63 mg ml^{-1}), which correlates with the total phenolic and flavonoid content of each extract (Jakhar and Dahiya 2017).

The administration of a tea made with dried leaves in boiled water (1 g 40 ml^{-1}) to patients with HIV each 2 days during 12 months decreased the plasmatic levels of oxidative stress biomarkers (AOPP and MDA) and increased those of T CD4 and CD8 lymphocytes without showing signs of hepatic or renal toxicity (Djohan et al. 2009).

The ethanolic extract from the entire plant (50–300 mg kg^{-1}) reduced glycemia and improved serum lipid profile (lowered cholesterol and triglyceride, restored HDL levels) in Wistar rats with alloxan-induced diabetes after 21 days of treatment, being 50 mg kg^{-1} the most effective dose (Owa et al. 2016).

The ethanolic extract (2.8%) from aerial parts of flowering plants produced significant increases of urinary volumetric excretion and sodium urinary excretion in Wistar rats (100 mg kg^{-1} p.o.) as did furosemide (Calderón et al. 1997).

Methanolic extracts of the whole plants tested on Swiss albino mice by the tail immersion method showed analgesic activity, showing the chloroform fraction the most potent analgesia. The chloroform fraction was rich in terpenoids, steroids, flavonoids, and phenolic compounds (Kalpana et al. 2018).

The aqueous, methanol, and n-hexane extracts (10, 20, 30 mg ml^{-1}) were tested against *Sitophilus oryzae*, *Callosobuchus chinensis*, and *Tribolium castaneum*. The hexane (30 mg ml^{-1}) and methanol (10 mg ml^{-1}) extracts resulted effective against *S. oryzae* (100% mortality after 2 and 7 days, respectively). The aqueous (30 mg ml^{-1}) and hexane (10 mg ml^{-1}) extracts were effective against *C. chinensis* (100% mortality after 6 and 8 days of treatment, respectively). The aqueous (30 mg ml^{-1}) and the hexane (10 mg ml^{-1}) extracts were effective against *T. castaneum* (100% mortality after 10 and 15 days (LD50 < 10 mg ml^{-1}after 24 hours) (Kazmi et al. 2017).

Toxicity *A. pungens* burrs can contaminate lucerne hay and other stock feeds and could cause digestive disturbances and skin ailment in cattle. Horses that graze on areas containing large amounts of this species have developed a form of staggers. The burrs can also contribute to vegetable fault in wool (Weeds of Australia 2016).

Legal status This species is inscribed in the FNA sixth edition.

References

Abedini W, Boeri P, Marunicci L, Ruscitti M, Scelzo L (2000) Biotécnicas aplicadas a especies forestales nativas. Invest Agr; Sist Recur For 9(1):31–43

Adrados H, Taira C, Wagner M, Gurni A (1997) Análisis de extractos de *Ligaria cuneifolia* y de su hospedante *Acacia caven*. II Congreso Mundial de Plantas Aromáticas y Medicinales. WOCMAP II. Mendoza. 10–15 Nov 1997. Abstract P-390

Ajaib M, Hussain T, Farooq S, Ashiq M (2016) Analysis of antimicrobial and antioxidant activities of *Chenopodium ambrosioides*: an ethnomedicinal plant. J Chem 2016:4827157., 11 pages. https://doi.org/10.1155/2016/4827157

Alban CJA (1984) Plantas Medicinales Usadas en Lima para los Trastornos Digestivos. Informe de Prácticas preprofesionales para Optar Título Profesional, Fac. Ciencias Biológicas, UNMSM, Lima

Alonso J, Desmarchelier CJ (2015) Plantas medicinales autóctonas de la Argentina. Bases científicas para su aplicación en atención primaria de la salud. 1a ed. CiudadAutónoma de Buenos Aires: Corpus Libros Médicos y Científicos

Álvarez C, Rodríguez P, Carvajal E (2011) Effect of extract paico (*Chenopodium ambrosioides*) gastrointestinal parasites of fighting cocks (*Gallus domesticus*). Cultura Científica 76–80

Amorín JL (1988) Guía taxonómica con plantas de interés farmacéutico - Bs. As Rev de Inf Fcia y Bioq 34:80 pp

Amorin JL, Rosow R (1988) Guia taxonómica con plantas de interés farmacéutico. Dominguezia 10(1):35–40

Arenas P (1981) Etnobotánica Lengua-Maskoy. Fundación para la Educación, Ciencia y Cultura, Buenos Aires, pp 168–169

Aronson J, NASH RJ (1989) Infrequent cyanogenesis in *Acacia caven* and general notes on cyanogenic glycoside data in subgenus Acacia. Bull Int Group Study Mimosoideae 17:97–110

Aronson J, Ovalle C (1989) Report on a study of the natural variability, biogeography and potential for genetic improvement of *Acacia caven*. Bull Int Group Study Mimosoideae 17:111–121

Asto-Guamán V, Escudero-Vilema ER, Andueza-Leal D, Toaquiza-Aguaguallo NC (2017) Actividad cicatrizante de extracto de hojas de llantén páramo (*Plantago australis*) en ratones (*Mus musculus*). VII Congreso Iberoamerican de Ciencias Farmacéuticas, Riobamba

Avila-Blanco ME, Rodríguez MG, Moreno Duque JL, Muñoz-Ortega M, Ven-tura-Juárez J (2014) Amoebicidal activity of essential oil of *Dysphania ambrosioides* (L.) Mosyakin & Clemants in an amoebic liver abscess hamster model. Evid Based Complement Alternat Med 2014:930208, 7 pages. https://doi.org/10.1155/2014/930208

Bakker CV, Blömeke B, Coenraads PJ, Schuttelaar ML (2011) Ascaridole, a sensitizing component of tea tree oil, patch tested at 1% and 5% in two series of patients. Contact Dermatitis 65(4):240–241

Bandoni AL, Retta D, Di Leo LP, van Baren C (2009) ¿Son realmente útiles los aceites esenciales? Bol Latinoam Caribe Plant Med Aromat 8(5):317–322

Barbieri CP, Fiuza LM (2004) Bioatividade de extratos vegetais à *Anticarsia gemmatalis* (Lepidoptera: Noctuidae). Resumos do XX Congresso Brasileiro de Entomologia, Gramado RS Brasil, 317 pp

Barbosa FS, Demolin Leite GL, Monteze Alves S, Fonseca Nascimento A, De Breu D'AV, Alves da Costa C (2011) *Chenopodium ambrosioides* against pests and natural enemies in commercial tomato plantation. Maringá 33(1):37–43

Barboza GE, Cantero JJ, Núñez C, Pacciaroni A, Ariza EL (2009) Medicinal plants: a general review and a phytochemical and ethnopharmacological screening of the native Argentine Flora. Kurtziana 34(12):7365

Barros L, Pereira E, Calhelha RC, Dueñas M, Carvalho AM, Santos-Buelga C, Ferreria ICFR (2013) Bioactivity and chemical characterization in hydrophilic and lipophilic compounds of *Chenopodium ambrosioides* L. J Funct Foods 5:1732–1740

Bayón ND, Colares MN, Am A, Viña SZ (2000) Anatomía de las Especies Medicinales de la Provincia Pampeana: Plantago L. (Plantaginaceae). Acta Farm Bonaer 19(4):263–272

Benedetti SR (2012) Monografía de espino, *Acacia caven* (Mol.) Mol. Instituto Forestal, Ministerio de Agricultura, Gobierno de Chile, Santiago de Chile

Bigoga JD, Saahkem PA, Ndindeng SA, Ngondi JL, Nyegue M, Oben JE, Leke RGF (2013) Larvicidal and repellent potential of *Chenopodium ambrosioides* Linn essential oil against *Anopheles gambiae* Giles (Diptera: Culicidae). Open Entomol J 7:16–22

Boelcke O (1989) Plantas vasculares de la Argentina - Bs.As., Ed. H. Sur, 2da. reimpresión, 132–369 pp

Bürger ME, Ghedini PC, Baldisserotto B, Palmeiro NMS, Almeida CE, Lenzi CF, Silva AC, Oliveira CQ (2002) Effect of *Plantago australis* leaves on different gastric ulcer models. Rev Bras 12(Suppl):113–114

Cafferata LFR, Jeandupeux R, Rimada RS (2005) Método Simple y Rápido para la Determinación de Ascaridol en Medio Acuoso Utilizando CLAE (RP-HPLC). Acta Farm Bonaer 24(4):567–571

Calado GP, Lopes AJO, Costa Junior LM, Lima FCA, Silva LA, Pereira WS, do Amaral FMM, García JBS, Cartágenes MSS, Nascimento FRF (2015) *Chenopodium ambrosioides* L. reduces synovial inflammation and pain in experimental osteoarthritis. PLoS ONE 10(11):e0141886. https://doi.org/10.1371/journal.pone.0141886

Calderón CP, García Aseff SB, Fuentes LB (1997) Evaluation of diuretic activity of *Alteranthera pungens* extract in rats. Phytother Res 11:606–608

Carballo MA, Cortada CM, Gadano AB (2005) Medicinal herbs: risks and benefits in their uses. Theoria 14(2):95–108

Carrere R (1990) El bosque natural uruguayo: utilización tradicional y usos alternativos. ANEXO No 2. Usos medicinales de especies del monte indígena.Serie Investigaciones N° 79 de CIEDUR Diciembre de 1990

Carrizo J, Isasmendi S (1998) Flora del valle de Lerma. Amaranthaceae Juss. Aportes Bot Salta-Ser. Flora 5(4):1–59

Cazorla D, Morales PM (2013) Prevalence of intestinal parasites in fighting cocks of Coro city, Falcon State, Venezuela. Rev Inv Vet Perú 24(4):489–502

Chekem MSG, Lunga PK, Tamokou JDD, Kuiate JR, Tane P, Vilarem G, Cerny M (2010) Antifungal properties of *Chenopodium ambrosioides* essential oil against *Candida* species. Pharmaceuticals 3:2900–2909

Chu SS, Feng Hu J, Liu ZL (2011) Composition of essential oil of Chinese *Chenopodium ambrosioides* and insecticidal activity against maize weevil, *Sitophilus zeamais*. Pest Manag Sci 67(6):714–718

Cialdella AM (1994) Acacia In: Kiesling R, Flora de San Juan 1:259–263

Cialdella AM (1996) Fabaceae Lindl. Tribu II. Acacieae Benth., nom. cons. Aportes Bot Salta-Ser. Flora - Ser. Flora 4(12):1–24

Clavijo-López F, Barrera Mosquera V, Rodríguez Iturralde L, Mosquera An-drade J, Yánez Ortiz I, Godoy Ortiz G, Grijalva Olmedo J (2016) Evaluation of Paico *Chenopodium ambrosioides* and lupine *Lupinus mutabilis* sweet as gastrointestinal antiparasitic in young cattle. La Granja: Revista de Ciencias de la Vida 24(2):95–110

Cruz GVB, Pereira PVS, Patrício FJ, Costa GC, Sousa SM, Frazao JB, Aragao-Filho WC, Maciel MCG, Silva LA, Amaral FMM, Barroqueiro ESB, Guerra RNM, Nascimento FRF (2007) Increase of cellular recruitment, phagocytosis ability and nitric oxide production induced by hydroalcoholic extract from *Chenopodium ambrosioides* leaves. J Ethnopharmacol 111(1):148–154

Cunha dos Santos FC, Flores Vogel FS, Buttow Roll VF, González Monteiro S (2013) *In vitro* effect of the association of citronela, Santa María herb (*Chenopodium ambrosioides*) and Quassia tincture on cattle tick *Rhipicephalus* (Boophilus) *microplus*. Ci Anim Bras, Goiânia 14(1):113–119

da Silva MG, Amorim RNL, Camara CC, Fontenele Neto JD, Soto-Blanco B (2014) Acute and sub-chronic toxicity of aqueous extracts of *Chenopodium ambrosioides* leaves in rats. J Med Food 17(9):979–984

de Andrade Santiago J, Cardoso MG, da Silva Figuereido AC, Campos de Moraes J, de Assis FA, Teixeira ML, Santiago WD, Sales TA, Camargo KC, Nelson DL (2014) Chemical characterization and application of the essential oils from *Chenopodium ambrosioides* and *Philodendron bipinnatifidum* in the control of Diabrotica speciosa (Coleoptera: Chrysomelidae). Am J Plant Sci 5:3994–4002

De la Sota ER, Mickel JT (1968) Sinopsis de las especies argentinas del género Anemia Swartz (Schizaeaceae). Rev Mus La Plata (n.s.), Bot 11:133–152

de Moura Sperotto ND, Steffens L, Veríssimo RM, Henn JG, Flores Péres V, Vianna P, Bogo Chhies JÁ, Roehe A, Saffi J, Moura DJ (2018) Wound healing and anti-inflammatory activities induced by a *Plantago australis* hydroethanolic extract standardized in verbascoside. J Ethnopharmacol 225:178–188

de Ruiz REL, Fusco M, Rapisarda AMP, Sosa A, Ruiz SO (1991) Estudio Fitoquímico de la "yerba del pollo" (*Alteranthera pungens* H.B.K., Amaranthaceae). Acta Farm Bonaer 10(1):25–27

Degenhardt RT, Farias IV, Grassi LT, Franchi GC Jr, Nowill AE, Bittencourt da S CM, Wagner TM, de Souza MM, Cruz AB, Malheiros A (2016) Characterization and evaluation of the cytotoxic potential of the essential oil of *Chenopodium ambrosioides*. Rev Bras 26:56–61

Demo M, Oliva MM, Lopez ML, Zunino MP, Zygadlo J (2005) Antimicrobial activity of essential oils obtained from aromatic plants of Argentina. Pharm Biol 43:129–134

Di Pasquo M, Rodríguez E, Nuñez Otaño N, Muñoz N, Silvestri L (2016) Esporas de helechos (monilofitas) y licofitas presentes en el Parque Nacional El Palmar (Entre Rios, Argentina). Bol Soc Argent Bot 51(2):269–298

Djohan YF, Camara C, Mondé AA, Koffi G, Niamke G, Dére L, Tiahou G, Djessou P, Sess D (2009) Intérêt des antioxydants dans la prise en charge des patients infectés par le VIH: apport de la consommation régulière de tisane d'*Alternanthera pungens*. Ann Biol Clin 67(5):563–568

Dogra J, Jha O, Mishra A (1977) Chemotaxonomy of Amaranthaceae. Study of triterpenes. Plant Biochem J 4:14–18

Domínguez JA (1928) Contribuciones a la Materia Médica Argentina, Bs. As., Ed. Peuser, 40 and 122–433 pp

Ebinger JE, Seigler DS, Clarke HD (2000) Taxonomic revision of South American species of the genus *Acacia* subgenus Acacia (Fabaceae: Mimosoideae). Syst Bot 25(4):588–617

Efferth T, Olbrich A, Sauerbrey A, Ross DD, Gebhart E, Neugebauer M (2002) Activity of ascaridol from the anthelmintic herb *Chenopodium anthelminticum* L. against sensitive and multidrug-resistant tumor cells. Anticancer Res 22(6C):4221–4224

el-Mofty MM, Khudoley VV, Sakr SA, Ganem NF (1992) Induction of neoplasms in Egyptian toads *Bufo regularis* by oil of Chenopodium. Oncology 49:253–255

Fenner R, Betti AH, Mentz LA, Rates SMK (2006) Plantas utilizadas na medicina popular brasileira com potencial atividade antifúngica. Rev Bras Cienc Farm 42:369–394

Filippa EM, Espinar LA (1993) Estudios morfohistológicos de la "yerba del pollo" (*Alternanthera pungens*) y su adulterante (*Guilleminea densa*). Acta Farrn Bonaerense 12(2):79–80

Flores CG, Pellegrini PA (2015) Traditional and expert knowledge around a plant medicinal use. The case of paico in Argentinian Patagonia. Revista 100-Cs 1(1):90–122

Flores IL, de Oliveira GT, Varella de Carvalho R, Guerra Lund R, Etges A (2016a) Experimental and clinical evaluation of *Plantago australis* extract as an anti-inflammatory agent to treat oral pathologies. Jentashapir J Health Res 7(5):e39105

Flores IL, de Oliveira GT, Varella de Carvalho R, Guerra Lund R, Etges A (2016b) In vitro cytotoxicity of *Plantago australis* ethanol extract used as an anti-inflammatory for the treatment of oral pathologies. Jentashapir J Health Res 7(1):e28515. https://doi.org/10.17795/jjhr-28515

França F, Lago EL, Marsden PD (1996) Plants used in the treatment of leishmanial ulcers due to *Leishmania* (viannia) *br*aziliensis in the endemic area of Bahia, Brazil. Rev Soc Bras Med Trop 29:229–232

Gadano A, Gurni A, López P, Ferraro G, Carballo M (2002) *In vitro* genotoxic evaluation of the medicinal plant *Chenopodium ambrosioides* L. J Ethnopharmacol 81:11–16

Gadano A, Gurni A, Carballo MA (2007) Herbal medicines: cytotoxic effects of chenopodiaceae species used in Argentinian folk medicine. Pharm Biol 45(3):217–222

Garcia D, Domingues MV, Rodrigues E (2010) Ethnopharmacological survey among migrants living in the Southeast Atlantic forest of Diadema, São Paulo, Brazil. J Ethnobiol Ethnomed 6:1–19

Gayathri J, Parvathi K, Chinthapalli B, Westhoff P, Raghvendra AS (2001) Immunological characteristics of PEP carboxylase from leaves of C3-, C4- and C3-C4 intermediate species of *Alternanthera*-comparison with selected C3- and C4- plants. Indian J Exp Biol 39(07):643–649

Gheno-Heredia YA, Nava-Bernal G, Martínez-Campo AR, Sánchez-Vera E (2011) Las plantas medicinales de la organización de parteras y médicos indígenas tradicionales de Ixhuatlancillo, Veracruz, México y su significancia cultural. Polibotánica 3:199–251

Gillij YG, Gleiser RM, Zygadlo JA (2008) Mosquito repellent activity of essential oils of aromatic plants growing in Argentina. Bioresour Technol 99:2507–2515

Giove NRA (1996) Tradicional medicine in the treatment of enteroparasitosis. Rev Gastroenterol Peru 16:197–202

Giusti L (1997) Chenopodiaceae. In: Hunziker AT (ed) Flora Fanerogámica Argentina, Programa ProFlora 40. Pugliese Siena, Córdoba, pp 1–52

Gómez Castellanos JR (2008) Epazote (*Chenopodium ambrosioides*). Revisión a sus características morfológicas, actividad farmacológica, y biogénesis de su principal principio activo, ascaridol. Bol Latinoam Caribe Plant Med Aromaticas 7(1):1–9

González SB, Huauquinao LN, González A, van Baren C, Di Leo LP, Bandoni A (2009) Uso popular del paico y composición química de su aceite esencial en la zona de Esquel (Chubut. Argentina). In: Pochettino ML, Ladio AH, Arenas PM (eds) Tradiciones & Transformaciones en Etnobotánica, CYTED, San Salvador de Jujuy, Argentina pp 303–307

Gupta MP (1995) 270 Plantas Medicinales iberoamericanas. Ed Andres Bello and CYTED pp 617

Harraz FM, Hammoda HM, El Ghazouly MG, Farag MA, El-Aswad AF, Bassam SM (2014) Chemical composition, antimicrobial and insecticidal activities of the essential oils of *Conyza linifolia* and *Chenopodium ambrosioides*. Nat Prod Res 29:879. https://doi.org/10.1080/14786419.2014.988714

Hausmann M, Obermeier F, Paper DH, Balan K, Dunger N, Menzel K, Falk W, Schoelmerich J, Herfarth H, Rogler G (2007) *In vivo* treatment with the herbal phenylethanoid acteoside ameliorates intestinal inflammation in dextran sulphate sodium induced colitis. Clin Exp Immunol 148:373–338

Henn JG, Steffens L, de Moura Sperotto ND, de Souza PB, Veríssimo RM, Menezes Boaretto FB, Hassemer G, Flores Péres V, Chirmer H, Picada JN, Saffi J, Moura DJ (2019) Toxicological evaluation of a standardized hydroethanolic extract from leaves of *Plantago australis* and its major compound, verbascoside. J Ethnopharmacol 229:145–156

Hernández F (1571) Historia natural de nueva España. Ed. UNAM. México. vol 3

Hieronymus G (1882) Plantae Diaphoricae Florae Argentinae - Bs. A.s, Ed. Kraft, 228–404 pp.

Hundiwale Jogendra C, Patil Avinash V, Kulkarni Mohan V, Patil DA, Mali Ravindra G (2012) A current update on phytopharmacology of the genus *Alternanthera*. J Pharm Res 5(4):1924–1929

Iamonico D, Sánchez del Pino I (2016) Taxonomic revision of the genus *Alternanthera* (Amaranthaceae) in Italy. Plant Biosystems 150(2):333–342. https://doi.org/10.1080/11263504.2015.1019588

Jain N, Alam MS, Kamil M, Ilyas M, Miwa M, Sakae A (1990) Two flavonol glycosides from *Chenopodium ambrosioides*. Phytochem 9(12):3988–3991

Jakhar S, Dahiya P (2017) Antimicrobial, antioxidant and phytochemical potential of *Alternanthera pungens* HB&K. Pharm Sci Res 9(8):1305–1311

Jardim CM, Jham GN, Dhngra OD, Freire MM (2010) Chemical composition and antifungal activity of the hexane extract of the Brazilian *Chenopodium ambrosioides* L. J Braz Chem Soc 21(10):1814–1818

Jesus RS, Piana M, Freitas RB, Brum TF, Alves CFS, Belke BV, Mossmann NJ, Cruz RC, Santos RCV, Dalmolin TV, Bianchini BV, Campos MMA, de Freitas BL (2018) *In vitro* antimicrobial and antimycobacterial activity and HPLC–DAD screening of phenolics from *Chenopodium ambrosioides* L. Braz J Microbiol 49:296–302

Jiménez Osornio FMVZJJ, Kumamoto J, Wasser C (1996) Allelopathic activity of *Chenopodium ambrosioides* L. Biochem Syst Ecol 24(3):195–206

Johnson MA, Croteau R (1984) Biosynthesis of ascaridole: iodide peroxidase-catalyzed synthesis of a monoterpene endoperoxide in soluble extracts of *Chenopodium ambrosioides* fruit. Arch Biochem Biophys 235:254–266

Joseph-Nathan P, Guimarães Leitão S, Corrêa Pinto S, Guimarães Leitão G, Ri-beiro Bizzo H, Paranhos Costa FL, Barbosa de Amorim M, Martínez N, Dellacassa E, Hernández-Barragán AM, Pérez-Hernández N (2010) Structure reassignment and absolute configuration of 9-epi-presilphiperfolan-1-ol. Tetrahedron Lett 51:1963–1965

Juliani HR, Zygadlo JA, Scrivanti R, Sota E, Simon JE (2004) The essential oil of *Anemia tomentosa* (Savigny) Sw. var. anthriscifolia (Schard.) Mickel. Flavour Fragr J 19:541–543

Kalpana G, Srutho K, Banu Z, Sumakanth M, Ravindernath A, Java Prakash D (2018) Screening of analgesic activity of methanolic extract and its fractions of *Alternanthera pungens*. IOSR J Pharm Biol Sci 13(4):53–58

Kazmi Z, Safdar N, Yasmin A (2017) Assessment of *Adiantum incisum*, *Alternanthera pungens* and *Trichodesma indicum* as bio-insecticides against stored grain pests. Proc Pak Acad Sci B Life Environm Sci 54(2):103–109

Kemabonta KA, Okogbue F (2002) *Chenopodium ambrosioides* (Chenopodiaceae) as a grain protectant for the control of the cowpea pest *Callosobruchus maculatus* Coleoptera: Bruchidae. J Fruit Ornam Plant Res 10:165–171

Kiuchi F, Itano Y, Uchiyama N, Honda G, Tsubouchi A, Nakajima-Shimada J, Aoki T (2002) Monoterpene hidroperoxides with trypanocidal activity from *Chenopodium ambrosioides*. J Nat Prod 65:509–512

Kliks MM (1985) Studies on the traditional herbal anthelmintic *Chenopodium ambrosioides* L.: ethnopharmacological evaluation and clinical field trials. Soc Sci Med 21(8):879–886

Korkina LG, Mikhal'Chik EV, Suprun MV, Pastore S, Dal Toso R (2007) Molecular mechanisms underlying wound healing and anti-infflamatory properties of naturally occurring biotechnologically produced phenylpropanoid glycosides. Cell Mol Biol 53(5):84–91

Kujawska M, Pieroni A (2015) Plants used as food and medicine by polish migrants in Misiones, Argentina. Ecol Food Nutr 00:1–25. https://doi.org/10.1080/03670244.2014.983498

Kumar R, Kumar Mishra A, Dubey NK, Tripathi YB (2007) Evaluation of *Chenopodium ambrosioides* oil as a potential source of antifungal, antiaflatoxigenic and antioxidant activity. Int J Food Microbiol 115:159–164

Lahitte HB, Hurrell JA, Valla JJ, Jankowski LS, Bazzano D, Hernández AJ (1999) Árboles urbanos. Biota Rioplatense IV. Inventario de la biota de la región del Delta del Paraná, Isla Martín García y Ribera Platense. Literature of Latin America. L.O.L.A. Buenos Aires, Argentina

Lall N, Meyer JJM (1999) *In vitro* inhibition of drug-resistant and drug-sensitive strains of *Mycobacterium tuberculosis* by ethnobotanically selected South African plants. J Ethnopharmacol 66:347–354

Lamarque AL, Maestri DM, Zygadlo JA, Grosso NR (1998) Volatile constituents from flowers of *Acacia caven* (Mol.) Mol. var. caven, Acacia aroma Gill. ex Hook., *Erythrina crista-galli* L. and *Calliandra tweedii* Benth. Flavour Fragr J 13:266–268

Lee KW, Kim HJ, Lee YS, Park HJ, Choi JW, Ha J, Lee KT (2007) Acteoside inhibits human promyelocytic HL-60 leukemia cell proliferation via inducing cell cycle arrest at G0/G1 phase and differentiation into monocyte. Carcinogenesis 28(9):1928–1936

Lorenzi H, Matos FJA (2002) Plantas medicinais no Brasil: nativas e exóticas. Editora Instituto Plantarum, Nova Odessa

Mabona U, Viljoen A, Shikanga E, Marston A, Van Vuuren S (2013) Antimicrobial activity of southern African medicinal plants with dermatological relevance: from an ethnopharmacological screening approach to combination studies and the isolation of a bioactive compound. J Ethnopharmacol 148:45–55.22

MacDonald D, VanCrey K, Harrison P, Rangachari PK, Rosenfeld J, Warren C et al (2004) Ascaridole-less infusions of *Chenopodium ambrosioides* contain a nematocide (s) that is(are) not toxic to mammalian smooth muscle. J Ethnopharmacol 92(2–3):215–221

Martínez GJ (2005) Recolección y Comercialización de Plantas Medicinales en el Departamento Santa María, Provincia de Córdoba, Argentina. Acta Farm Bonaer 24:575–584

Martínez G (2007) Farmacopea natural y tratamiento de afecciones de la piel en la medicina tradicional de los campesinos de las sierras de Córdoba (República Argentina). Dominguezia 24(1):27–46

Martínez Crovetto RN (1981a) Las plantas utilizadas en medicina popular en el Noroeste de Corrientes (Argentina). Miscelánea 69:1–140. Inst M Lillo, Tucumán

Martínez Crovetto R (1981b) Plantas utilizadas en la medicina popular en el NO de Corrientes. Tucumán, Min. de Cultura y Educ., Fundación Miguel Lillo 42–139 pp

Martínez OG, de la Sota ER, Narváez PL (2003) Flora del valle de Lerma. Schizaceae Kaulf. Aportes Bot Salta-Ser. Flora 7(2):1–7

Marzocca A (1997) Vademécum de malezas medicinales de la Argentina, indígenas y exóticas. Orientación Gráfica Editora, Buenos Aires

Medeiros CAM, Boica-Junior AL, Leite-Torres A (2005) Efeito de extratos aquosos de plantas na oviposição da traça-das-crucíferas, em couve. Bragantia, Campinas 64:227–232

Mello AFS, Machado ACZ, Inomoto MM (2006) Potencial de Controle da Erva-de-Santa-Maria sobre *Pratylenchus brachyurus*. Fitopatol Bras 31(5):513–516

Mickel JT (1962) A monographic study of the fern genus Anemia, subgenus Coptophyllum. Iowa State J Sci 36:349–482

Montoya-Cabrera MA, Escalante-Galindo P, Meckes-Fisher M, Sánchez-Vaca G, Flores-Alvarez E, Reynoso-Garcia M (1996) Fatal poisoning caused by epazote oil: *Chenopodium graveolus*. Gac Méd Méx 132:433–437

Monzote L, Montalvo A, Scull R, Miranda M, Abreu J (2007) Activity, toxicity and analysis of resistance of essential oil from *Chenopodium ambrosioides* after intraperitoneal, oral and intralesional administration in BALB/c mice infected with *Leishmania amazonensis*: a preliminary study. Biomed Pharmacother 61:148–153

Monzote L, Stamberg W, Staniek K, Gille L (2009) Toxic effects of carvacrol, caryophyllene oxide, and ascaridole from essential oil of *Chenopodium ambrosioides* on mitochondria. Toxicol Appl Pharmacol 240:337–347

Monzote L, Nance MR, García M, Scull R, Setzer WN (2011) Comparative chemical, cytotoxicity and Antileishmanial properties of essential oils from *Chenopodium ambrosioides*. Nat Prod Commun 6(2):281–286

Monzote L, García M, Pastor J, Gil L, Scull R, Maes L, Cos P, Gille L (2014) Essential oil from *Chenopodium ambrosioides* and main components: activity against Leishmania, their mitochondria and other microorganisms. Exp Parasitol 136:20–26

Monzote L, Geroldinger G, Tonner M, Scull R, De Sarkar S, Bergmann S, Bacher M, Staniek K, Chatterjee M, Rosenau T, Gille L (2018) Interaction of ascaridole, carvacrol, and caryophyllene oxide from essential oil of *Chenopodium ambrosioides* L. with mitochondria in Leishmania and other eukaryotes. Phytother Res 32(9):1729-1740

Morais SM, Dantas JDP, Silva ARA, Magalhães EF (2005) Plantas medicinais usadas pelos índios Tapebas do Ceará. Rev Bras Farmacogn 15:169–177

Moreno Mendoza MA, Parada Palacios EA, Mejía Valencia JG, Espinoza Madrid PA (2013) Subchronic toxicology of *Chenopodium ambrosioides* (epazote) infusion orally administered to NIH mice. Rev Cubana Planta Med 18(1):157–170

Moraes Monteiro JN, Barros Archanjo A, Passos GP, Vidal Costa A, Cardoso Porfirio L, Freire Martins IV (2017) *Chenopodium ambrosioides* L. essential oil and etanol extract on control of canine Ancylostoma spp. Semina Cienc Agrar 38(4):1947–1954

Moroni P, Hassemer G, O'Leary N (2018) Typification of Plantago names (Plantagineae, Plantaginaceae) linked to the Flora of Argentina. Novon 6(4):364–377

Moya MA, Escudero VG (2015) Las plantas medicinales en el control de nemátodos gastrointestinales en cabras: potencial de las plantas que crecen en la región de Coquimbo, Chile. Rev Bras Pl Med Campinas 17(3):480–494

Murashige T, Skoog F (1962) A revised medium for rapid growth and bio assays with tobacco tissue cultures. Physiol Plant 15(3):473–497

Nascimento FRF, Cruz GVB, Pereira PVS, Maciel MCG, Silva LA, Azevedo APS, Barroqueiro ESB, Guerra RNM (2006) Ascitic and solid Ehrlich tumor inhibition by *Chenopodium ambrosioides* L. treatment. Life Sci 78:2650–2653

Navone GT, Zonta ML, Gamboa MI (2014) Phyttotherapy mbyá-guaraní in the control of intestinal parasitosis. An exploratory study with *Chenopodium ambrosioides* L. var. anthelminitcum in five communities in Misiones, Argentina. Polibotánica 37:135–151

Neiva VA, Ribeiro MNS, Cartágenes MSS, Moraes-Coutuinho DF, Nascimento FRF, Reys AS, Amaral FMM (2011) Estudos pré-clínicos de atividade giardicida de *Chenopodium ambrosioides* L. e a padronizacão dos extratos na pesquisae desenvolvimento de fitoterápicos. Rev Ciên Saúde 13:155–165

Neiva VA, Ribeiro MNS, Nascimento FRF, Cartágenes MSS, Coutinho-Moraes DF, do Amaral FMM (2014) Plant species used in giardiasis treatment: ethnopharmacology and *in vitro* evaluation of anti-Giardia activity. Rev Bras Farmacogn 24:215–224

Nunes Cysne D, Silva Fortes T, Silva Reis A, de Paulo RB, dos Santos FA, Mendonça do Amaral FM, Meireles Guerra RN, Farias Marinho CR, Nicolete R, Fernandes Nascimento FR (2016) Antimalarial potential of leaves of *Chenopodium ambrosioides* L. Parasitol Res 115(11):4327–4334

Orfila EN (1971) Estudio morfológico y anatómico de la "yerba del pollo" *Alternanthera pungens* Humbold, Bonpland et Kunth (Amaranthaceae), para su reconocimiento farmacobotánico. Rev Fac Agron UNLP 47(2):153–168

Ortiz J (1966) Algunos forestales chilenos de la estepa septentrional. Boletín Técnico N°23. Ministerio de Agricultura. Santiago, Chile. 45 p

Owa SO, Oyewale DB, Taiwo AA, Edewor-Ikupoyi TI, Okosun JA, Otohiniyi DA, Akujobi YO, Nwonuma CO, Adeyemi OS (2016) Antidiabetic and antidyslipidemic effect of ethanolic extract of *Alternanathera pungens* on alloxan-induced diabetic rats. Asian J Biochem 11:82–89

Palmeiro NS, Esteves Almeida C, Ghedini PC, Goularts LS, Baldisserotto B (2002) Analgesic and anti-inflammatory properties of *Plantago australis* hydroalcoholic extract. Acta Farm Bonaer 21(2):89–92

Palmeiro NS, Almeida CE, Ghedini PC, Goulart LS, Pereira MCF, Huber S, da Silva JEP, Lopes S (2003) Oral subchronic toxicity of aqueous crude extract of *Plantago australis* leaves. J Ethnopharmacol 88:15–18

Pandey AK, Palni UT, Tripathi NN (2014) Repellent activity of some essential oils against two stored product beetles *Callosobruchus chinensis* L. and *C. maculatus* F. (Coleoptera: Bruchidae) with reference to *Chenopodium ambrosioides* L. oil for the safety of pigeon pea seeds. J Food Sci Technol 51(12):4066–4071

Patel DK (2017) A phytochemical of modern medicinal perspective. J Coastal Life Med 5(2):88–94

Peña Rivera A, Ferrer-Gallego PP (2016) Sobre la presencia de *Alternanthera pungens* Kunth (Amaranthaceae) en la Flora Valenciana. Flora Montib 62:31–36

Pereira W, Ribeiro BP, Sousa AIP, Serra ICPB, Mattar NS, Fortes TS, Reis AS, Silva LA, Barroqueiro ESB, Guerra RNM, Nascimento FRF (2010) Evaluation of the subchronic toxicity of oral treatment with *Chenopodium ambrosioides* in mice. J Ethnopharmacol 127:602–605

Pereira-Cenci T, Nunes T, Barwaldt C, Moraes A, Sarkis-Onofre R, Boscato N (2014) Influence of cleansers and phytotherapic against *Candida albicans* on the properties of denture liners. RFO, Passo Fundo 19(2):205–211. https://doi.org/10.5335/rfo.v19i2.3898

Pinheiro Neto VF, Ribeiro RM, Morais CS, Campos MB, Vieira DA, Guerra PC, Abreu-Silva AL, Silva Junior JR, Nascimento FRF, Borges MOR, Borges ACR (2017) *Chenopodium ambrosioides* as a bone graft substitute in rabbits radius fracture. BMC Complement Altern Med 17:350–360

Pinto SC, Leitão GG, Bizzo HR, Martinez N, Dellacassa E, Santos FM Jr, Cos-ta FLP, de Amorim MB, Leitão SG (2009) (-) epi-presilphiperfolan-1-ol, a new tri-quinane sesquiterpene from the essential oil of *Anemia tomentosa* var. anthriscifolia (Pteridophyta). Tetrahedron Lett 50:4785–4787

Pinto SC, Leitão GG, Castellar A, Bizzo HR, Leitão SG (2013) Chemical composition of the volatile fractions from wild and *in vitro* plants of *Anemia tomentosa* var. anthriscifolia (Pteridophyta). J Essent Oil Res 25:198–202

Pollack Y, Segal R, Golenser J (1990) The effect of ascaridole on the *in vitro* development of *Plasmodium falciparum*. Parasitol Res 76(7):570–572

Pometti CL, Cialdella AM, Vilardi JC, Saidman BO (2007) Morphometric analysis of varieties of *Acacia caven*: (Leguminosae, Mimosoideae): taxonomic inferences in the context of other Argentinean species. Plant Syst Evol 264:239–249

Potawale SE, Luniya KP, Mantri RA, Mehta UK, Md. Sadiq MW, Vetal YD, Deshmukh RS (2008) *Chenopodium ambrosioides*: an ethnopharmacological review. Pharmacol Online 2:272–286

Quiroga E, Sgroi N, Selis A, Vattuone M (2004) Antifungal activity of ethanolic extracts of argentine plants. Abstract FT-36. Congresso Iberoamericano de Plantas Medicinais. 20–23 maio 2004. Angra dos Reis

Rahn K (1996) A phylogenetic study of the Plantaginaceae. Bot J Linn Soc 120:145–198

Raju M, Varakumar S, Lakshminarayana R, Krishnakantha TP, Baskaran V (2007) Carotenoid composition and vitamin A activity of medicinally important green leafy vegetables. Food Chem 101:1598–1605

Raju M, Varakumar LR, Krishnakantha T, Baskaran V (2007) Carotenoid composition and vitamin a activity of medicinally important green leafy vegetables. Food Chem 101(4):1598–1605

Ratera EL, Ratera MO (1980) Plantas de la flora argentina empleadas en medicina popular. Buenos Aires, Ed. Hemisferio sur S. A.

Roig JT (1992) Plantas medicinales, aromáticas o venenosas de Cuba - La Habana, Cuba, Ed. Científico-Técnica, Tomo II - 2da. ed., 3ra. reimpresión. 836:837 - 1125 pp

Rojas J, Solís H, Palacios O (2010) *In vitro* anti-*Trypanosoma cruzi* activity of essential oils of ten medicinal plants. An Fac Med 71(3):161–165

Rondina RVD, Coussio JD (1969) Estudio fitoquímico de plantas medicinales argentinas. RIA INTA 6:351–366

Rondina RVD, Coussio JD (1981) Ensayo Fitoquímico Orientativo de Plantas con Actividad Farmacológica. Facultad de Farmacia y Bioquímica, Universidad de Buenos Aires, Argentina

Ronsted N, Göbel E, Franzyk H, Rosendal Jensen S, Olsen CE (2000) Chemo-taxonomy of Plantago. Iridoid glucosides and caffeoyl phenylethanoid glycosides. Phytochem 55:337–348

Ruffa MJ, Ferraro G, Wagner ML, Calcagno ML, Campos RH, Cavallero L (2002) Cytotoxic effect of Argentine medicinal plant extracts on human hepatocellular carcinoma cell line. J Ethnopharmacol 79:335–339

Sá D, Santana ASCO, Silva FCL, Soares LAL, Randau KP (2016) Anatomical and histochemical analysis of *Dysphania ambrosioides* supported by light and electron microscopy. Rev Bras 26:533–543

Saggese D (1959) Yerbas medicinales argentinas, Rosario, págs 1–189

Sangeeta NS, Batish DR, Singh HP (2017) Comparative phytotoxic effect of the essential oil of *Chenopodium ambrosoides* and *Mentha longifolia* on *Avena fatua*. Intl J Adv Tech Eng Sci 5(1):691–697

Santayana MP, Blanco E, Morales R (2005) Plants known as té in Spain: an ethno-pharmaco-botanical review. J Ethnopharmacol 98:1–19

Santos MG, Rocha LM, Carvalho ES, Kelecom A (2003) Isoafricanol um sesquiterpeno in comun encontrado em *Anemia tomentosa* var. anthriscifolia (Pteridófita). Docs Instit Agronôm Campinas 74:58

Santos MG, Rocha LM, Carvalho ES, Kelecom A (2006) Isoafricanlo um sesquiterpeno in comun encontrado em *Anemia tomentosa* var. anthriscifolia (Pteridófita). Rev Bras Pl Med 8(4):71–75

Scarpa GF, Cassá LA (2015) Etnobotánica de los helechos (Ophioglossidae, Equisetidae y Polypodiidae) en Argentina: recopilación y análisis entre grupos criollos e indígenas. Rev Museo Arg Cs Nat n.s. 17(1):1–12

Shah H (2014) Antibacterial and antifungal activities of the crude extracts from the stem of *Chenopodium ambrosioides* Linn, an indiginous medicinal plant. Afr J Pharm Pharmacol 7(48):1560–1567

Shah H, Khan AA (2017) Phytochemical characterisation of an important medicinal plant, *Chenopodium ambrosioides* Linn. Nat Prod Res 31(19):2321–2324

Shameem SA, Khan KZ, Waza AA, Shah AH, Qadri H, Ganai BA (2019) Bio-activities and chemoprofiling comparisons of *Chenopodium ambrosioides* L and *Chenopodium botrys* L. growing in Kashmir, India. Asian J Pharm Clin Res 12(1):124–129

Sharry S, Adema M, Basiglio M, Villareal B, Nikoloff N, Briones V, Abedini W (2010) Técnicas *in vitro* y *ex vitro* para la conservación y propagación de especies de interés medicinal de la flora Bonaerense. XIX° SILAE Congress "Fernando Cabieses Molina". Villasimius, Cagliari, Italy pp. 94

Silva G, Orrego O, Hepp R, Tapia M (2005) Búsqueda de plantas con propiedades insecticidas para el control de *Sitophilus zeamais* en maíz almacenado. Pesq Agrop Bras 40(1):11–17

Smillie WG, Pessoa SB (1924) A study of the anthelmintic properties of the constituents of the oil of *Chenopodium*. J Pharmacol Exp Ther 24:359–370

Smith GH, Roberts JM, Pope TW (2018) Terpene based biopesticides as potential alternatives to synthetic insecticides for control of aphid pests on protected ornamentals. Crop Prot 110:125–130

Soraru SB, Bandoni AL (1978) Plantas de la medicina popular Argentina - Bs.As., Ed. Albatros, 1ra. ed., 17:18 - 153 pp

Sousa ZL, de Oliveira FF, Oliveira da Conceiçao A, Mattos Silva LA, Rossi MH, da Silva SJ, Andrioli JL (2012) Biological activities of extracts from *Chenopodium ambrosioides* Lineu and Kielmeyera neglecta Saddi. Ann Clin Microbiol Antimicrob 11:20

Spegazzini G (1924) Acacieas Argentinas. Bol Acad Nac Cienc 26:161–334

Tolaba JA (2006) Flora del Valle de Lerma, Chenopodiaceae Vent. Aportes Bot Salta-Ser. Flora 7(18):1–60

Toursarkissian M (1980) Plantas medicinales de la Argentina - Buenos Aires Ed. Hemisferio Sur, 2–178 pp

Trillo C (2016) Traditional management practices of plant resources in units of cultural landscapes on the west of the province of Cordoba, Argentina. Z Áridas 16(1):86–111

Trivellato Grassi L, Malheiros A, Meyre-Silva C, da Silva BZ, Dalmarco Monguilhott E, Fröde TS, da Silva KABS, de Souza MM (2013) From popular use to pharmacological validation: a study of the anti-inflammatory, anti-nociceptive and healing effects of *Chenopodium ambrosioides* extract. J Ethnopharmacol 145:127–138

Vázquez-Covarrubias DA, Jiménez-Pérez A, Castrejón-Ayala F, Figueroa-Brito R, Montes Belmont R (2015) Effects of five species of *Chenopodiaceae* on the development and reproductive potential of *Copitarsia decolora* (Lepidoptera: Noctuidae). Fla Entomol 98(1):80–85

Velasco-Castilho CV, Ferra Neto JF, Guimarães Leitão S, Barreto CS, Corrêa Pinto S, Barboza da Silva NC (2018) *Anemia tomentosa* var. anthriscifolia i*n vitro* culture: sporophyte development and volatile compound profile of an aromatic fern. Plant Cell Tiss Org Cult 133(3):311–323

Vertuani S, Beghelli E, Scalambra E, Malisardi G, Copetti S, dal Toso R, Baldis-serotto A, Manfredini S (2011) Activity and stability studies of verbascoside, a novel antioxidant, in dermo-cosmetic and pharmaceutical topical formulations. Molecules 16:7068–7080

Villalobos-Delgado LH, González-Mondragón EG, Salazar Govea AY, Ramírez Andrade J, Santiago-Castro AJT (2017) Potential application of epazote (*Chenopodium ambrosioides* L.) as natural antioxidant in raw ground pork. LWT Food Sci Technol 84:306–313

Weeds of Australia (2016) Alternanthera pungens. Biosecurity Queensland Edition Fact Sheet

Wei H, Liu J, Li B, Zhan Z, Chen Y, Tian H, Lin S, Gu X (2015) The toxicity and physiological effect of essential oil from *Chenopodium ambrosioides* against the diamondback moth, Plutella xylostella (Lepidoptera: Plutellidae). Crop Prot 76:68–74

Yoou M-S, Kim H-M, Jeong H-J (2015) Acteoside attenuates TSLP-induced mast cell proliferation via down-regulating MDM2. Int Immunopharmacol 26:23–29

Zhu WX, Zhao K, Chu SS, Liu ZL (2012) Evaluation of essential oil and its three main active ingredients of Chinese *Chenopodium ambrosioides* (family: Chenopodiaceae) against Blattella germanica. J Arthropod Borne Dis 6(2):90–97

Chapter 7
Yungas

7.1 Introduction: Yungas, the Green Face of the Andes

The Yungas are mountain jungles of Northwest Argentina that have just been included by UNESCO in the World Network of Biosphere Reserves (2002). They have different characteristics according to the altitude, with a dense and humid forest vegetation and high biodiversity in the lower regions and predominantly shrubs and grasslands in the highlands (Fig. 7.1). The climate is warm and humid to subhumid. The conditions of temperature and humidity vary according to the altitude, latitude, and exposure of the slopes. The mountain ranges form a barrier that condenses the humid winds from the South Atlantic anticyclone creating the conditions for a thick wooded development. Rains are mainly during summer with annual precipitation oscillating between 900 and 1300 mm. During the colder months, the characteristic mists from the cloudy jungles compensate the lack of rain. It has a well-organized fluvial network, and the soils have abundant organic matter, with frequent local landslides.

The fauna has typical species such as *Amazona tucumana* (loro alisero), *Antigone vipio* (white-naped crane), *Cinclus schulzi* (mirlo de agua), *Eriocnemis glaucopoides* (picaflor frente azul), *Psittacidae* spp. (macaws), *Pyrrhura molinae* (chiripepé de la Yunga), and *Synallaxis scutata* (pijuí anaranjado). Among the mammals there are the *Anoura caudifer* (snout bat), *Cavia tschudii* (cuis serrano), *Dasyprocta punctata* (reddish agouti), *Hippocamelus antisensis* (taruca), and *Scirus ignites* (red squirrel). Among the amphibians, there is the marsupial frog (*Gastrotheca gracilis*), which is at risk of extinction and was included by the ONG Global Wildlife in the initiative *The Search for Lost Species*. There are also *Leopardos wiedii* (gato salvaje), *Mazama americana* (corzuela roja), *Panthera onca* (yaguareté), *Puma concolor concolor*, *Pecari tajacu*, and *Tapirus terrestres* (tapir), among others.

The flora varies according to the altitude, and four regions could be distinguished: *Pedemontane jungle*, hot and humid with predominance of epiphytes, bryophytes,

M. A. Alvarez, *Pharmacological Properties of Native Plants from Argentina*,
https://doi.org/10.1007/978-3-030-20198-2_7

Fig. 7.1 Yungas, province of Jujuy, Argentina. (Personal collection PG Alvarez)

ferns, and arboreal species such as *Calycophyllum multiflorum Griseb.*, *Chrysophyllum marginatum* (Hook. & Arn.) Radlk., *Enterolobium contortisiliquum* (pacará), *Handroanthus impetiginosus, Jacaranda mimosifolia*, *Parapiptadenia excelsa* (cebil), and *Tipuana tipu*, among others; *Montana jungle*, temperate-warm and humid, where *Blepharocalyx gigantea* (horco-molle), *Cinnamomum porphyrium* (Griseb.) Kosterm., *Enterolobium contortisiliquum, Juglans australis* (nogal criollo), *Parapiptadenia excelsa*, myrtles, and cedars, among others, predominate; *Montana forest*, temperate and humid with frequent winter frosts, where alder, *Podocarpus parlatorei* (pino del cerro), and walnut predominate; and *temperate-cold and subhumid high-altitude meadows*, with patches of montane forest alternating shrubland and, at a higher altitude, pure herbaceous communities (De la Sota 1972; Cabrera 1976; Morales et al. 1995; Brown et al. 2001; Grau 2006; Coll-Aráoz et al. 2016).

Among the species with medical use are *Acacia macracantha* Humb. & Bonpl. Ex Willd (Fabaceae), *Achyrocline alata* (Kunth) DC (Asteraceae), *Achyrocline flaccida* (Weinm) DC (Asterace), *Ageratum conyzoides* L. (Asteraceae), *Aloysia polystachya* (Griseb.) Mold. (Lamiaceae), *Amaranthus quitensis* Kunth. (Amaranthaceae), *Anredera cordifolia* (Ten.) Steenis (Basellaceae), *Clinopodium gilliesii* (Benth.) Kunze (Lamiaceae), *Equisetum giganteum* L. (Equisetaceae), *Buddleja brasiliensis*

Jacq. Ex Spreng. (Buddlejaceae), *Fagara coco* (Gill.) Engler (Rutaceae), *Erythroxylum coca* Lam. var. coca (Erythroxylaceae), *Lepechinia vesiculosa* (Benth.) Epling. (Lamiaceae), *Lepidium didymus* (L.) Brassicaceae, *Maytenus cuezzoi* Legname (Celastraceae), *Myroxylon peruiferum* L. f. (Fabaceae), *Pavonia sepium* St. Hil. (Malvaceae), *Petiveria alliaceae* L. (Phytolaccaceae), *Pluchea sagittalis* (Lam.) Cabrera, (Asteraceae), *Sambucus nigra* subsp. peruviana (Kunth) R. Bolli (Adoxaceae), *Satureja boliviana* (Benth.) Briquet (Lamiaceae), *Satureja parvifolia* (Phil.) Epling (Lamiaceae), *Schinus molle* L. (Anacardiaceae), *Senecio crepidifolius* DC. (Asteraceae), *Smallanthus macroscyphus* (Baker) A. Grau ex A.A. Sáenz (Asteraceae), *Smilax campestris* Griseb. (Smilacaceae), *Solanum sisymbriifolium* Lam.var. sisymbriifolium (Solanaceae), *Tabebuia impetiginosa* (Mart. ex DC.) Standl. (Bignoniaceae), *Tagetes campanulata* Griseb. (Asteraceae), *Tillandsia usneoides* (L.) L. (Bromeliaceae), *Trichocereus arborícola* Kimnach (Cactaceae), *Urera baccifera* (L.) Gaudich. (Urticaceae), and *Xanthium ambrosioides* Hook. & Arn. (Asteraceae).

7.2 *Lepidium didymum* L. (Brassicaceae)

Synonymy *Coronopus leptocarpus* Boelcke, *Coronopus leptocarpus* Boelcke var. microcarpus, *Coronopus didymus* (L.) Sm., *Coronopus didymus* L. var. procumbens, *Coronopus didymus* L. var. macrocarpus, *Senebiera didyma* (L.) Pers., *Senebiera pinnatifica* DC., *Lepicochlea americana* Rojas Acosta.

Vernacular name Mastuerzo, quimpe, quimpi, calachi, yerba del ciervo, mastruço, erva-de-santa-maria, mentruz, mentruz-rasteiro (Portuguese), swinecress (English).

General information *L. didymium* is a stinky weed that spontaneously grows in parks, gardens, orchards, and pasture, preferring shady places. It is also considered an edible species, being added to salads or to beverages. In Argentina it could be found in the provinces of BAI, CAT, CHA, COR, COS, DFE, ERI, FOR, JUJ, LPA, LRI, MEN, MIS, SAL, SCR, SDE, SFE, SJU, SLU, TDF, and TUC. It is also used to feed cattle, mainly sheep. Blooming is in spring (Amorin and Rosow 1988; Lorenzi and Matos 2002; Alonso and Desmarchelier 2015).

Botanical characteristics *L. didymum* is an annual or biannual herb whose height ranges between 10 and 20 cm. It has a procumbent stem with leafy, glabrous, or pubescent branches. Leaves are basal and bipinnate with whole or barely dentate lobes. The white flowers have glabrous sepals on the back side. Nectaries are tongue-shaped. The fruit is an indehiscent, reticulate, bi-globose silicle, with a rough surface and a seed in each compartment. Pedicels are cylindrical with scarce antrorse hairs (Prina 1995; Ferreira Jorge et al. 1986).

Ethnobotanical information The tisane made with the whole plant is used as an expectorant and digestive and to treat intermittent fever. The juice is used as vulnerary; to treat hemorrhoids, gangrene, as antitussive, antimalarial, antiscorbutic, for hernia and prolapse, menstrual diseases, neonatal and pediatrics care, in cases of gingivitis. It also cited its use in pregnancy and to facilitate birth and in cancerous processes (Soraru and Bandoni 1978; Ratera and Ratera 1980; Toursarkissian 1980; Scarpa 2004; Barboza et al. 2009). The tisane made with the aerial parts is taken as antiscorbutic, vulnerary, anti-inflammatory, antiseptic, hemostatic, astringent for humoral and blood diseases and to prevent pyorrhea. Decoction of flowers is used in cases of vaginal fungus (Barboza et al. 2009).

Chemical data In the dried entire plant, there were identified benzyl cyanide, glucotrapaeolin (benzyl isothiocyanate glycoside), chrysoeriol, chrysoeriol-6"(OAc)-4'β-D-glucoside, cymaroside, rhamnetin, β-sitosterol, sinapine, 1,8-dihydroxy anthraquinone, and myristic acid. The EO from aerial parts contains benzyl cyanide, benzylmercaptan, benzene, acetonitrile, benzyl disulfide, benzyltrisulfide, and hydrocarbons of high molecular weight (Fig. 7.2). In the dried seed, there are benzylnitrile and glucosinolates (Haque 1989; Ruiz et al. 1989, 1993; Prabhakar et al. 2002; Mantena et al. 2005; Barboza et al. 2009; Moismann Busnardo et al. 2010).

Biological activities The aqueous extract of the whole plant (200–400 mg kg^{-1}) was reported as effective in cases of allergies, inflammation, wound healing, hepatotoxicity, and hyperglycemia (Mantena et al. 2005). The aqueous extract was more active as anti-inflammatory and wound healer than the ethanolic extract (Prabhakar et al. 2002). The hydroalcoholic extract (200, 400, 600 mg kg^{-1}) from the whole plant showed a dose-dependent decrease of paw inflammation (decrease of leukocyte migration, inhibition of neutrophil influx, and exudation) on the model of pleurisy and paw edema induced by carrageenan in male Swiss mice. Similar results were obtained on edema induced by other phlogogens (substance P, bradykinin, and histamine) (Moismann Busnardo et al. 2010).

The aqueous extract (1 drop during 7 days) was effective to accelerate skin scars of Wistar rats, showing an increase of fibroblasts and collagen fibers (Nitz et al. 2006). The aqueous and ethanolic extracts (200 mg kg^{-1}) showed similar activity (Prabhakar et al. 2002).

The crude methanol extract had antimicrobial activity against *Staphylococcus aureus* ATCC 25923, *Staphylococcus epidermidis* ATCC 12228, *Bacillus subtilis* ATCC 6633 (Gram- positive), and *Klebsiella pneumonia* ATCC 13883 (Gram-negative). Against the fungi *Aspergillus niger* ATCC 16404, *Aspergillus flavus* ATCC 32611, and *Alternaria solani* ATCC 11078, the inhibitory activity was lower (Uddin et al. 2014). On the other hand, the methanolic extract (10 mg ml^{-1}) did not show *in vitro* antiviral activity against herpes simplex virus type 1 and 2, bovine viral diarrhea virus type 1, influenza virus type A, and human immunodeficiency virus type 1 (HIV-1) (Ruffa et al. 2004). The biological activities were attributed to the antioxidant activity of flavonoids and tannins (Mantena et al. 2005; Prabhakar et al. 2006; Moismann Busnardo et al. 2010).

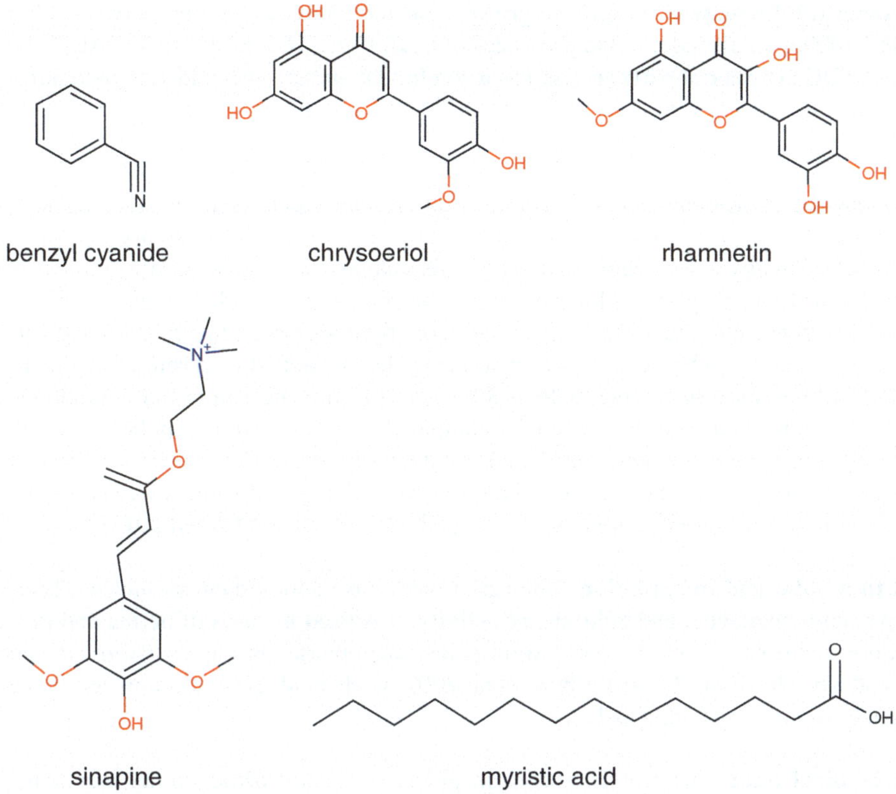

Fig. 7.2 Chemical compounds from *Lepidium didymum* L.

Toxicity The hydroethanolic extract (0.1–100 mg ml^{-1}) showed moderate toxicity against *Eisenia andrei* (DL50 = 3.498 mg ml^{-1}) and *Artemia salina* (CE50 = 773.69 ppm) (Ponce et al. 2007).

7.3 *Clinopodium gilliesii* (Benth.) Kunze (Lamiaceae)

Synonymy *Satureja parvifolia* (Phil.) Epling, *Micromeria eugenioides* (Griseb.) Hieron., *Xenopoma eugenioides* Griseb., *Bystropogon minutus* Briq., *Satureja eugenioides* Griseb., *Satureja oligantha* Briq., *Micromeria gilliesii* Benth., *Oreosphacus parvifolia* Phil., *Satureja gilliesii* (Benth.) Briq.

Vernacular name Muña-muña, salvilori, muña, hillencoma, mulla-mulla, yerba del pájaro, hierba del amor.

General information *C. gilliesii* grows in the mountainous regions, between 1200 and 4000 m of altitude, of the provinces of CAT, COR, JUJ, LRI, MEN, SAL, SJU, and TUC. The aerial parts are used as a condiment and are also added to *mate* infusion. The branches are used as fuel (Orfila and Farina 1997; Brodkiewicz et al. 2017).

Botanical characteristics *C. gilliesii* is an aromatic undershrub up to 2 m high, with stiff, erect branches; quadrangular reddish-brown pubescent stems; and large hairs, eventually glabrescent. Leaves are imbricate and subsessile or with petioles up to 1 mm long; they are oblong or lanceolate, obtuse to rounded at the apex, and broadly cuneate by the base, with an entire or slightly serrate margin. Both leaf surfaces are shortly pubescent, the lower surface paler and dotted with yellowish glands. The inflorescence is a cyme, with 1–3 flowers. Flowers are solitary or pauciflorus, shortly pedicelled. The shortly tubular-campanulate calyx is slightly pubescent, with shortly curly hairs, teethed, ribbed, and glandular on the exterior, with a basal ring of hairs on the interior. The white corolla is externally hairy, and anthers are subsessile. The fruit is a trigonous, oblong-ovoid-shaped, glabrous nutlet (Wood 2011).

Ethnobotanical information The aerial parts have antioxidant, stomachic, laxative, emmenagogue, and aphrodisiac activity. It is used in cases of hernia and prolapse, menstrual diseases, postpartum pains, menopause, and against *soroche* and headache (Ruthsatz 1974; Barboza et al. 2009; Dadé et al. 2009; Simoni and Perea 2016; Bektas and Cilkiz 2016).

Chemical data The leaf EOs contain pulegone, cis-dihydrocarvone, p-cymene, 1,8-cineol (eucalyptol), E-isocitral, carvacrol, carvacryl acetate, carvone, cis-dihydrocarvone, piperitone oxide, piperitenone, piperitenone oxide, lippione, dihydrolippione, linalool, isopulegol, α- and β-pinene, limonene, pulegone, sabinene, menthol, menthone, neoisomenthol, and α-terpinene (Fig. 7.3) (Zygaldo et al. 1993; Muschietti et al. 1996; Viturro et al. 2000; van Baren et al. 2006; Luna et al. 2008; Barboza et al. 2009; Vázquez et al. 2016; Reynoso et al. 2018). The composition of the EOs varies according to the chemotype, characteristics of soil, altitude, and weather (Viturro et al. 2000).

Biological activities The decoction yielded the highest concentration of phenolic compounds and consequently the highest antioxidant activity, than the hydrolate and the ethanolic extract (Cabana et al. 2013). The EOs obtained by hydrodistillation from the aerial parts have antioxidant activity, being pulegone (74.5%) and piperitenone oxide the most abundant chemicals (Dadé et al. 2009; Cabana et al. 2013; Barbieri et al. 2015). The anti-inflammatory activity from the aqueous and ethanolic extracts was tested in male Wistar rats through the carrageenan-induced edema paw and cotton pellet-induced granuloma formation models. Rats pretreated with the ethanol extract (500 mg kg^{-1}) and the topical application (100 μl ml^{-1}) of the EO produced a significant reduction of edema after 2 h of treatment. Both treatments reduced transudative and granuloma weights (Reynoso et al. 2018).

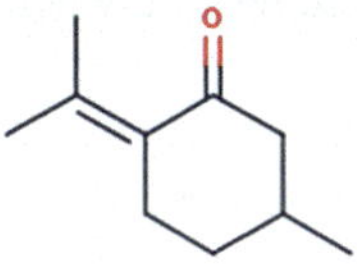

pulegone

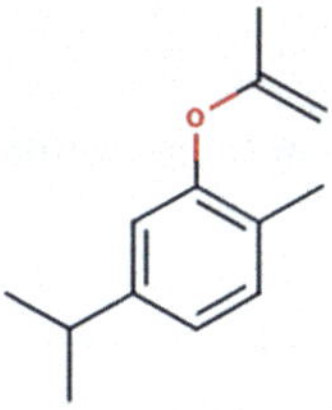

carvacryl acetate

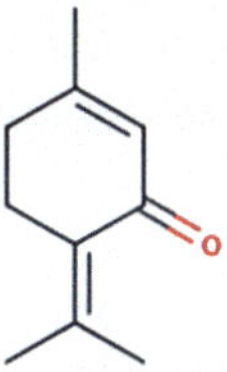

piperitenone

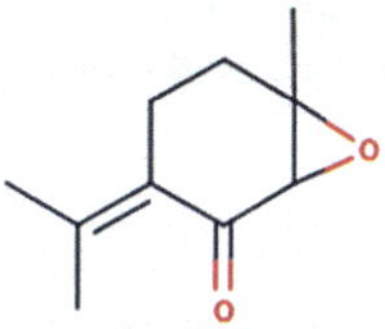

lippione

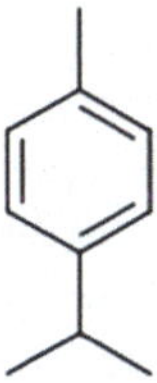

p-cymene

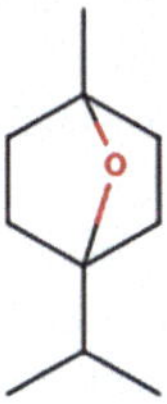

1,8-cineol

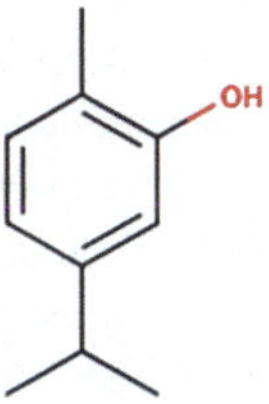

carvacrol

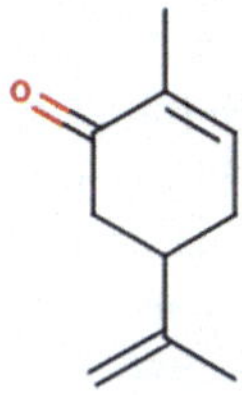

carvone

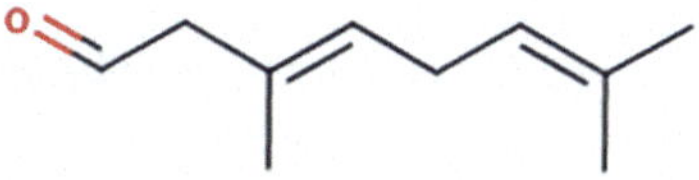

isocitral

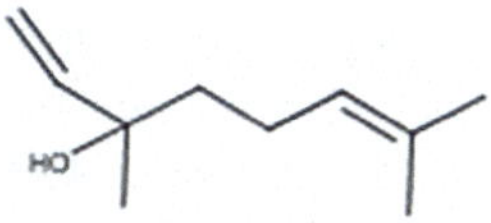

linalool

sabinene

Fig. 7.3 Chemical compounds from *Clinopodium gilliesii* (Benth.) Kunze

The hexane, dichloromethane, and methanol extracts were active against *Escherichia coli* and *Trichophyton mentagrophytes* (MIC 1 mg ml^{-1}) (Feresin et al. 2001). The total flavonoid extract (methanol 50% and 80%) from aerial parts tested against Gram-negative rods (*Escherichia coli* ATCC 25922, *Pseudomonas aeruginosa* ATCC27853, *Klebsiella pneumoneae*, *Salmonella enteriditis*, and *Shigella* sp.), Gram-positive cocci (*Staphylococcus aureus* ATCC 6538, *Staphylococcus*, and *Streptococcus β-hemoliticus* A Group), and Gram-positive spore-forming rods (*Bacillus subtilis* BSGC A1) inhibited bacterial growth, except in the case of *S. epidermidis*, *Klebsiella*, and *Salmonella* (Hernández et al. 2000). Methanol and dichloromethane extracts tested against *Staphylococcus aureus* F7 methicillin-resistant and *P. aeruginosa* ATCC 6738 growing as biofilms were inhibited by both extracts (Brodkiewicz et al. 2017). The EO (24% w/w) obtained by hydrodistillation of aerial parts (leaves and flowers) (0.25–2.00 mg ml^{-1}) was active against *Yersinia enterocolítica*-PI, *Salmonella enteritidis*-MI, *Escherichia coli* ATCC 25922, *E. coli* isolated from clinical strains, and *Salmonella* sp-LM. Moderate activity was found against *Staphylococcus aureus* methicillin-sensitive and *Staphylococcus aureus* LM (Luna et al. 2008). A reversible inhibition of spore germination was attained between 0.1% and 1.0% EO concentration for the fungi *Alternaria solani* (Ell. Et Mart.), *Sclerotium cepivorum* Berk., and *Colletotrichum coccodes* (Wallr.) Hughes. At 1.0% EO concentration, mycelial growth of fungi was totally inhibited, but no effect on mycelia was shown at 0.01% (Zygaldo and Grosso 1995).

The EOs have insect repellent, antifungal, and antibacterial activities (Zygaldo et al. 1995). The *in vitro* effect of the EO (50% and 100%) and the corresponding hydrolate (residue of hydrodistillation) showed activity against *Streptococcus mutans* (ATCC 25175). After 72 h of exposition, the inhibition of growth was similar to that obtained with chlorhexidine glycolate (0.12%) with an increasing effect with time (Rivadeneira-Cajas and Álvarez-Velasco 2015).

Insect repellant activity against *Triatoma infestans* (Klug) was also reported (Lima et al. 2011). EOs obtained from leaves resulted active against permethrin-resistant head lice (50% knock down, 32.06 min after fumigation) being the most effective EO from plant chemotypes with more piperitone content (Toloza et al. 2010).

Methanol and water extracts of leaves showed significant activity against *Plasmodium falciparum* K1 (IC50 3.0 and 8.5 μg ml^{-1}, respectively) that was attributed to the content on eriodyctiol (IC 17.2 μg ml^{-1}), luteolin (IC 6.4 μg ml^{-1}), ursolic (IC 4.9 μg ml^{-1}), and oleanolic acid (IC 9.3 μg ml^{-1}) (Debenedetti et al. 2002). That extracts also have activity against *P. falciparum* 3D7 strain and *Trypanosoma brucei* rhodesiense (van Baren et al. 2006).

The aqueous extracts (100 mg ml^{-1}) showed allelopathic activity (67% inhibition of growth) on wheat (Mongelli et al. 1997).

Toxicity The aqueous extract (100, 1000, and 2000 μg ml^{-1}) was cytotoxic against *Artemia salina* (Mongelli et al. 1995). Cytotoxicity of eriodictyol on mammalian KB cell was relatively low (IC50 174.2 μg ml^{-1}) (van Baren et al. 2006). The administration of a single dose (of 4000 and 8000 mg kg^{-1}) of ethanolic extract of aerial parts of *C. gilliesii* on shaved skin of male rats did not show death or toxic symptoms after 24–48 h of treatment (Reynoso et al. 2018).

7.4 *Schinus molle* L. Var. Molle (Anacardiaceae)

Synonymy *Schinus molle* var. argentifolius Marchand, *Schinus molle* var. hyungan angustifolium Sessé & Moc., *Schinus biluminosus* Salisb., *Schinus molle* var. huigan (Molina) Marchand (Alonso and Desmarchelier 2015).

Common name Aguaribay, árbol de la pimienta, curanguay, chichita sauce, molle, pimentero, chichita péndula, falso pimentero, anacahuita, pirul, aroeira (Portuguese), Califorina pepper tree (Eng.).

General information *S. molle* is an evergreen tree, very resistant to high temperatures and drought, and native from Central and South America that is also worldwide found, mainly in arid and saline soils. In Argentina, *S. molle* grows in the provinces of BAI, COR, COS, ERI, JUJ, MIS, SAL, and SFE. Leaves are used as dye, and its reddish fruits are edible having a taste similar to pepper; they are also used to produce fermented beverages. It is also considered as an ornamental tree (Goldstein and Coleman 2004; Albán-Castillo et al. 2018). It was proposed as an indicator of heavy metal contamination (Al, As, Co, Cu, Cd, Cr, Ni, Pb, Ti, V, and Zn) (Alcalá-Jáuregui et al. 2015).

Botanical characteristics *S. molle* is a tree, 3–15 m height, with dark brown and intensely fissured bark. The leaves are imparipinnate, persistent, alternate, and aromatic, with a winged rachis and 20–40 leaflets. Leaflets are linear-lanceolate and entire or toothed by the margins. The fruit is a drupe, 4–6 mm in diameter, pink to reddish, and bright (Rodríguez et al. 2005; Machado et al. 2018).

Ethnobotanical information The decoction of the entire plant is used as antipyretic, anti-inflammatory, emmenagogue, antifungal, antibacterial, antirheumatic, and anti-conjunctivitis, to treat skin affections, ulcer, urinary tract diseases, and respiratory diseases. The infusion made with the bark is taken as purgative and stimulant. The decoction of the leaves is considered as analgesic, antitussive, antidepressant, anti-inflammatory, and hypotensive, for neonatal and pediatrics care and for hernia and prolapse. Leaves and fruits are taken as emmenagogue, anti-inflammatory, and antiseptic. The fruit is used as diuretic, antirheumatic, and anti-inflammatory for amenorrhea and dysmenorrhea. The EO is prescribed as antibacterial, antifungal, and cytotoxic, and for venereal diseases. The infusion made with the oleoresin is considered antirheumatic, wound healing, and purgative (Bernhard et al. 1983; Yueqin et al. 2003; Goldstein and Coleman 2004; Díaz et al. 2008; Barboza et al. 2009; Guerra-Boone et al. 2013; López et al. 2014; Martins et al. 2014; Maema et al. 2016; Pellegrini et al. 2017).

Chemical data *S. molle* contains the pentacyclic triterpenoid α-amyrin. The EO from leaves contains croweacin, limonene, phellandrene, carvacrol, cismen-2-en-1-ol, trans-piperitol, and laccase. In the leaves there are isoquercitrin 6”-O-phydroxybenzoate, 2”-*O*-α-l-rhamnopyranosyl-hyperin-6“-*O*-gallate, gallic acid, methyl gallate, chlorogenic acid, 2”-O-α-l-rhamnopyranosyl-hyperin, quercetin 3-O-β-d-neohesperidoside, miquelianin, quercetin, rutin, quercetin 3-O-β-d galacturonopyranoside, isoquercitrin, hyperin, isoquercitrin 6“-gallate, hyperin 6”-O-gallate, and (+) catechin. In the EO from fruits, there are α-bergamont-*trans*, α-cadinol, γ-cadinene, bourbonene, α- and β-phellandrene, β-spathulene, D-limonene, silvestrene, α-pinene, β-pinene, carvacrol, myrcene, camphene, o-ethylphenol, p-cymene, p-cymol, γ-cubebene, α- and γ-muurolene (Fig. 7.4). In the fruits there are the triterpenoid isomasticadienonic, isomasticadienonalic, masticadienonic, and epi-isomasticadienolic acids and the flavonoid chamaejasmin

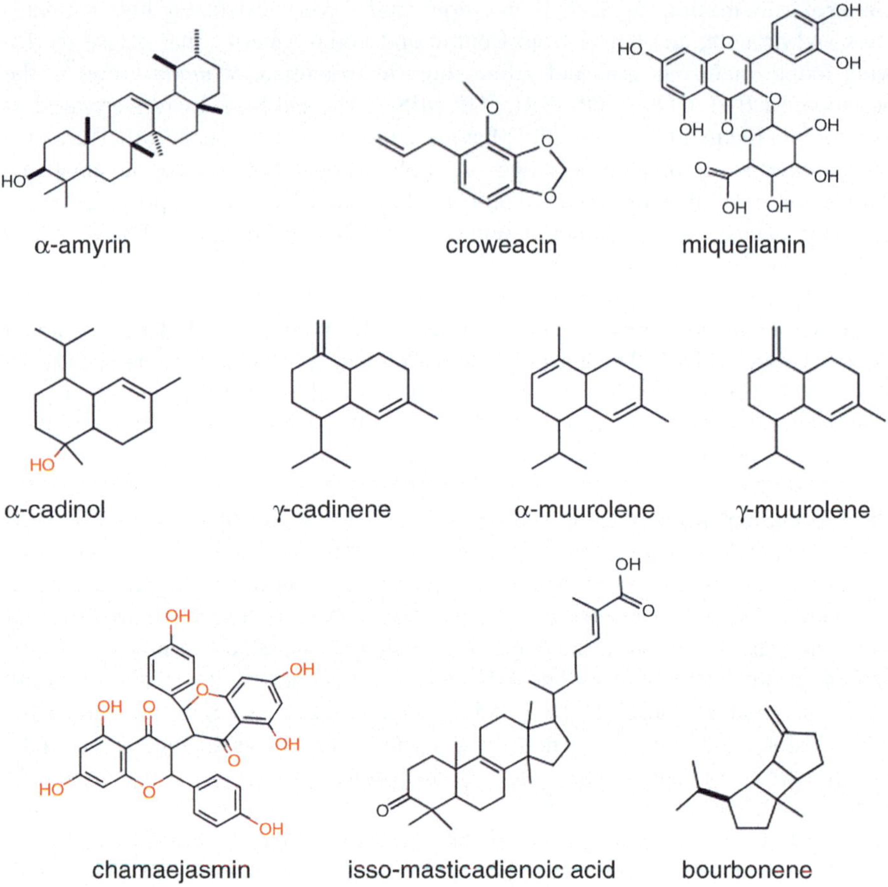

Fig. 7.4 Chemical compounds from *Schinus molle* L. var. molle

(Fig. 7.4) (Díaz et al. 2008; Barboza et al. 2009; Deveci et al. 2010; González et al. 2011; Simionatto et al. 2011; de Mendonça Rocha et al. 2012; Caballero et al. 2014; Doleski Muhd et al. 2015; Dos Santos et al. 2015; Malca-García et al. 2017; Machado et al. 2018).

Biological activities Light, high temperature, and high pressure produced a decrease of the biological activity of the EOs due to degradation or volatilization (Beyki et al. 2014; de Oliveira et al. 2014).

The hexane extract from leaves showed antidepressant-like activity in the tail suspension test in mice, a predictive model of depression. The extract (600–1000 mg kg^{-1}) significantly reduced the locomotor activity of mice in the open-field test. A similar effect was produced by the flavonoid rutin (0.3, 1.0, and 3.0 mg kg^{-1}) isolated from the ethanolic extract (Machado et al. 2007; Machado et al. 2008).

An ointment containing the EOs (3.5, 2.0, or 1.75% final concentration) and vaseline showed wound healing activity in cattle and Balb C53 mice, being 2% the most effective EO concentration (Alba González et al. 2009).

The aqueous extract of fruits showed inhibitory effect on the angiotensin-converting enzyme in dogs and rabbits (Olaffson et al. 1997).

The hexane extract and the EO from fruits had activity against *Streptococcus pneumoniae* (MIC 62.5 mg ml^{-1}), attributed to γ-cadinene (Molina-Salinas et al. 2007; Pérez-López et al. 2011). The hexane extract was also effective against *H. influenzae* and *Mycobacterium tuberculosis* (Molina-Salinas et al. 2007). The methanolic extract was active against *Proteus* spp., *Streptococcus* spp., *Escherichia coli*, and *Entereobacter* spp. (Matela et al. 2018). The EO and the hexane extract (30 μg ml^{-1}) have antimicrobial activity against *E. coli* 0157:H7 and *Bacillus cereus* (Deveci et al. 2010). The ethanolic extract of fruits was effective against *Staphylococcus aureus* (Cruz-Carrillo et al. 2010), *S. aureus* ATCC 25923, and *E. coli* ATCC25922 and has a weak activity against *P. aeruginosa* ATCC 27853 (de Mendonça Rocha et al. 2012). The ethanol extract (25 μg ml^{-1}) was active against *Candida albicans* ATCC 10231 (inhibition halo ≥20 mm) that colonize dental prosthesis, but at a lesser extent than fluconazole (inhibition halo ≥31 mm) at the same concentration (Saravia-León and Guillinta-Vallejos 2012). The EO from leaves showed a low activity against *C. albicans* (MIC 2.5 mg ml^{-1}), but it was effective against *Fusarium solani, Aspergillus flavus, A. japonicus, A. niger, A. oryzae, A. fumigatus, A. parasiticus* ATCC 16992, *Fusarium moniliforme* ATCC 52539, *Alternaria alternata* (Fries), and *Penicillium italicum* Wehmer (Dikshit et al. 1986; Alanís-Garza et al. 2007; Rhouma et al. 2009; Martins et al. 2014; López-Meneses et al. 2015).

The EOs are used as food preservative for their antioxidant and antimicrobial activity (Guala et al. 2016), for example, against *Salmonella* (103 CFU g^{-1}) inoculated in minced beef meat (Hayouni et al. 2008). Chitosan nanoparticles loaded with *S. mole* leaf extract (0.05% v/v) were effective against *Aspergillus parasiticus* spores, reducing the viability of fungal spores as the concentration of the chitosan: *S. molle* bionanocomposites raised (Luque-Alcaraz et al. 2016).

S. molle ethanol extracts and rutin extracted from leaves showed anti-lymphoma potential on induced lymphoma mice models (Male CD-1 nu/nu nude), cytotoxic activity on U-937 cells, and strong lethality on brine shrimp (BS) larvae. The anti-cancer property was connected with the presence of α-phellandrene and rutin (Calzada et al. 2018). It also showed antitumor and cytotoxic activity (IC50 50 ± 7 μg ml^{-1}) against the human hepatocellular carcinoma cell line, Hep G, and cytotoxicity on breast carcinoma and leukemic cell lines (Ruffa et al. 2002; Díaz et al. 2008).

The aqueous extract had a discrete inhibitory activity of trypsin and elastase that was postulated as partially responsible of the insecticidal activity (Macció et al. 2014). The crude hexane extract from fruits was effective against *Cydia pomonella* (Linneaeus) larvae, a phytopathogen of fruit trees, with 60% mortality at 5 g kg^{-1} concentration; from the surviving adults, a great proportion showed malformations (Chirino et al. 2001). The hexane extract from leaves was active against oriental adult cockroach (*Blatta orientalis* L.) with a 83.33% repellency at a concentration of 0.075 mg l^{-1} (Deveci et al. 2010). The EO from leaves (0.78% w/v) was active against bed bugs (*Cimex lectularis*), reaching a 100% of efficiency after a 24-hr treatment (250 μg 125 cm^{-2}); that activity was attributed to γ-pinene, muurolol, and limonene content (Machado et al. 2018). The aqueous and ethanol extracts had larvicidal and pupicidal activity against immature *Culex quinquefasciatus* with an increased mortality as exposure time and concentration rise (100% mortality of third instar after 72 h exposure at 100 mg l^{-1}) (Girmay et al. 2014). Also, the aqueous and methanolic extract (4%) of leaves was active against the adult *Bophilus decoloratus* and *Rhipicephalus pulchellus* cattle ticks (Feyera and Abdisa 2016). Activity of the n-hexane extract against eggs and adults of *Ctenocephalides felis felis*, a flea that infests dogs and cats in Brazil, was *in vitro* demonstrated. Efficacy was 100% at 800 μg cm^{-2} (LD50 524.80 μg cm^{-2} after 48 h). EOs from fruits and leaves showed 100% efficacy against adult fleas at 800 μg cm^{-2} (LD5 353.95 μg cm^{-2}) and at 50 μg cm^{-2} (LD50 12.02 μg cm^{-2}), but not against flea eggs (Batistia et al. 2016). The EOs from seeds produced 77–100% mortality of *Drosophila melanogaster* Meigen (Diptera: Drosophilidae) and had a repellent effect on *Tribolium confusum* Jacquelin du Val (Coleoptera: Tenebrionidae) (Steimbauer 1995).

In the case of *Trypanosoma cruzi*, the methanolic extract (150 μg ml^{-1}) resulted effective against epimastigotes (IC50 16.31 μg ml^{-1}) after 96 h of treatment (Molina-Garza et al. 2014); the crude hydro-alcoholic extract (250 μg ml^{-1}) produced a higher inhibition (IC50 50 μg ml^{-1}) (Pizzolatti et al. 2002). The differences were attributed to variation in bioactive concentration, geographical origin, etc. of the extracts. The extract from leaves was repellent and insecticidal against *Triatoma infestans* (vinchuca, winchuka) a blood-sucking bug that is the main vector of the Chagas-Mazza disease (Ferrero et al. 2006). An acaricidal effect against varroa (*Varroa destructor*) parasitizing hive of bees and against tick (*Rhipicephalus microplus*) was also reported (Torres et al. 2012; Guala et al. 2014; Guala et al. 2016).

On the other hand, a slight activity of the aqueous extract (50% and 100% w/v) from leaves on *Helicoverpa zea* (Boddie) (Lepidoptera: Noctuidae) showed 16% mortality (Guevara et al. 2018). On the contrary, the antifeeding effect of the aqueous extract (2.5, 3.0, 4.3, and 5.6% w/v) on the leaf beetle *Xanthogaleruca luteola* (Muller) (Coleoptera: Chrysomelidae) produced a 98.32% antifeeding effect. The ethanolic extract produced more larval mortality (LC50 1.28% w/v at day 7) than the aqueous extract at the same concentration (Huerta et al. 2010; Chiffelle et al. 2013). An insecticidal effect was also reported against *Sitophilus zeamais* Motschulsky (Coleoptera: Curculinoideae) (Silva et al. 2005; Benzi et al. 2009; Awoke et al. 2014; Arias et al. 2017) and *Sitophilus oryzae* L. (Benzi et al. 2009). Crude leaf methanolic extract had antihelmintic activity against *Haemonchus contortus*, a parasite of small ruminants (Zenebe et al. 2017). Leaf EO has allelopathic activity on *Allium cepa* and *Lactuca sativa* (Simionatto et al. 2011).

Toxicity The EO was reported not to be toxic for humans (Carrasco 1998; Guba 2008) nor carcinogenic for animals (Guba 2008). The EO was highly toxic (LC50 47 μg ml^{-1}) to *Artemia salina* (Martins et al. 2014). Also, it was cytotoxic to breast carcinoma, leukemia cell line K562 (0.89 μg ml^{-1}), and NCI-ADR/RES (ovarian expressing the resistance phenotype for adryamycin) (3.40 μg ml^{-1}), attributed to caryophyllene oxide (Díaz et al. 2008). Immunotoxicity to human lymphocytes (LD50 30.07 μg ml^{-}1) and human macrophages (42.07 μg ml^{-1}) was also reported. A reduction in IL-10 production at low concentrations was correlated with the increase in TNF-ũFC; production (Apollo Duarte et al. 2018).

7.5 *Smilax campestris* Griseb. (Smilacaceae)

Synonymy *Smilax campestris* Griseb. var. rubiginosa, *Smilax rubiginosa* Griseb., Smilax campestris Griseb. var. marginulata, *Smilax marginulata* Mart. ex Griseb., *Smilax montana* Griseb., *Smilax scalaris* Griseb., *Smilax viminea* Griseb.

Vernacular name Zarzaparrilla, zarzaparrilla negra, zarzaparrilla blanca, japicanga, japecanga (Portuguese).

General information *S. campestris* grows in temperate thickets and grasslands near rivers and streams, in mountains, and fields at high altitude (1800 m.a.s.l.) in the provinces of BAI, CHA, COR, COS, ERI, FOR, JUJ, MIS, SAL, SDE, SFE, and TUC. The phenotypic plasticity of leaves and rhizomes is due to ecological factors (Andreata 1997). Blooming is from July to September. A tonic, bitter, and digestive beverage is prepared from the leaves and aerial stems (Rugna et al. 2003, 2011). It contains steroid saponins which are used as precursors in the semi-synthesis of cortisone.

Botanical characteristics *S. campestris* is a dioecious climbing plant about 2–4 m high with lignified rhizome and tuberose hickening. The underground stem runs horizontally near the ground surface (10–15 cm depth). The woody, brown stem, approximately 5–10 cm length, has swollen nodes that can produce negative geotropic stems and adventitious roots. Leaves are variable both in shape and in size (ovate, oblong-elliptic, elliptical, oblong, rarely cordate) with obtuse, acute, or emarginated coriaceous apex. Basal acrodrome venation has three major midribs, with spiniform emergences in midrib extensions. Inflorescences are umbels with 30–45 flowers, terete peduncle, and 2–20-mm-long globose receptacle. Perigone is purple-red, with elliptic tepals and papillae at the apex. The globular ovary contains hemotrope ovules, 1 per locule. The fruit is a black-violet globose berry with purple pulp, and subglobose seeds, wrapped by a layer of hyaline, elastic tissue (Guaglianone and Gattuso 2006; Martins et al. 2010; Martins et al. 2013; Colares et al. 2017).

Ethnobotanical information The roots are considered as antisyphilitic, diaphoretic, and diuretic, for infections, hernia and prolapse, and menstrual diseases. Rhizomes are considered as antisyphilitic, diuretic, sudorific, aperitive, refrigerant, anti-gout, hypolipemic, and laxative. The aerial parts are used as anti-inflammatory, antidiarrheal, and for healing wounds. The leaves are considered as antisyphilitic and antirheumatic. Finally, flowers are used to treat vaginal fungus. It is also prescribed to treat impotence, rheumatism, gout, and skin conditions; it is considered diuretic and diaphoretic. The roots and the basal parts of the stems are used as diuretics and laxative for infants (Mandrile and Bongiorno de Pfirter 1991; Toursarkissian 1980; Rodrigues and Carvalho 2001; Barboza et al. 2009; Silva et al. 2013).

Chemical data Leaves contain kaempferol, kaempferol-3-O-rutinoside, quercetin, quercetin-3-O-glucoside, quercetin-3-O-rhamnogalactoglucoside, isorhamnetin, isorhamnetin-3-O-rutinoside, and rutin. In rhizomes there are quercetin-3-glucoside, 3-rutinoside, procyanidin, and propelargonidin (Fig. 7.5) (Rugna et al. 2002a, 2004, 2008, 2013; Barboza et al. 2009; Morais et al. 2014; Salas-Coronado et al. 2017).

Biological activity The antioxidant activity from rhizomes, which was partly attributed to the proanthocyanidin content, follows a circannual pattern. The highest antioxidant activity is reached in summer (February) with two more peaks during pre-flowering and fructification (Rugna et al. 2002b, 2003). The ethanol and butanol fractions showed a strong antioxidant activity.

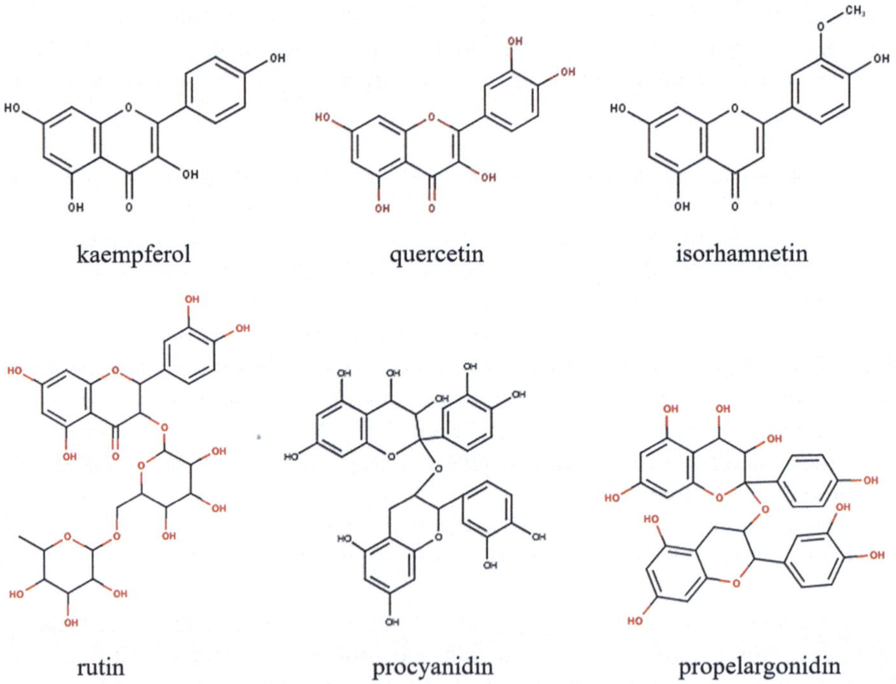

Fig. 7.5 Chemical compounds from *Smilax campestris* Griseb

The methanolic extracts (80%) from rhizome showed antifungal activity against *Candida albicans* and *Aspergillus niger* (Battista et al. 2007). The ethanol extract and its fractions were active against *C. albicans*, *C. glabrata*, *C. krusei*, *C. parapsilosis*, *C. tropicalis*, and *Cryptococcus gattii*. The ethyl acetate fraction exhibited the highest activity against *C. krusei*, (MIC 62.5 mg ml^{-1}) and was also active against the *C. gattii* (MIC 1000 mg ml^{-1}). The hexane fraction was active against *C. glabrata* (MIC 125 mg ml^{-1}) and *C. krusei* (MIC 500 mg ml^{-1}). The ethanol extract was active against *C. krusei*, *C. parapsilosis* and *C. gattii* (MIC 1000 mg ml^{-1}), and *C. albicans*, *C. glabrata*, and *C. tropicalis* (MIC 2000 mg ml^{-1}). The dichloromethane fraction displayed activity against *C. krusei* and *C. tropicalis* (MIC 2000 mg ml^{-1}). Based on those results, a new therapeutic strategy could be implemented specifically against *C. krusei*, which presents intrinsic resistance to fluconazole, and against *C. gattii*, which is implicated in HIV-associated infections (Morais et al. 2014). The ethanolic extract of leaves showed inhibitory activity against *Staphylococcus epidermidis* (MIC 1000 mg ml^{-1}) and a worthy antioxidant activity (IC50 13.1 ± 0.1 mg ml^{-1}) (Ferreira de Lima 2016).

***In vitro* culture** *In vitro* cultures were initiated from seeds in MS solid culture medium without plant growth regulators. Explants transferred to MS medium with the addition of different plant growth regulator relationships (BAP, IAA) were tested to regenerate the whole plant. A methanolic extract (50%) of leaves from regenerated plantles rendered kaempferol, quercetin, and isorhamnetin. The highest total phenolics content was obtained in MS medium with BAP (2 mg ml^{-1}) and IAA (0.4 mg ml^{-1}) (Rugna et al. 2010).

Toxicity Mutagenic activity on *Salmonella*/microsome assay was reported. *S. campestris* infusions had direct mutagenic activity on TA98 strain (Fernández de Sá Ferreira and Vargas 1999).

7.6 *Smallanthus macroscyphus (Baker)* A. Grau ex A.A. Sáenz (Asteraceae)

Synonymy *Polymnia macroscypha* Baker, *Smallanthus riograndensis* C.A. Mondin, *Smallanthus araucariophilus* C.A. Mondin.

Vernacular name Yacón del campo, wild yacon.

General information *S. macroscyphus* belongs to the yacon group whose main representative is *S. sonchifolius*. *Smallanthus* species show a clear preference for disturbed habitats, such as riverbanks, landslides, and roadsides. The distribution is affected by precipitation seasonality (Grau and Rea 1997; Vitali and Katinas 2015). This specie is an invader of abandoned sugar cane fields as well as of the free space between land plots (Serra-Barcellona et al. 2014). In Argentina, *S. macroscyphus* grows in the provinces of JUJ, SAL, and TUC. It has been consumed as a fruit since pre-Columbian times. Lately it was revalued as a functional or nutraceutical food (Grau and Rea 1997; Aybar et al. 2001). Under modern nutritional and agronomic requirements, some traits like carbohydrate high polymerization degree and short growth cycle may be very useful in breeding programs (Grau and Rea 1997; Coll Aráoz et al. 2008, 2014). Blooming is from December to June.

Botanical characteristics *S. macroscyphus* is a perennial herbaceous plant, 1–3 m tall, with a caulescent rhizome or rhizophore that gives rise to tuberose fusiform adventitious roots and aerial caulescent branches. The stem is pubescent at the apex and almost glabrous by maturity with opposite decussate opposite simple, green-yellowish leaves. The leaf blade is membranous, pubescent, of triangular symmetry, and deltoid ovate, with an acute apex and margins irregularly scalloped. Terminal inflorescences are capitula clusters arranged in corymbiform or paniculiform cymes, with yellow tubular corolla and ovate, subtriangular anthers (Coll Aráoz et al. 2008; Vitali et al. 2015).

Ethnobotanical information The roots are used to treat urinary and hepatic disorders; it is also used as hypoglycemic, antitumoral, antimicrobial, antifungal, and anti-inflammatory (Hilgert 2000; Cabrera et al. 2007; Barboza et al. 2009; Serra-Barcellona et al. 2012; Vitali 2017).

Chemical data Kaurenoic acid methyl ester is the predominant diterpene in the root cortex; also there are fructooligosaccharides with high polymerization degree. In the aerial parts, there are melampolides, melampolide-type sesquiterpene lactones (STLs), (6R∗,7S∗,8S∗,9S∗)-8-hydroxy-9-methacryloxy-10-carbomethoxymelampo-1(10),4,11(13)-trien-6,12-olide, (6R∗,7S∗,8S∗, 9S∗)-8-Hydroxy-9-angeloxy-10-carbomethoxymelampo-1(10), 4,11(13)-trien-6,12-olide, and (4S∗,5S∗,6R∗,7R∗,8S∗,9S∗)-4,5-epoxy-8-hydroxy-9-angeloxy-10-carbomethoxymelampo-1(10),11(13)-dien-6,12-olide (de Pedro et al. 2003; Barboza et al. 2009; Coll Aráoz et al. 2010, 2014). The decoction of leaves contains caffeic acid, 3-caffeoylquinic acid (ac. chlorogenic), three dicaffeoyl quinic acids (3,4-dicaffeoylquinic acid, 3,5-dicaffeoylquinic acid, and 4,5-dicaffeoylquinic acid), and the lactone polymatin A (Fig. 7.6) (Serra-Barcellona et al. 2016). The sesquiterpene lactone content varies according to the latitude. In lower latitudes the sesquiterpene lactone profile is more complex containing mainly uvedaline, fluctuanine, and enhydrin. In higher latitudes there are mainly polymatin A and a few other lactones as isopolymatin A and sonchifolin (Coll-Aráoz et al. 2016).

Biological activities The sesquiterpene lactones are produced as a mechanism of defense in glandular trichomes of the foliar surface. They accumulate underneath the cuticle and are released when the glandular trichomes are mechanically disrupted. The biological activity of those sesquiterpene lactones is deterrent to herbivores, toxic to pathogens like fungi and bacteria, and allelopathic (Coll-Aráoz et al. 2016).

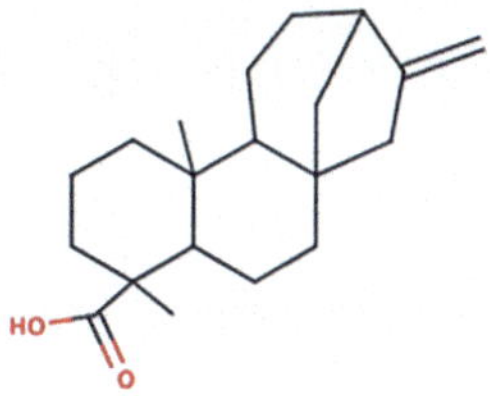

kaurenoic acid

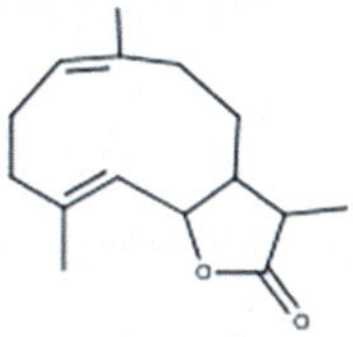

melampolide basic structure

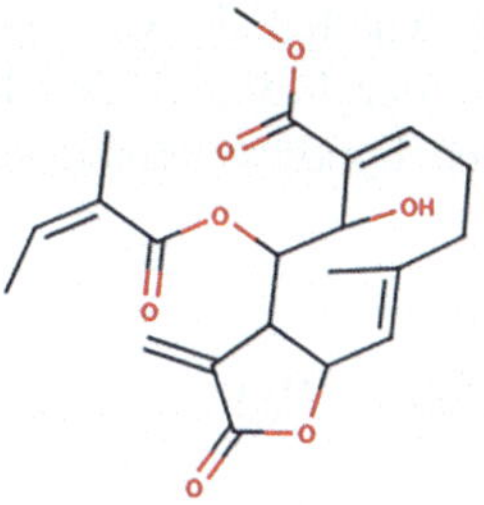

polymatin A

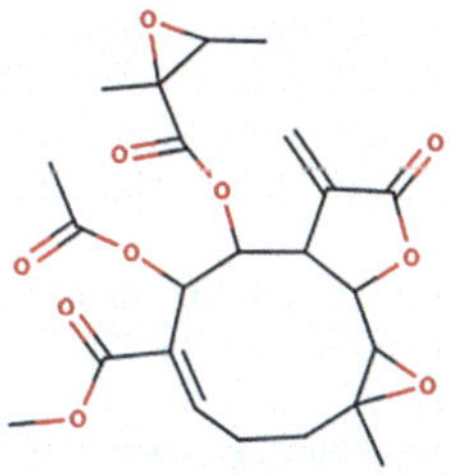

enhydrin

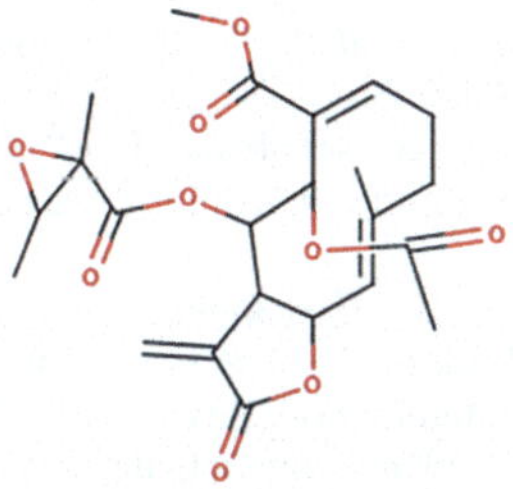

uvedalin

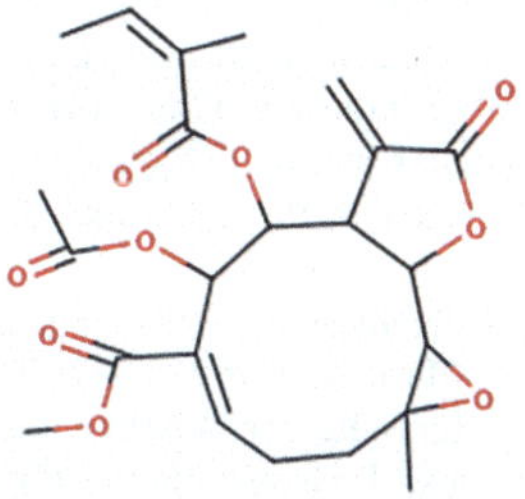

fluctuanin

Fig. 7.6 Chemical compounds from *Smallanthus macroscyphus* (Baker) A. Grau ex A.A. Sáenz

The aqueous extract of dried *S. macroscyphus* leaves has high hypoglycemic effect on rats. Glucose tolerance tests on Wistar rats showed that both the leaf decoction and polymatin A significantly diminished glycemia. Daily administration of the decoction (70 mg polymatin A 100 ml^{-1}, dose 8 ml kg^{-1}) or a water solution of polymatin A (dose 7 ml kg^{-1}) during 30 days produced a strong decrease of glycemia in diabetic rats. On the other hand, the aqueous extracts (10% decoction, 140 ml kg^{-1}) of dried *S. macroscyphus* leaves produce a hypoglycemic effect in a dose-dependent manner on adult male Wistar rats with induced diabetes. Polymatin A has an effective inhibition of postprandial blood glucose peak and a hypoglycemic activity in diabetic animals. Probably, the glucose-lowering effect of the decoction could be mediated by an improvement in the glycemic control mechanisms, including insulin secretion and/or extrapancreatic pathways without risk of hypoglycemia, which is one of the most undesirable side effects of diabetes treatments. Moreover, the treated animals reduced urinary glucose excretion. Polymatin A was identified as the major hypoglycemic principle of *S. macroscyphus* (Cabrera et al. 2007; Coll Aráoz et al. 2008; Serra-Barcellona et al. 2014).

Toxicity The 10% decoction and pure polymatin A have a safety margin in rats at the doses that led to the hypoglycemic effect. The aqueous extract of leaves orally administered is nontoxic, at least up to the maximum level assayed (280 mg kg^{-1} day^{-1}). Similarly, polymatin A has no toxic effect even at the highest dose tested (28 mg kg^{-1} day^{-1}) (Serra-Barcellona et al. 2016).

The lethal dose 50 (LD50) of the 10% decoction of leaves and of polymatin A (above 14.0 and 2.8 g kg^{-1} b.w., respectively) is significantly higher than the effective hypoglycemic doses.

References

Alanís-Garza BA, González-González GM, Salazar-Aranda R, Waksman de Torres N, Rivas-Galindo VM (2007) Screening of antifungal activity of plants from the northeast of Mexico. J Ethnopharmacol 114(3):468–471

Alba González A, Bonilla Rivera P, Arroyo Acevedo J (2009) Actividad cicatrizante de una pomada con aceite esencial de *Schinus molle* L. "molle " en ganado vacuno con heridas infectadas y en ratones. Cienc Invest 12(1):29–36

Albán-Castillo J, Espinoza G, Rojas R, Santibáñez D (2018) Color in the memory: plant dyes used in the tradition of Andean and Amazonian Preuvian communities. Ecología Aplicada 17(1):85–96

Alcalá-Jáuregui J, Rodríguez Ortiz JC, Hernández Montoya A, Díaz Flores PE, Filippini MF, Martínez Carretero E (2015) Bark of *P. laevigata* (Fabaceae) and S. molle (Anacardiaceae) as bioindicator of heavy metal contamination. Rev Fca UNCUYO 47(2):83–95

Alonso J, Desmarchelier CJ (2015) Plantas medicinales autóctonas de la Argentina. Bases científicas para su aplicación en atención primaria de la salud, 1a ed. Corpus Libros Médicos y Científicos, Ciudad Autónoma de Buenos Aires

Amorin JL, Rosow R (1988) Guia taxonómica con plantas de interés farmacéutico. Dominguezia 10(1):35–40

Andreata RHP (1997) Revisão das espécies brasileiras do gênero Smilax Linnaeus (Smilacaceae). Pesqui Bot 47:7–244

Apollo Duarte J, de Bairros Zambrano LA, Dias Quintana L, Balhego Rocha M, Gomes Schmitt E, Agusti Boligon A, Anraku de Campos MM, Souza de Oliveira LF, Mansur Machado M (2018) Immunotoxicological Evaluation of *Schinus molle* L. (Anacardiaceae) Essential Oil in Lymphocytes and Macrophages. Evid Based Complement Alternat Med 2018:6541583. https://doi.org/10.1155/2018/6541583

Arias OJ, Silva AG, Figueroa CI, Fischer GS, Robles-Bermúdez A, Rodríguez-Maciel JC, Lagunes-Tejeda A (2017) Insecticidal, repellent and antifeeding activity of powder and essential oil of *Schinus molle* L. fruits against *Sitophilus zeamais* (Motschulsky). Chilean J Agric Anim Sci ex Agro-Ciencia 33(2):93–104

Awoke Y, Asayew G, Melaku G, Derbew M, Kedir S, Raja N (2014) Evaluation of certain plant leaf powders and aqueous extracts against maize weevil, *Sitophilus zeamais* Motsch. (Coleoptera: Curculionidae). Asian J Agric Sci 6(3):83–88

Aybar MJ, Sánchez Riera AN, Grau A, Sánchez SS (2001) Hypoglycemic effect of the water extract of *Smallanthus sonchifolius* (yacón) leaves in normal and diabetic rats. J Ethnopharmacol 74:125–132

Barbieri N, Costamagna M, Gilabert M, Perotti M, Schuff C, Isla MI, Benavente A (2015) Antioxidant activity and chemical composition of essential oils of three aromatic plants from La Rioja province. Pharm Biol 54:168–173. https://doi.org/10.3109/13880209.2015.1028077

Barboza GE, Cantero JJ, Núñez C, Pacciaroni A, Ariza Espinar L (2009) Medicinal plants: a general review and a phytochemical and ethnopharmacological screening of the native Argentine Flora. Kurtziana 34(12):7365

Batistia LC, Cid YP, de Almeida AP, Prudêncio ER, Riger CJ, de Souza MAA, Coumendouros K, Chaves DS (2016) *In vitro* efficacy of essential oils and extracts of *Schinus molle* L. against *Ctenocephalides felis felis*. Parasitology 143(5):627–638. https://doi.org/10.1017/S0031182016000081

Battista S, García GF, Rugna AZ, Wagner ML, Gurni AA (2007) Actividad antimicótica en extractos de diferentes órganos de *Smilax campestris* Griseb-Smilacaceae. Boletín Latinoamericano y del Caribe de Plantas Medicinales y Aromáticas 6(6):330–331

Bektas T, Cilkiz M (2016) A pharmacological and phytochemical overview on *Satureja*. Pharm Biol 54(3):375–412. https://doi.org/10.3109/13880209.2015.1043560

Benzi V, Stefanazzi N, Ferrero A (2009) Biological activity of essential oils from leaves and fruits of pepper tree (*Schinus molle* L.) to control rice weevil (*Sitophilus oryzae* L.). Chilean J Agric Res 69(2):154–159

Bernhard RA, Shibamoto T, Yamagucho K, White E (1983) The volatile constituents of *Schinus molle* L. J Agric Food Chem 31(2):463–466

Beyki M, Zhaveh S, Khalili TS, Rahmani-Cherati T, Abollahi A, Mansour B, Bayat Tabatabaei M, Mohsenifarc A (2014) Encapsulation of *Mentha piperita* essential oils in chitosan–cinnamic acid nanogel with enhanced antimicrobial activity against *Aspergillus flavus*. Ind Crop Prod 54:310–319

Brodkiewicz IY, Molina RDI, Vega López G, Alberto MR, Vera NR, Arena ME (2017) Anti-pathogenic activity of herbs used in Argentinean traditional infusion. Glob J Agric Innov Res Dev 4:36–46

Brown AD, Grau HR, Malizia LR, Grau A (2001) Argentina. In: Kappelle M, Brown AD (eds) Bosques nublados del Neotropico. INBio, Costa Rica, pp 623–659

Caballero GM, Padin EV, Pollio ML (2014) Chemical composition of oleoresins from berries of *Schinus molle* L. grown in Buenos Aires Province (Argentina). J Food Technol 12(3):73–77

Cabana R, Silva LR, Valentao P, Viturro CI, Andrade PB (2013) Effect of different extraction methodologies on the recovery of bioactive metabolites from *Satureja parvifolia* (Phil.) Epling (Lamiaceae). Ind Crop Prod 48:49–56

Cabrera AL (1976) Regiones fitogeográficas argentinas. In: W F Kugler (ed.). Enciclopedia Argentina de Agricultura y Jardinería, 2nd. Acme, Buenos Aires 2 (1): 1–85

Calzada F, Solares-Pascasio JI, Valdes M, Garcia-Hernandez N, Velázquez C, Ordóñez-Razo RM, Barbosa E (2018) Antilymphoma Potential of the Ethanol Extract and Rutin Obtained of the Leaves from *Schinus molle* Linn. Pharm Res 10:119–123

Cabrera W, de Pedro A, Perotti ME, Grau A, Catalán C, Genta S, Sánchez SS (2007) Hypoglycemic effect of *Smallanthus macroscyphus* leaves. Identification of the active principle. Biocell 31(2):311

Carrasco RE (1998) Estudio de los aceites y determinación de la actividad antimicrobiana del fruto de Schinus molle L. Tesis de Maestría en Recursos Vegetales y Terapéuticos. UNMSM. Lima

Chiffelle I, Huerta A, Celis M, Araya JE (2013) Proximal analysis and insecticidal effects of extracts from pepper tree (*Schinus molle*) leaves on elm leaf beetle (*Xanthogaleruca luteola*) larvae. Ind Crop Prod 43:523–528

Chirino M, Cariac M, Ferrero AA (2001) Actividad insecticida de extractos crudos de drupas de *Schinus molle* L. (Anacardiaceae) sobre larvas neonatas de *Cydya pomonela* L. (Lepidoptera: Tortricidae). Bol San Veg Plagas 27:305–314

Colares MN, Martínez Alonso SM, Spegazzini E, Hernández MP, Perrotta VG, Nova MC, Arambarri AM (2017) Anatomy and Histochemistry of the Underground Organs of Climbing Medicinal Plants from the Shore of Rio de la Plata (Argentina). Lilloa 54(2):123–140

Coll Aráoz MV, Mercado MI, Grau A, Ponessa GI (2008) Morfología y anatomía foliar, caulinar y radicular de *Smallanthus macroscyphus* (Asteraceae). Lilloa 45(1–2):23–33

Coll Aráoz MV, Mercado MI, Grau A, Catalán CAN (2010) Ent-kaurane derivatives from the root cortex of yacon and other three *Smallanthus* species (Heliantheae, Asteraceae). Biochem Syst Ecol 38:1042–1048

Coll Aráoz MV, Kortsarz González AM, Mercado MI, Ponessa GI, Grau A, Catalán C (2014) Ontogeny and total sugar content of yacon tuberous roots and other three *Smallanthus* species (Heliantheae, Asteraceae), insights on the development of a semi-domesticated crop. Genet Resour Crop Evol 61:163–172

Coll-Aráoz MV, Mercado MI, Grau A, Catalán CAN (2016) Intraspecific variation of sesquiterpene lactones associated to a latitudinal gradient in *Smallanthus mac-roscyphus* (Heliantheae: Asteraceae). Chemoecology 26:143. https://doi.org/10.1007/s00049-016-0213-1

Cruz-Carrillo A, Rodríguez NN, Rodríguez CA (2010) *In vitro* evaluation of the antibacterial effect of *Bidens pilosa*, *Lantana camara*, *Schinus molle* and *Silybum marianum*. Rev UDCA Act Div Cient 13(2):117–124

Dadé MM, Fioravanti D, Schinella GR, Tournier HA (2009) Total antioxidant capacity and polyphenol content of 21 aqueous extracts obtained from native plants of Traslasierra valley (Argentina). Boletín Latinoamericano y del Caribe de Plantas Medicinales y Aromáticas 8(6):529–539

Debenedetti S, Muschietti L, van Baren C, Clavin M, Broussalis A, Martino V, Houghton P, Warhurst D, Steele J (2002) *In vitro* antiplasmodial activity of Argentinian plants. J Etnopharmacol 80:163–166

De la Sota ER (1972) Sinopsis de las Pteridofitas del Noroeste de Argentina, I. Darwiniana 17:11–103

de Oliveira JL, Campos EVR, Bakshi M, Abhilash PC, Fraceto LF (2014) Application of nanotechnology for the encapsulation of botanical insecticides for sustainable agriculture: prospects and promises. Biotechnol Adv 32(8):1550–1561

de Pedro A, Cuenca MR, Grau A, CAN C, Gedris TE, Herz W (2003) Melampolides from *Smallanthus macroscyphus*. Biochem Syst Ecol 31:1067–1071

Deveci O, Sukan A, Tuzun N, Kocabas EEH (2010) Chemical composition, repellent and antimicrobial activity of *Schinus molle* L. J Med Plant Res 4(21):2211–2216

Díaz C, Quesada S, Brenes O, Aguilar G, Cicció JF (2008) Chemical composition of *Schinus molle* essential oil and its cytotoxic activity on tumour cell lines. Nat Prod Res 22:1521–1534

Dikshit A, Nanqvi AA, Husain A (1986) *Schinus molle*: a new source of natural fungitoxicant. Appl Environ Microbiol 51(5):1085–1088

Doleski Muhd PS, Ferreira Cuelho CH, Calil Brondani J, Palermo Manfron M (2015) Chemical composition of the *Schinus molle* L. essential oil and their biological activities. Rev Cuba Farm 49(1):132–143

Dos Santos Calvancanti A, de Souza Alvez M, da Silva LCP, dos Santos Patrocínio D, Nalesso Sanchez M, Siquiera de Almeida Chavez D, de Souza A (2015) Volatiles composition and extraction kinetics from *Schinus terebinthifolius* and *Schinus molle* leaves and fruit. Rev Bras 25:356–362

Feresin GE, Tapia A, López SN, Zacchino SA (2001) Antimicrobial activity of plants used in traditional medicine of San Juan province, Argentine. J Ethnopharmacol 78:103–101

Fernandes de Sá Ferreira IC, Ferrão Vargas VM (1999) Mutagenicity of medicinal plant extracts in Salmonella/microsome assay. Phytother Res 13(5):397–400

Ferreira Jorge LI, de Oliveira FV, Koschtschak MRV (1986) Diagnose comparativa das espécies Chenopodium ambrosioides L. (erva-de-santa-maria) e *Coronopus didymus* (L.) Sm (mastruço): principais características morfo-histológicas e químicas. Rev Bras Farmacognosia 1(2):143–153

Ferreira de Lima T (2016) Propiedade antimicrobiana e atividade antioxidante do extracto etanólico das folhas de *Smilax campestris* Grisebach. Thesis Grau de bacharela em Biotecnologia. Universidade Federale de Grande Dourados, Brazil

Ferrero AA, González JOW, Chopa CS (2006) Biological activity of *Schinus molle* on *Triatoma infestans*. Fitoterapia 77:381–383

Feyera T, Abdisa E (2016) *In vitro* acaricidal activity of crude extracts of *Schinus molle* (L.) leaves against field population of *Bophilus decoloratus* and *Rhipicephalus pulchellus* ticks. Afr J Pharm Pharmacol 10(36):772–777

Girmay K, Fikre B, Asmelash A, Getachew B, Tekle E, Raja N (2014) Evaluation of water and ethanol extracts of *Schinus molle* Linn. Against immature *Culex quinquefasciatus* Say (Diptera: Culicidae). J Coast Life Med 2(6):471–477

Goldstein DJ, Coleman RC (2004) *Schinus molle* L. (Anacardiaceae) Chicha production in the Central Andes. Econ Bot 58:523–529

Grau A, Rea J (1997) Yacón – *Smallanthus sonchifolius* (Poepp. & Endl.) H. Robinson: 199-242 En Andean roots and tubers: ahipa, arracacha, maca and yacón. In: Hermann M, Heller J (eds). IPGRI (International Plant Genetic Resources Institute). . Gatlesben, Germany

Grau HR (2006) Dinámica de bosques en el gradiente altitudinal de las Yungas Argentinas. In: Arturi M, Frangi JL, Goya JF (eds) Ecología y Manejo de los bosques de Argentina. Editorial de la Universidad Nacional de La Plata, pp 1–30

Guaglianone ER, Gattuso S (2006) Flora del valle de Lerma- Smilacaceae Vent. Aportes Botánicos de Salta-Ser Flora 7(16):1–6

Guala MS, Lapissonde MO, Elder HV, Pérez GA (2014) Efecto Acaricida del Aceite Esencial de Aguaribay (*Schinus molle* L.) y sus Fracciones en Colmenares de Abejas (Apis mellifera) en Relación con la Composición Química. Información Tecnológica 25(2):151–156

Guala MS, Lapissonde MO, Elder HV, van Baren CM, Bandoni AL, Dellacassa E (2016) Rose pepper (*Schinus molle* L.) oil. In: Preedy V (ed) Essential oils in food preservation, flavor and safety. Academic Press, Elsevier Inc., San Diego, pp 689–695

Guba R (2008) Toxicity myths–essential oils and their carcinogenic potential. Center for Aromatic Medicine, Australia

Guerra-Boone L, Alvarez-Román R, Salazar Aranda R, Torres-Cirio A, Rivas-Galindo VM, Waksman de Torres N, González González GM, Pérez-López LA (2013) Chemical compositions and antimicrobial and antioxidant activities of the essential oils from *Magnolia grandiflora*, *Chrysactinia mexicana*, and *Schinus molle* found in northeast Mexico. Nat Prod Commun 8:135–138

Guevara DJ, Araya JE, Huerta A, Chiffelle I (2018) Extracts of *Schinus molle* and *Artemisia absinthium* against *Helicoverpa zea* on fresh ear corn in Ecuador. Chilean J Agric Anim Sci 34(3):216–225

Haque I (1989) Composition of essential oil of *Coronopus didymus* seeds. J Chem Soc Pak 11:80–81

Hayouni EA, Chraief I, Abedrabba M, Bouix M, Leveau J-Y, Mohammed H, Hamdi M (2008) Tunisian *Salvia officinalis* L. and *Schinus molle* L. essential oils: their chemical compositions and their preservative effects against *Salmonella* inoculated in minced beef meat. Int J Food Microbiol 125:242–251

Hernández NE, Tereschuk ML, Abdala LR (2000) Antimicrobial activity of flavonoids in medicinal plants from Tafí del Valle (Tucumán, Argentina). J Ethnopharmacol 73:317–322

Hilgert N (2000) Especies vegetales empleadas en la insalivación de hojas de "coca" (*Erythroxylum coca* var. coca, Erythroxylaceae). Darwiniana 38(3–4):241–252

Huerta A, Chiffelle I, Puga K, Azúa F, Araya J (2010) Toxicity and repellence of aqueous and ethanolic extracts from *Schinus molle* on the elm leaf beetle *Xanthogaleruca luteola*. Crop Protect 29:1118–1123

Lima B, Lopez S, Luna L, Agüero MB, Aragón L, Tapia A, Zacchino S, López ML, Zygadlo J, Feresin GE (2011) Essential oils of medicinal plants from the Central Andes of Argentina: chemical composition, and antifungal, antibacterial, and insect-repellent activities. Chem Biodivers 8:924–936

López A, Castro S, Andina MJ, Ures X, Munguía B, Llabot JM, El-der H, Dellacassa E, Palma S, Domínguez L (2014) Insecticidal activity of micro-encapsulated *Schinus molle* essential oil. Ind Crop Prod 53:209–216

López-Meneses AK, Plascencia-Jatomea M, Lizardi-Mendoza J, Rosas-Burgos EC, Luque-Alcaraz AG, Cortez-Rocha MO (2015) Antifungal and antimycotoxigen-ic activity of essential oils from *Eucalyptus globulus*, *Thymus capitatus* and *Schinus molle*. Food Sci Technol Campinas 35(4):664–671

Lorenzi H, Matos FJA (2002) Plantas Medicinais no Brasil: nativas e exóticas. Instituto Plantarum, São Paulo, pp 189–190

Luque-Alcaraz AG, Cortez-Rocha MO, Velázquez-Contreras CA, Acosta-Silva AL, Santacruz-Ortega HC, Burgos-Hernández A, Arguelles-Monal WM, Plascencia-Jatomea M (2016) Enhanced antifungal effect of chitosan/pepper tree (*Schinus molle*) essential oil bionanocomposites on the viability of Aspergillus parasiticus spores. J Nanomat 2016:6060137. https://doi.org/10.1155/2016/6060137

Luna L, Lima B, Tapia A, Feresin GE, Duschatzky C, Possetto M, de Lampasona MP, Schuff C (2008) Chemical composition and antibacterial activity of *Satureja parvifolia* (Phil.) Epling essential oil. J Essent Oil Bear Pl 11:106–111

Macció L, Vallés D, Cantera AM (2014) Search of inhibitory activity of trypsinand elastase in extracts of native plants. INNOTEC 9:84–90

Machado CD, Raman V, Rehman JU, Maia BHLNS, Meneghetti EK, Almeida VP, Silva RZ, Farago PV, Khan IA, Budel JM (2018) *Schinus molle*: anatomy of leaves and stems, chemical composition and insecticidal activities of volatile oil against bed bug (*Cimex lectularius*). Revista Brasileira de Farmacognosia 29:1. https://doi.org/10.1016/j.bjp.2018.10.005

Machado DG, Kaster MP, Binfaré RW, Dias M, Santos ARS, Pizzolatti MG, Brighente IMC, Rodrigues ALS (2007) Antidepressant-like effect of the extract from leaves of *Schinus molle* L. in mice: evidence for the involvement of the monoaminergic system. Prog Neuro-Psychopharmacol Biol Psychiatry 31:421–428

Machado DG, Bettio LEB, Cunha MP, Santos ARS, Pizolatti MG, Brighente IMC, Rodrigues ALS (2008) Antidepressant-like effect of rutin isolated from the ethanolic extract from *Schinus molle* L. in mice: evidence for the involvement of the serotonergic and noradrenergic systems. Eur J Pharmacol 587:163–168

Maema LP, Potgieter M, Mahlo SM (2016) Invasive alien plant species used for the treatment of various diseases in Limpopo Province, South Africa. Afr J Tradit Complement Altern Med 13:223–231

Malca-García GR, Hennig L, Ganoza-Yupanqui ML, Piña-Iturbe A, Bussmann RW (2017) Constituents from the bark resin of *Schinus molle*. Rev Bras 27:67–69

Mandrile EL, Bongiorno de Pfirter G (1991) Zarzaparrilla, *Smilax campestris* Frisebach (Cmilacaceae) o (Liliaceae). Bifase 6(4)

Mantena SK, Mutalik S, Srinivasa H, Subramanian GS, Prabhakar KR, Reddy KR, Srinivasan KK, Unnikrishnan MK (2005) Antiallergic, antipyretic, hypoglycemic hypoglycemic and hepatoprotective effects of aqueous extract of *Coronopus didymus* Linn. Biol Pharm Bull 28:468–472

Martins AR, Bombo AB, Soares AN, Appezzato-da-Glória (2013) Revista Brasileira de Farmacognosia. Brazilian Journal of Pharmacognosy 23(4):576–584

Martins MDR, Arantes S, Candeias F, Tinoco MT, Cruz-Morais JC (2014) Antioxidant, antimicrobial and toxicological properties of *Schinus molle* L. essential oils. J Ethnopharmacol 151:485–492

Martins AR, Pütz N, Soares AN, Bertolsi Bombo A, Appezzato da Glória B (2010) New approaches to underground systems in Brazilian Smilax species (Smilacaceae). J Torrey Bot Soc 137(2–3):220

Matela KS, Mekbib SB, Pillai MK (2018) Antimicrobial activities of extracts from *Gleditsia triacanthos* L. and *Schinus molle* L. Pharmacol Online 2:85–92

de Mendonça Rocha PM, Rodilla JM, Diez D, Elder H, Guala MS, Silva LA, Pombo EB (2012) Synergistic antibacterial activity of the essential oil of Aguaribay (*Schinus molle* L.). Molecules 7:12023–12036

Moismann Busnardo TCP, Padoani C, Mora TC, Weber Bivatti M, Fröde TS, Bürger C, Claudino VD, Dalmarco EM, de Souza MM (2010) Anti-inflammatory evaluation of *Coronopus didymus* in the pleurisy and paw oedema models in mice. J Ethnopharmacol 128:519–525

Molina-Garza ZJ, Bazaldúa-Rodríguez AF, Quintanilla-Licea R, Galaviz-Silva L (2014) Anti-*Trypanosoma cruzi* activity of 10 medicinal plants used innortheast Mexico. Acta Trop 136:14–18

Molina-Salinas GM, Pérez-López A, Beceril-Montes P, Salazar-Aranda R, Said-Fernández S, Waksman de Torres N (2007) Evaluation of the flora of Northern Mexico for *in vitro* antimicrobial and anti-tuberculosis activity. J Ethnopharmacol 109:435–441

Mongelli ER, Coussio JD, Ciccia GN (1995) Estudio de toxicidad aguda de plantas medicinales argentinas mediante el bioensayo de *Artemia salina* Leach. Dominguezia 12(1):35–42

Mongelli E, Desmarchelier C, Coussio J, Ciccia GN (1997) The potential effects of allelopathic mechanisms on plant species diversity and distribution determined by the wheat rootlet growth inhibition bioassay in South American plants. Rev Chil Hist Nat 70:83–89

Morais MI, Amaral Pinto ME, Guimarães Araújo S, Fonsêca Castro H, Duarte-Almeida JM, Rosa LH, Rosa CA, Johann S, Rodrigues dos Santos Lima L (2014) Antioxidant and antifungal activities of *Smilax campestris* Griseb. (Smilacaceae). Nat Prod Res 28:1275. https://doi.org/10.1080/14786419.2014.895728

Morales JM, Sirombra J, Brown AD (1995) Riqueza de árboles en las Yungas argentinas. In: Brown AD, Grau HR (eds) Investigación, conservación y desarrollo en Selvas Subtropicales de Montaña. Laboratorio de Investigaciones Ecológicas de las Yungas (UNT) y Proyecto de Desarrollo Agroforestal, Tucumán, pp 157–162

Muschietti L, van Baren C, Vila R, Clos M, Cañigueral S, Adzet T, Coussio J (1996) Chemical composition of the leaf oil of *Satureja odora* and *Satureja parvifolia*. J Essent Oil Res 8:681–684

Nitz AC, Ely JB, Acampora AJ, Tames DR, Corrêa BP (2006) Estudo morfométrico no processo de cicatrização de feridas cutâneas em ratos, usando: *Coronopu didymus* e *Calendula officinalis*. Arquivos Catarinenses de Medicina 35(4):74–79

Olaffson K, Jaroszewski JW, Smitt UW, Nyman U (1997) Isolaton of Angiotensin Converting Enzyme (ACA) inhiiting Triterpenes from *Schinus molle*. Planta Médica 63:352–353

Orfila EN, Farina EL (1997) Flora del Valle de Lerma. Lamiaceae Barnh. (Labiatea Juss.). Aportes Botánicos de Salta- Ser Flora 4(1):1–62

Pellegrini MC, Alonso-Salces RM, Umpierrez ML, Rossini C, Fuselli SR (2017) Chemical composition, antimicrobial activity, and mode of action of essential oils against *Paenibacillus larvae*, etiological agent of American foulbrood on *Apis mellifera*. Chem Biodivers 14:1–44

Pérez-López A, Torres Cirio A, Rivas-Galindo VM, Salazar Aranda R, Waksman de Torres N (2011) Activity against Streptococcus pneumoniae of the Essential Oil and γ-Cadinene Isolated from *Schinus molle* Fruit. J Essent Oil Res 23:25–28

Pizzolatti MG, Koga AH, Grisard EC, Steindel M (2002) Trypanocidal activityof extracts from Brazilian Atlantic rain forest plant species. Phytomedicine 9:422–426

Ponce RI, Cabana M, Vaira M, Giunta SA (2007) Bioassays to determine toxicity and bioactivity of Alcaloids in *Coronopus dydimus*. Bol Latinoam Caribe Plant Med Aromaticas 6(5):268–269

Prabhakar KR, Srinivasan KK, Rao PGM (2002) Chemical investigation, anti-inflammatory and wound healing properties of *Coronopus didymus*. Pharm Biol 40:490–493

Prabhakar KR, Veeresh VP, Vipan K, Sudheer M, Priyadarsini KI, Satish RBSS, Unnikrishnan MK (2006) Bioactivity-guided fractionation of *Coronopus didymus*: a free radical scavenging perspective. Phytomedicine 13:591–595

Prina AO (1995) Las Crucíferas de la provincial de La Pampa, República Argentina. Rev Fac Agronomía-UNLPam 8(2):29–67

Ratera EL, Ratera MO (1980) Plantas de la flora argentina empleadas en medicina popular. Ed. Hemisferio sur S.A.

Reynoso M, Balderrama Coca ME, Brodkiewicz IY, Jaime G, Perotti M, Schuff C, Vera NR (2018) Anti-inflammatory effects and safety of extracts and essential oil from *Clinopodium gilliesii* (muña muña). Int J Pharm Sci Drug Res 10(4):306–314

Rhouma A, Ben Daoud H, Ghanmi S, ben Salah H, Romdhane M, Demak M (2009) Antimicrobial activities of leaf extracts of *Pistacia* and *Schinus* species against some plant pathogenic fungi and bacteria. J Plant Pathol 91(2):339–345

Rivadeneira-Cajas D, Álvarez-Velasco P (2015) Essential oil *Schinus molle* L. (Molle) as an antimicrobial potential on *Streptococcus mutans*. Study *in vitro*. Kiru 12(2):8–14

Rodríguez R, Ruiz E, Elissetche J (2005) Árboles en Chile. Universidad de Concepción, Concepción, Chile

Rodrigues VEG, Carvalho DA (2001) Plantas medicinais no domínio dos Cerrados. Lavras, UFLA

Ruffa MJ, Ferraro G, Wagner ML, Calcagno ML, Campos RH, Cavallaro L (2002) Cytotoxic effect of Argentine medicinal plant extracts on human hepatocellular carcinoma cell line. J Ethnopharmacol 79:335–339

Ruffa MJ, Wagner ML, Suriano M, Vicente C, Nadinic J, Pampuro S, Salomón H, Campos RH, Cavallaro L (2004) Inhibitory effect of medicinal herbs against RNA and DNA viroses. Antivir Chem Chemother 15:153–159

Rugna A, Gurni AA, Wagner ML (2002a) Ritmo de producción de flavonoides en rizomas de *Smilax campestris* Griseb.-Smilacaceae. XI Congreso Italo-Latinoamericano di Etnomedicine "Alberto Di Capua", p 42

Rugna A, Gurni AA, Wagner ML (2002b) Estudio variacional de flavonoides en ejemplares masculinos y femeninos de *Smilax campestris* Griseb.-Smilacaceae. Acta Farmacéutica Bonaerense 21(2):119–121

Rugna A, Polo J, Evelson P, Gurni AA, Llesuy S, Wagner ML (2003) Antioxidant activity in rhizomes from *Smilax campestris* Griseb. Smilacaeae. Mol Med Chem 1:21–25

Rugna AZ, Gurni AA, Waagner ML (2004) Estudio Comparativo de Flavonoides en los Órganos Aéreos de *Smilax campestris* Griseb. -Smilacaceae- Acta Farm. Bonaerense 23(3):379–382

Rugna A, Ricco R, Gurni A, Wagner M (2008) Variaciones en el Contenido de los Polifenoles Foliares en *Smilax campestris* Griseb. - Smilacaceae - según su Grado de Desarrollo. Lat Am J Pharm 27(2):247–249

Rugna AZ, Romero O, Mazzeo M, Santamaria JM, Gurni AA, Wagner ML (2010) Producción de polifenoles en ejemplares de *Smilax campestris* Griseb. (Smilacaceae) cultivadas en condiciones controladas de cultivo. Dominguezia 26(2):31–36

Rugna AA, Ricco RA, Gurni AA, Wagner ML (2011) *Smilax campestris* Griseb. –Smilacaceae–: variaciones en la producción de polifenoles en hojas sanas y atacadas por la oruga de la mariposa *Agraulis vanillae* L. –Heliconidae. Dominguezia 27(2):27–33

Rugna AZ, Gurni AA, Wagner ML (2013) Phenological variations of polyphenols in *Smilax campestris* (Smilacaceae). Turk J Bot 37:350–354

Ruiz REL, Fusco M, Rapisarda AMP, Sosa A, Ruiz SO (1989) Estudio Fitoquímico del "mastuerzo" (*Coronopus didymus* Sm. Brassicaceae). Acta Farm Bonaer 8(3):171–173

Ruiz REL, Fusco M, Sosa A, Ruiz SO (1993) Aislamiento de un Glicósido Flavonoide y de una Saponina del "Mastuerzo" (*Coronopus didymus* Sm., Brassicaceae). Acta Farm Bonaer 12(2):101–104

Ruthsatz B (1974) Los arbustos de las estepas andinas del noroeste argentino y su uso actual. Boletín de la Sociedad Argentina de Botánica 16(1–2):27–45

Salas-Coronado R, Hernández-Carlos B, Llaguno-Guilberto J, Santos-Sánchez NF (2017) Phenolic compounds in genus smilax (Sarsaparilla). In: Phenolic compounds (Ed. Marcos Soto-Hernández) InTech Open. https://doi.org/10.5772/66896

Saravia-León N, Guillinta-Vallejos G (2012) Antifungal activity *Schinus molle* ethanol extract and fluconazole against *Candida albicans*. KIRU 9(1):39–41

Scarpa GF (2004) Medicinal plants used by the Criollos of Northwestern Argentine Chaco. J Ethnopharmacol 91(1):115–135. https://doi.org/10.1016/j.jep.2003.12.003

Serra-Barcellona C, Cabrera WM, Honoré SM, Mercado MI, Sánchez SS, Genta SB (2012) Safety assessment of aqueous extract from leaf *Smallanthus sonchifolius* and its main active lactone, enhydrin. J Ethnopharmacol 144:362–370

Serra-Barcellona C, Coll Aráoz MV, Cabrera WM, Habib NC, Honoré SM, Catalán CA, Grau A, Genta SB, Sánchez SS (2014) *Smallanthus macroscyphus*: a new source of antidiabetic compounds. Chem Biol Interact 209:35–47

Serra-Barcellona C, Honoré SM, Cabrera WM, Habib NC, Sb G, Sánchez SS (2016) Polymatin A from *Smallanthus macroscyphus* leaves: a safe and promising antidiabetic compound. J Appl Toxicol 36:1516. https://doi.org/10.1002/jat.3312

Silva G, Orrego O, Hepp R, Tapia M (2005) Búsqueda de plantas con propiedades insecticidas para el control de *Sitophilus zeamais* en maíz almacenado. Pesq Agropec Bras (Brasilia) 40(1):11–17

Silva AF, Silva Ferreira L, Silva CG, Campana PRV, Enoque MM, Rabelo MFR, Almeida VL (2013) Espécies do cerrado I: Etnobotánica, fitoquímica e atividade antimicrobiana. 64° Congresso Nacional de Botânica, Belo Horizonte, 10-15 de Novembro de 2013

Simionatto E, Chagas MO, Peres MTLP, Hess SC, da Silva CB, Ré-Poppi N, Gebara SS, Corsino J, Farias Morel A, Stuker CZ, Matos MFC, de Carvalho JE (2011) Chemical composition and biological activities of leaves essential oil from *Schinus molle* (Anacardiaceae). J Essent Oil Bear Plants 14:590. https://doi.org/10.1080/0972060X.2011.10643976

Simoni AA, Perea MC (2016) Las plantas que curan: el lugar que ocupan las plantas medicinales desde la cosmovisión de los escolares de la Comunidad India Quilmes (Tucumán-Argentina). Mundo de antes 10:143–172

Sorarú SB, Bandoni AL (1978) Plantas de la Medicina Popular. Ed. Albatros S.R.L., Buenos Aires, págs. 48–49

Steimbauer M (1995) The insecticidal and repellent activity of *Schinus molle* L. (Anacardiaceae) against *Drosophila melanogaster* Meign (Diptera: Drosophilidae) and *Tribolium confusum* Jaquelin du Val (Coleoptera: Tenebrionidae). Gen Appl Ent 26:13–18

Toloza AC, Zygaldo J, Biurrun F, Rotman A, Picollo MI (2010) Bioactivity of Argentinean essential oils against permethrin resistant head lice, *Pediculus humanus* capitis. J Insect Sci 10:185. Available online: insectscience.org/10.185

Torres FC, Machado Lucas A, Sardá Ribeiro VL, Martins JR, von Poser G, Guala MS, Elder HV, Cassel E (2012) Influence of essential oil fractionation by vacuum distillation on acaricidal activity against the cattle tick. Braz Arch Biol Technol 55:613–621

Toursarkissian M (1980) Plantas Medicinales de la Argentina. Ed. Hemisferio Sur. Buenos Aires, Argentina

Uddin G, Ali J, Feroz D, Rauf A (2014) Phytochemical and pharmacological study of *Coronopus didymus* L. Topclass. J Herb Med 3(1):1–4

van Baren C, Anao I, Lira PD et al (2006) Triterpenic acids and flavonoids from *Satureja parvifolia*. Evaluation of their antiprotozoal activity. Z Naturforsch C 61:189–192

Vázquez AM, Aimar ML, Decarlini MF, Demmel GI, Cantero JJ, Ruiz GM (2016) Volatile Organic Constituents of *Clinopodium gilliesii* (Benth.) Kuntze: analysis by HS-SPME and classic hydrodistillation. Rev Colomb investing agroindustriales, p. 91–100

Vitali MS, Sancho G, Katinas L (2015) A revision of *Smallanthus* (Asteraceae, Milleriaeae), the "yacón" genus. Phytotaxa 2014(1):1–84

Vitali MS, Katinas L (2015) Modelado de distribución de las especies argentinas de *Smallanthus* (Asteraceae), el género del "yacón": un cultivo potencial para la agricultura familiar. Rev Fac Agron La Plata 114(Núm. Esp.1):110–121

Vitali MS (2017) Anatomia foliar del género *Smallanthus* (Asteraceae, Millerieae). Bol Soc Argent Bot 52(3):463–472

Viturro CI, Molina A, Guy I, Charles B, Guinaudeau H, Fournet A (2000) Essential oils of *Satureja boliviana* and *S. parvifolia* growing in the region of Jujuy, Argentina. Flavour Frag J 15:377–382

Wood JRI (2011) *Clinopodium* L. (Lamiaceae) in Bolivia. Kew Bull 66:199–226

Yueqin R, Recio M, Manez S, Giner R, Cerda N, Rios J (2003) Isolation of two triterpenoides and a biflavanone with anti-inflammatory activity from *Schinus molle* fruits. Planta Med 69:893–898

Zenebe S, Feyera T, Assefa S (2017) *In vitro* Anthelmintic Activity of Crude Extracts of Aerial Parts of *Cissus quadrangularis* L. and Leaves of *Schinus molle* L. against *Haemonchus contortus*. Biomed Res Int 2017:1905987. https://doi.org/10.1155/2017/1905987

Zygaldo JA, Lamarque AL, Maestri DM, Grosso NR (1995) Use of essential oils as natural antioxidants. Grasas Aceites 46:285–288

Zygaldo JA, Grosso NR (1995) Comparative study of the antifungal activity of essential oils from aromatic plants growing wild in the central region of Argentina. Flavour Fragr J 10:113–118

Zygaldo J, Merino F, Maestri M, Guzmán C (1993) The essential oil composition of *Satureja odora* and *S. parvifolia* from Argentina. J Essent Oil Res 5:549–551

Chapter 8
Chaco and Espinal

8.1 Introduction

8.1.1 Chaco

The Chaco region extends through the province of FOR, CHA, eastern SAL, JUJ, TUC, CAT and LRI; all of SDE; north of SLU, COR and SFE Fe; and northwest of COR (Cabrera 1971). It covers plains and mountains of low elevation with mostly subtropical climate, warm and humid summers, and temperate and dry winters (Fig 8.1a). Rains are concentrated in summer, and occasionally a wave of polar cold is perceived. As for the fauna, large emblematic animals such as the *Panthera onca* (yaguareté) and the *Priodontes maximus* (tatú carreta, giant armadillo) are in critical danger of extinction. The *Myrmecophaga tridactyla* (giant anteater) and the *Tapirus terrestris* (tapir) are also among the typical fauna of the eco-region, as well as monkeys, coatis, foxes, and weasels.

In the Dry Chaco, the characteristic vegetation is the xerophilous forest, with trees that become lower and thinner towards the Arid Chaco. The deciduous xerophilous forest predominates, having also palmares, savannahs, and halophilic steppes. The highest areas have xeric forests (quebrachales) of red quebracho and white quebracho, with mistol, itin, yuchán, pitch, several cacti, duraznillo, and bushes of the *Acacia* (tusca, teatín) and *Capparis* (*sacha-quince*, *sacha*-watermelon) genus. In some low areas, salinity and restrictions in drainage condition the floristic composition, giving rise to communities of palo santo (*Bulnesia sarmientoi*, the *palosantales*), carob trees (*Prosopis* spp.), and chañar (*Geoffroea decorticans*) and, in the salt flats, to vegetation with a predominance of halophytic species. The forests alternate with pastures ("Pampas") of aibé (*Stipa paposa*), associated with old clogged channels. There are also horco-quebracho, visco, churqui (espinillo), molle, molle de coco, and coconut. A particular formation of the Córdoba and Puntana mountain ranges is the palmares of carandilla (*Trithrinax* spp.). Above these forests within the altitudinal gradient are graminoid steppes, which form biogeographic islands with

M. A. Alvarez, *Pharmacological Properties of Native Plants from Argentina*,
https://doi.org/10.1007/978-3-030-20198-2_8

Fig. 8.1 Chaco (**a**) and Espinal (**b**) eco-regions from Argentina. MA Alvarez personal collection

features of the Puna. Tabaquillo (*Polylepis australis*) groves appear in these tall grasslands.

In the Humid Chaco, the high forests of the albardones are mainly composed of quebracho colorado chaqueño (*Schinopsis balansae*), quebracho blanco (*Aspidosperma quebracho-blanco*), guayaibí (*Patagonula americana*), urunday (*Astronium balansae*), lapacho (*Tabebuia* sp.), guayacán (*Handroanthus chrysanthus*), viraró (*Ruprechtia salicifolia*), palo amarillo (*Phyllostylon rhamnoides*, ibirácatú), and mistol (*Ziziphus mistol*). There are dense or open forests, bordering estuaries and ravines, mainly composed of carob (*C. siliqua*), tatané (*Chloroleucon tenuiflorum*), guaraniná (*Sideroxylon obtusifolium*), churqui (*A. caven*, espinillo del Espinal), and *Prosopis* sp.; grasslands, which are composed of wire grass as dominant species, accompanied by many other herbaceous and often tree-like elements (forming savannahs) of low-forest species; cañadas, constituted by pure scrublands (*pajonales*) or by palm tree forests (*palmares*), with white palm (*Trithrinax campestris* or *caranday*); and estuaries, without vegetation cover and with diverse aquatic species, such as pirí, peguajó (*Thalia geniculata*), totoras (*Schoenoplectus californicus*), cucharero (*Echinodorus grandiflorus*), and camalote (*Eichhornia crassipes*), among others.

8.1.2 Espinal

The Espinal extends in the form of an irregular arch around the LPA province through the center of SFE and COR, a large part of SLU, the center of LPA to the south of BAI. Towards the south it extends along the ravines of the Paraná river and through the banks of conchilla and dead dunes of the northeast of BAI, and towards the West it penetrates through the depressions and along the rivers until finally reaching the area of Sierra de los Padres (Cabrera 1971). This is a region of plains, low mountains areas, and dunes (Fig 8.1b). The climate is, due to the extension of the region, extremely variable, warm and humid in the north, and temperate and dry with marked water deficits, towards the west and south.

Many animals from the Pampa, affected by hunting and habitat transformation, are more frequent or are only associated with the Espinal forests. The most emblematic case is the pampas deer, since two of its last populations are located in the grasslands with *calden* (*Prosopis caldenia*), in San Luis, and in the pastures with *talas* (*Celtis spinosa*), in the northeast of Buenos Aires. Other examples are the wildcat, the gray fox, and the rhea. Much of the Espinal is located in regions of high-agricultural and urban development, which is why Espinal has been greatly reduced during the last decades. The creation of more protected natural areas and a sustainable management of livestock are the challenges the Espinal is currently facing. The Espinal can be divided into three subregions, namely: *Espinal del Ñandubay*, the Mesopotamian sector, with a humid climate; the *Espinal de los Algarrobos*, the central, subhumid sector of transition between the Pampa and the Chaco; finally, *Espinal del Caldén*, semiarid sector, transition between the Pampa and the Monte. In Espinal predominates the xerophilous forest vegetation that is quite similar to the Chaco region but more impoverished. There are also palm groves, grass savannahs, and steppes, and predominant trees are *Prosopis* spp. In the northeast of the *Espinal del Ñandubay*, algarrobos (*Prosopis* sp.), molle (*Schinus molle*), and espinillo (*A. caven*) forests predominate, with other species, among which the palm trees yatay (*Butia yatay*) and carandilla, which are associated with the forest or palm trees. In the center of the eco-region, the dominant *Prosopis* species are the white and black algarrobo (carob) that are accompanied by *A. caven*, *G. decorticans*, and *C. spinosa*. The herbaceous plants are mostly *pampeanas* species, with predominance of grasses characteristic of temperate grasslands and, in part, *chaqueñas*.

8.2 *Adiantum raddianum* C. Presl (Pteridaceae)

Synonymy *Adiantum cuneatum* Langsd. & Fisch., hom. illeg., *Adiantum aemulum* T. Moore, *Adiantum colpodes* T. Moore, *Adiantum moorei* Baker, *Adiantum amabile* T. Moore, hom. illeg.

Vernacular name Culantrillo, culandrillo, curatú, tipichá, avença (Portuguese), maidenhair ferns (English).

General information *A. raddianum* is a widely distributed species in Brazil, Peru, Uruguay, and Argentina. In Argentina it is located in the provinces of BAI, CAT, CHA, COR, COS, ERI, JUJ, LRI, MIS, SAL, SFE, SLU, and TUC. It has tolerance to different light conditions, and it usually grows isolated or forming bushes in the riverbanks, rock walls, waterfalls, slope forests, swamps, and ravines. It is cultivated as ornamental (Fujino et al. 1983; Silva et al. 2007; de la Sota et al. 2009; Arana and Bianco 2011; Arana et al. 2013).

Botanical characteristics *A. raddianum* is a fern with a terrestrial or saxicolous perennial sporophyte, creeping suberect rhizomes without clathrate scales, exceptionally with hairs. Petiole is shiny brown, and deltoid pinna is glabrous, several times divided, with scales or wax and free to reticulated veins (Fig. 8.2). They have polysaccharide mucilage in the epidermis that would function as protection from solar radiation. The sporangium (reproductive structure) is located alongside and at the end of the veins forming marginal-clustered sores or covering the entire surface of the blade or segment, without indusium or with a false glabrous kidney-shaped indusium, or protected by the modified leaf margin. Spores are trilete, with or without perisporium (Arana and Bianco 2011; Santos-Silva et al. 2011; Winter et al. 2011).

Ethnobotanical information In folk medicine, leaves are considered abortive, antitussive, and emmenagogue, with the ability to dissolve the mucus for respiratory disorders, as diaphoretic, emollient, diuretic, and expectorant, for treating pain, coughs, urinary disorders, alopecia, and menstrual difficulties (Martínez Crovetto 1981; De Feo 1992; Bresciani et al. 2003; Silva et al. 2007; Barboza et al. 2009).

Chemical Data The compounds identified in the spores are the alkaloids spermidine and N(1)-N(10)-caffeoyl-feruloyl spermidine (Fig. 8.3).

The whole plant contains the flavonoids, querciturone, kaempferol, astragalin, the terpenoids filicene (filic-3-ene), filicenal, adiantol (adian-5-en-3α-ol) and isoadiantone, secofilicanes, hydroxyhopane, neohopene, trisnorhopane, 17 αH-trisnorhopan-21-one, glaucanol B acetate, 21-hydroxy-30-norhopan-22-one, isoadiantol B, fern-9(11)-en-25-ol, fernane, fern-7-en-25-ol, 25-norfern-7-en-10b-yl formate, 7α,8α-epoxyfernan-25-ol, 7β,25-epoxyfern-8-ene, 7β,25-epoxyfern-9(11)-en-8α-ol (Fig. 8.4) (Shiojima et al. 1997; Bresciani et al. 2003; Barboza et al. 2009; Guerra Santos et al. 2010; Lai and Lim 2011; Pan et al. 2011; Thomas 2014).

Biological activity The main biological activities reported were antioxidant, antinociceptive, and antimicrobial. The antioxidant activity was attributed to its content in phenolic compounds and particularly flavonoids (Lai and Lim 2011; Shrivastava et al. 2014).

The hexane fraction obtained after a methanolic extraction from raw material, as well as pure filicene and filicenal, showed potent analgesic activity in two models of

Fig. 8.2 *Adiantum raddianum* C. Presl (Pteridaceae). Buenos Aires Botanical Garden. M A Alvarez personal collection

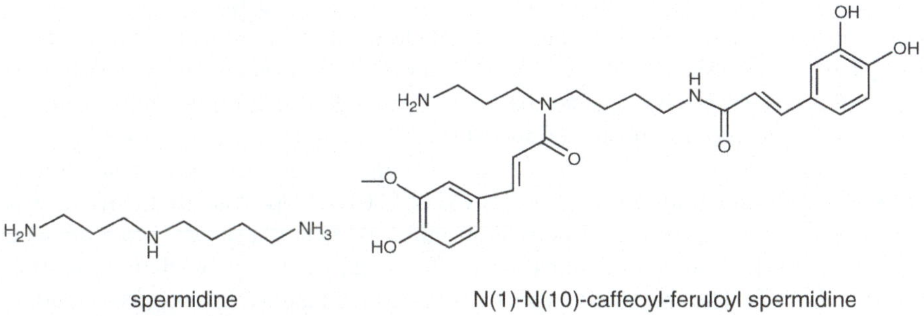

Fig. 8.3 Chemical compounds found in *Adiantum raddianum* C. Presl spores

pain in mice (writhing test and formalin-induced pain) (Bresciani et al. 2003; Pan et al. 2011). Filicene had antinociceptive effect in male Swiss mice challenged with acetic acid-, capsaicin-, and glutamate-induced nociception tests and hot-plate and open-field tests, each test involving different systems. Apparently, the mechanism of action involves interactions with the cholinergic, dopaminergic, glutamatergic, GABAergic, and tachykinergic systems. Filicene had higher antinociceptive activity of inflammatory pain (acetic acid-induced abdominal constrictions) in the animal model and was less effective in models of acute pain (glutamate and capsaicin tests). It was not found interacting with the opioid, serotoninergic, or adrenergic systems (de Souza et al. 2009).

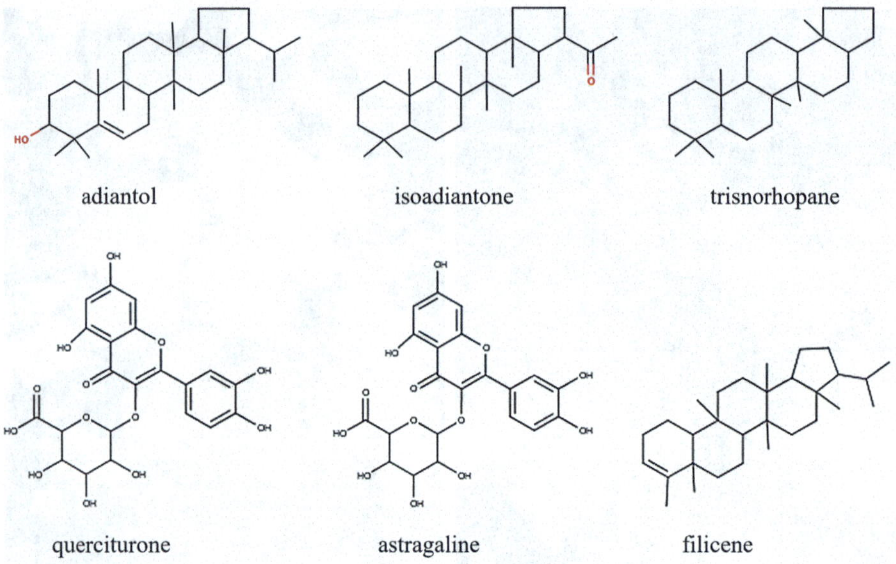

Fig. 8.4 Some chemical compounds found in *Adiantum raddianum* C. Presl

For testing the antimicrobial activity, successive extractions with petroleum ether, acetone, methanol, and water were made from air-dried fronds with sori. A sample of each extract (0.46%, 2.8%, 5.1%, and 0.70%, respectively) was tested for its antibacterial activity against bacteria involved in skin infections such as *Staphylococcus aureus* subsp. aureus (MTCC 96), *Escherichia coli* (MTCC 443), *Pseudomonas aeruginosa* (MTCC 741), *Klebsiella pneumoniae* subsp. pneumoniae (MTCC-109), and *Serratia marcescens* (MTCC 6164). Antibacterial activity was not found in the aqueous and petroleum ether extracts. Acetone extract showed moderate level of inhibition towards *S. aureus* and a lower level of inhibition towards *E. coli*. The methanol extract was effective against *P. aeruginosa* and *S. marcescens* and *S. aureus* (MIC 50 mg ml^{-1}, MBC 25 mg ml^{-1}) (Thomas 2014). Moreover, silver nanoparticles (AgNPs) were elaborated with aqueous extracts (10 ppm) from *A. raddianum* fronds. The antibacterial activity of those particles (1 nmol ml^{-1}) was tested against *S. aureus*, *Proteus* spp., *K. pneumoniae*, and *E. coli* resulting in significant inhibition of both *P. aeruginosa* (inhibition zone 10 mm) and *K. pneumoniae* (inhibition zone 8 mm), a moderate activity against *S. aureus*, and no activity against *Proteus* spp. The potential of *A. raddianum* aqueous extracts for developing novel eco-friendly biocidal formulations was hypothesized (Dev et al. 2018).

Toxicity Toxicity analysis is not available (Rastogi et al. 2018).

8.3 *Aloysia citriodora* Palau (Verbenaceae)

Synonymy *Aloysia citriodora* Ortega ex Pers., hom. illeg.; *Aloysia sleumeri* Moldenke; *Aloysia triphylla* (L'Hér.) Britton, hom. illeg.; *Lippia citriodora* (Ortega ex Pers.) Kunth, comb. illeg.; *Lippia triphylla* (L'Hér.) Kuntze; *Verbena citriodora* Cav., hom. illeg.; *Verbena triphylla* L'Hér.; *Zappania citriodora* Lam., hom. illeg.; *Aloysia triphylla* (L'Hér.) Britton f. serrulata.

Vernacular name Chaquesayasaya, cedrón, verbena citronela, limón verbena, haurippampacay, verbena olorosa, hierba de la princesa, yerba Luisa, vervein odorante.

General information *A. citriodora* is a herbal species distributed from the South of the United States down to South America and cultivated worldwide. It grows in light and well-drained soils and in temperate-warm climate resisting frost quite well. Their leaves are very aromatic; hence, it is frequently used as a flavor essence and odor corrector in food industry and pharmacy, as well as in perfumery and cosmetics. It is also employed in the elaboration of beverages, sweets, cookies, desserts, and pastry. In Argentina *A. citriodora* grows in the provinces of CAT, JUJ, LRI, and SAL (Rzedowski and Calderón de Rzedowski 2002; Bandoni et al. 2003; Santos-Gomes et al. 2005).

Botanical characteristics *A. citriodora* is a woody, profusely branched shrubby plant, 3–3.5 m tall, with woody stem, and simple, thin, elongated, rough leaves with a very characteristic sweet and sour lemon smell. Leaves are opposite, subsessile to shortly petiolate, with an entire to serrated margin. The small flowers are white or pale violet in the outside and violet in the inside, usually arranged in terminal spikes. Mericarps were glabrous and rough on the outside (Muñoz López de Bustamante 1996; Rzedowski and Calderón de Rzedowski 2002; Gattuso et al. 2008; Múlgara de Romero et al. 2012; Calzada-Sánchez et al. 2014).

Ethnobotanical information There are reports about the traditional use of this species dating back to the seventeenth century; it was considered as a medicinal plant by the Inca culture. The infusion and decoction are traditionally used as digestive, eupeptic, antispasmodic, mild sedative, expectorant, carminative, and tonic. The infusion and decoction made with leaves are taken in cases of stomachache, as cardiotonic, febrifuge, diuretic, antispasmodic, in odontalgia, and against hemorrhoids. The aerial parts are used as antiemetic, for heart problems and backache. There are also references about its addition to the traditional beverage mate (Gupta 1995; Barboza et al. 2009; Elechosa et al. 2017; Bahramsoltani et al. 2018).

Chemical data The leaf essential oil has neral, geranial, nerol, geraniol, bicyclosesquiphellandrene, spathulenol, nerolidol, ß-bourbonene, trans-ß-caryophyllene, caryophyllene-2,6-β-oxide, limonene, geranyl acetate, neryl acetate, trans calamenene, β-citronellene, β-pinene, neryl acetate, α-thujone, cis-carveol, carvone, 1,8-cineole, caryophyllene oxide, and linalool. The leaves have vervain, geniposidic acid, geniposide, mussaenoside, luteolin 7-diglucuronide, salvigenin, eupatorin, eupafolin, verbascoside (or acteoside), hispidulin, 6-hydroxyluteolin, luteolin 7-O-β-glucoside, luteolin, cismaritin, diosmetin, cirsiliol, chrysoeriol, apigenin, pectolinarigenin, p-cymene, and ß-bourbonene. The flower oil has 1-octene, 1-octen-3-ol, p-cymene, (Z)-β-ocimene, and trans-carveol (Fig. 8.5) (Lamaison and Petitjean 1993; Carnat et al. 1999; Barboza et al. 2009; Ricco et al. 2011). The content and composition of the EOs differ according to the region and the phenological state of

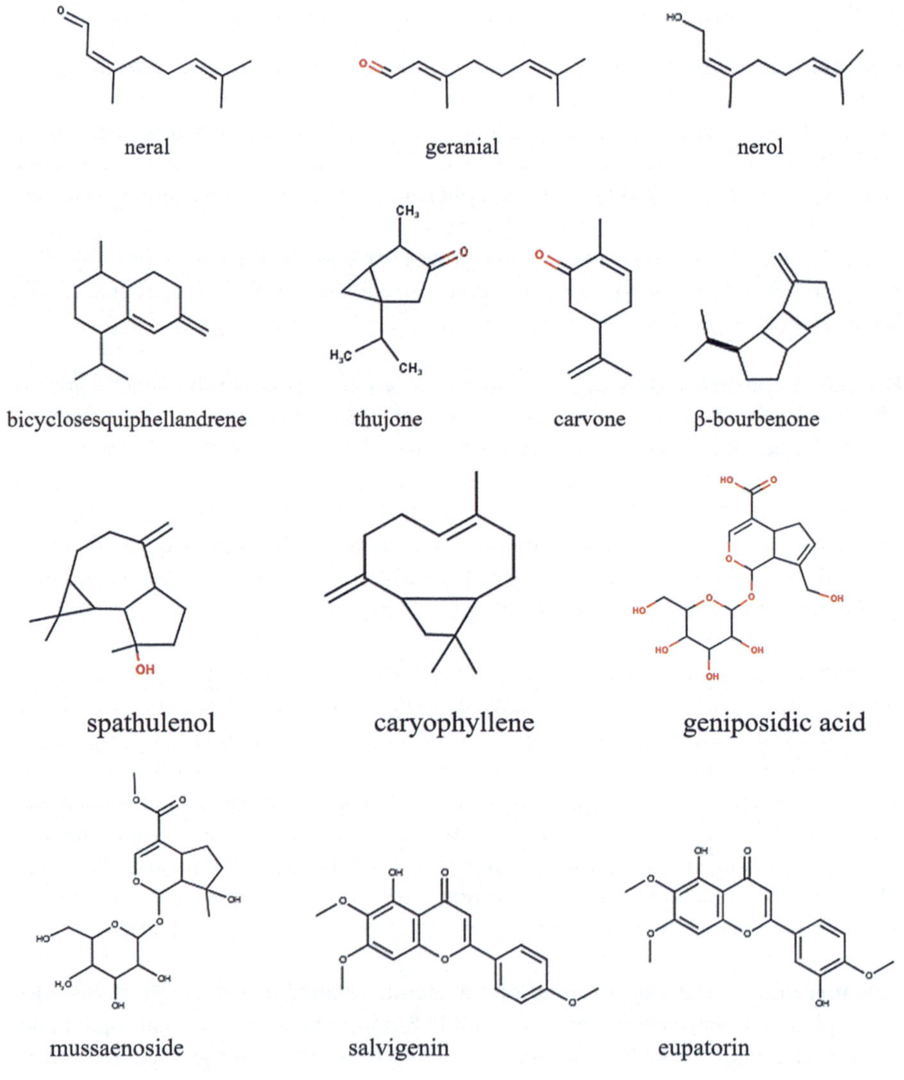

Fig. 8.5 Chemical compounds from *Aloysia citriodora* Palau (Verbenaceae)

the plant (Molina et al. 2003; Di Leo et al. 2007; Oliva et al. 2010; Ricco et al. 2011). According to Di Leo et al. (2007), in Argentina *A. citriodora* leaves contain more than 0.2% of oil (v/w of dried material). Four different chemotypes, thujones, citronellal, carvone, or citral, were identified in northwestern Argentina depending on the predominating component from the EO (Elechosa et al. 2017).

Biological activities The biological activities reported for *A. citriodora* aerial parts are giardicidal, analgesic, anti-inflammatory, antipyretic, sedative, and proconvulsant. The EO has antidiarrheal, nematicidal, protective, antimicrobial, antigenotoxic, and spasmolytic activities (Barboza et al. 2009).

The infusion made with crushed leaves had high phenylpropanoid, glucuronidated flavonoid, and iridoid glycoside content that correlates with its high antioxidant activity (Rodríguez Vaquero et al. 2010). One of the chemical compounds found in *A. citriodora*, verbascoside, was reported to have antioxidant, anti-inflammatory, and chemopreventive activities (Alipieva et al. 2014). There are numerous reports about the antimicrobial activity of *A. citriodora* EO and its main components (Bagher Hashemi et al. 2018; Pérez Zamora et al. 2018). The acetone extract from aerial parts has antimicrobial activity against *Bacillus subtilis*, *Staphylococcus aureus*, *Escherichia coli*, *Klebsiella pneumoniae*, and *Proteus vulgaris* (Kumar et al. 2008), and the ethanolic extract (20%) of leaves was active against *Listeria monocytogenes* ATCC 19117 (Bayoub et al. 2010).

The EO (900 mg ml^{-1}) was active against *S. aureus*, *S. epidermidis*, *B. cereus*, *E. coli*, *K. pneumoniae*, *Proteus mirabilis*, *Enterobacter aerogenes*, *K. ozaenae*, *Enterococcus* sp., *B. subtilis*, and *C. albicans* (Oliva et al. 2010; Oliva et al. 2015). Other assays have reported that the EO had inhibition values between 0.05 and >2 mg ml^{-1} against Gram-positive bacteria being not active against Gram-negative bacteria (Sartoratto et al. 2004). However, it showed a moderate activity on 12 of the *E. coli* serotypes tested with MIC values between 400 and 1000 μg ml^{-1} (Duarte et al. 2007).

On the other hand, *A. citriodora* inhibited *in vitro* replication of four dengue virus serotypes (DENV-1, DENV-2, DENV-3, and DENV-4 isolated from Colombian patients), attributed to the EO geranial and neral content (Ocazionez et al. 2010).

Treating *Trypanosoma cruzi* epimastigotes with *A. citriodora* EO at different concentrations (50, 100, 250, and 500 μg ml^{-1}) resulted in the inhibition of parasite growth (IC50 96.49 mg ml^{-1}) (Rojas et al. 2010). *In vivo* analysis of anti-*T. cruzi* activity in mice has shown that parasitemia, amastigote amount, inflammatory infiltration in the heart, and kinase-MB plasma levels were significantly reduced at the 28th postinfection day (Rojas et al. 2012). At an EO dose of 250 mg kg^{-1}, parasitemia diminished (85.4%) in 5-day-old male BALB/c mice, while at a dose of 300 mg kg^{-1} of citral, the parasitemia reduction is lower (43.7%). Additionally, the inflammatory infiltrates diminished 51.7% at 300 mg kg^{-1} citral (Rojas Armas et al. 2015).

Extracts and verbascoside were tested on an insulin-resistant hypertrophic 3T3-L1-adipocyte model to study their effects on triglyceride accumulation, inflammation, and oxidative stress, while a murine model of diet-induced obesity was studied to assess the *in vivo* metabolic response. It was found a dose-dependent decrease of triglyceride accumulation. The extract showed higher activity than the verbascoside, probably do to the synergistic action of the EO components (Herranz-López et al. 2015).

Verbascoside reduced the severity of intestinal macroscopic and microscopic lesions and decreased pro-inflammatory cytokine levels in various models of murine colitis (Hausmann et al. 2007; Mazzon et al. 2009), colonic injury, and inflammation in mice (Nishitani et al. 2013). The infusion offered beneficial effects against dextran sulfate sodium-induced colonic inflammation in rats, prevented epithelial destruction, glandular dilation and edema, and reduced the severity of colonic alterations (Lenoir et al. 2012). The bioavailability and intestinal absorption of polyphenols derived from lemon verbena infusion (1268 μM verbascoside, 270 μM isoverbascoside, 217 μM luteolin, 88 μM apigenin, 674 μM diosmetin) in healthy and colitic rats were not altered by colonic inflammation; on the contrary, urinary excretion of hydroxycinnamic acids derived from verbascoside was affected (Felgines et al. 2014).

The spasmolytic action of the decoction (15% w/w) was tested on isolated rat duodenum. The aqueous extract relaxed the intestinal smooth muscle and non-competitively inhibited the dose-response acetylcholine curve. Vitexin, but not isovitexin, was reported to contribute to the effect (Ragone et al. 2007). The hexane extract and citral were *in vitro* tested for their effect on uterine contractions, *in vivo* anti-inflammatory properties, and gastric wound healing capabilities (3, 10, 30, and 56 μg ml^{-1}). Both inhibited the contractions induced by prostaglandine-PGF2α and had anti-inflammatory activity without any injury even at the highest dose (100–800 mg kg^{-1}). These results supported the popular use of *A. citriodora* infusion to treat menstrual pain and primary dysmenorrhea (Ponce-Monter et al. 2010).

Preliminary analysis of the hydroalcoholic extract (600 mg kg^{-1}) effect on the tail suspension test in Swiss albino mice showed an antidepressive effect (Barrios et al. 2016).

The anesthetic effect in silver catfish and the recovery times from anesthesia of subadult and postlarvae white shrimp were also tested. The EO was effective in both cases, inducing anesthesia at the concentrations of 150 and 300 μl l^{-1} being the shortest recovery time from anesthesia (10 min) in subadult shrimp reached at 300 μl l^{-1}. In postlarvae shrimp, the shortest induction and recovery times (less than 10 min) were both seen at 300 μl l^{-1}. The EO induced deep anesthesia without mortality (Parodi et al. 2013).

The chemoprotective effect of leaf infusions (5%) against genetic damage induced by acrylamide (50 mg kg^{-1}) was tested by the alkaline version of the comet assay technique (Singh et al. 1988) in BALB/c mice. The administration of the infusion before acrylamide administration reduced the capacity of acrylamide to induce genetic damage (Zamorano-Ponce et al. 2006). A similar model was used to analyze the protection of the *A. citriodora* infusion (5%) from the genetic damage induced by cisplatin (6 mg kg^{-1}) in mouse bone marrow cells. A significant prevention of cisplatin-induced genetic damage was observed in animals previously treated with the infusion. The authors attributed those results to the high-superoxide and hydroxyl radical scavenging activity related to the content of flavonoid and polyphenols (Valentao et al. 2002; Zamorano-Ponce et al. 2004).

A repellence test against *Aedes aegypti* was made by an arm-in-cage assay with the EO (90%). Differences were seen depending on the geographic origin of *A. citriodora*, and the EO from plants from *Traslasierra* showed more repellence with a significantly longer repellency time (90 min) than those from Córdoba, probably due to the highest limonene content in the former one (Gillij et al. 2008). Also, the acute larvicidal toxicity of the EO against *Culex quinquefasciatus* Say (Diptera: Culicidae) was reported. *C. quinquefasciatus* is a vector of lymphatic filariasis and of diseases, such as West Nile and St. Louis encephalitis. Early 3rd larvae of mosquito were treated with EO (5–100 μl l^{-1}) being LC50 65.6 μl l^{-1}after 24 h. A combination of *A. citriodora* and *Satureja montana* EOs (1:1) was highly effective (LC50 18.3 μl l^{-1}, LD90 28.7 μl l^{-1}) against *C. quinquefasciatus* larvae (Benelli et al. 2017). Activity from pulegone and citral against head lice (*Pediculus humanus capitis*) was also demonstrated. Knockdown and mortality activities were 47–53% and 42–55%, respectively (Gonzalez-Audino et al. 2011).

Toxicity Cytotoxicity on RAW 264.7 ECACC (European collection of cell cultures) mammal cells at a concentration of 4 × 105 cells ml^{-1} was evaluated with 20, 50, and 100 μg ml^{-1} EO. The CI50 was superior to 100 μg ml^{-1}; therefore, the EO was not considered as toxic (Rojas et al. 2010). Oral acute toxicity was tested in albino female mice with a single dose of *A. citriodora* EO (50, 300 or 2000 mg kg^{-1}). No signs of toxicity or death were observed (Rojas Armas et al. 2015).

In vitro culture Explants (pieces of leaves and nodal segments) were cultivated in MS, ½ MS, and ¼ MS culture medium (Murashige and Skoog 1962) with sucrose 3%, 0.7% agar, and different concentrations of BAP and NAA. Calli appeared in all media and plant growth regulators relationship after 14 days of culture. Rhizogenesis appeared after 60 days in MS medium with 2 mg l^{-1} BAP and 1 mg l^{-1} NAA. Stems were obtained from nodal segments in ½ MS medium without plant growth regulators (Severin et al. 2005).

Legal status *A. citriodora* figures in the FNA VI ed. (1978) and in the Argentina Food Code (1969). It is included in the pharmacopoeias of France, Spain, Mexico, and Europe (Bandoni 2000). It figures in the GRAS (generally regarded as safe) list, and it was considered as a dietary supplement by the US Food and Drug Administration (FDA) (Barboza et al. 2001).

8.4 *Bauhinia forficata* subsp. pruinosa (Vogel) Fortunato & Wunderlin (Fabaceae)

Synonymy *Bauhinia pruinosa* Vogel, *Bauhinia candicans* Benth., *Bauhinia forficata* Link var. pruinosa, *Bauhinia forficata* Link var. candicans, *Pauletia candicans* (Benth.) A. Schmitz, *Pauletia pruinosa* (Vogel) A. Schmitz, *Bauhinia forficata* auct. non Link, *Bauhinia forficata* auct. non Hook. & Arn.

Vernacular name Pezuña de vaca, pata de vaca, caobá, mahogani, pata de buey, pata de chivo, pata de toro, falsa caoba, caoba falsa, caoba del país, palo buey, torop, mororó do sertão (Portuguese).

General information The natural habitat of this plant is South of Brazil, Uruguay, and North of Argentina, in soils rich in organic matter, being originally from Asia. *B. forficata* is a quite-resistant tree, highly tolerant to drought. Optimal growth is attained in sunny temperate climate. It is cultivated as ornamental. In Argentina it grows in the provinces of CAT, JUJ, LRI, MEN, SAL, SJU, and TUC. Blooming is in summer.

Botanical characteristics *B. forficate* is an annual tree, 4–8 m high, and trunk has a fissured brown-grayish bark. The crown is globular and irregular, with hairy and thorny pubescent branches. Leaves are alternate, light green, and divided in two oblong lobes resembling the hoof of a cow. The white flowers are hermaphrodite, solitary, or arranged in axillary clusters (Fig. 8.6). The fruit is a legume with several flattened, oval, blackish seeds (Juárez de Varela 2001; Garibotti Lusa and Bona 2009).

Ethnobotanical information In Argentinean folk medicine, *B. forficata* has several uses. The leaf is considered as antidiabetic, astringent, diuretic, expectorant, antinephritic, antitussive, antihemorroidal, abluent, and blood depurative. The flowers are considered as antihemorroidal and digestive, and they are externally used against dandruff (Barboza et al. 2009).

Chemical data The dried leaf have 3,7-di-O-a-rhamnopyranosylquercetin, 3,7-di-O-a-rhamnopyranosyl kaempferol (kaempferitrin), astragalin, bauhinoside, guanidine, quercitrin, rhamnose, rutin, saponins, and trigonelline. Leaves and flowers have sitosterol, campesterol, stigmasterol, cholesterol, stigmasta-3,5-dien-7-one, choline, trigonelline, sitosterol 3-O-β-glucoside, sitosterol 3-O-β-D–xylopyranoside, and

Fig. 8.6 *Bauhinia forficata* subsp. pruinosa (Vogel) Fortunato & Wunderlin (Fabaceae). Flower (**a**) personal collection M C Fernández and leaves (**b**) MA Alvarez personal collection

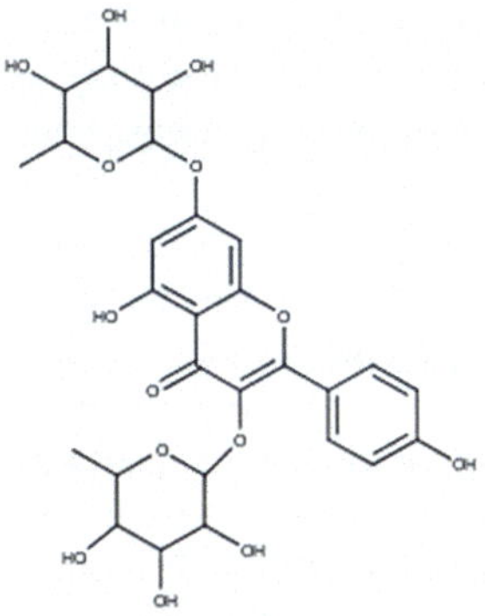

kaempferitrin

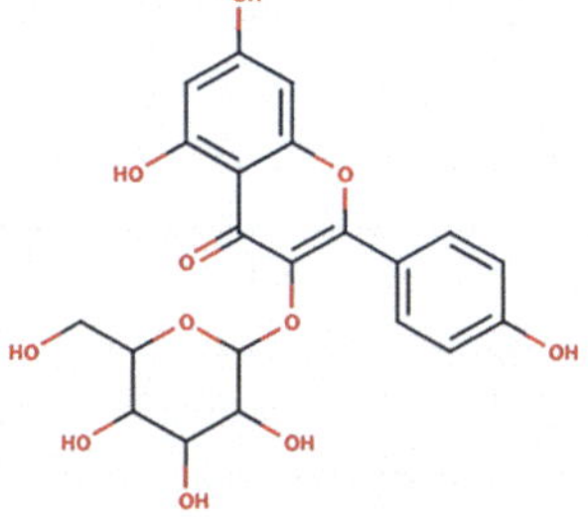

astragalin

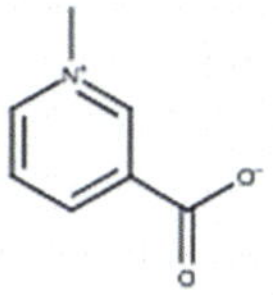

trigonelline

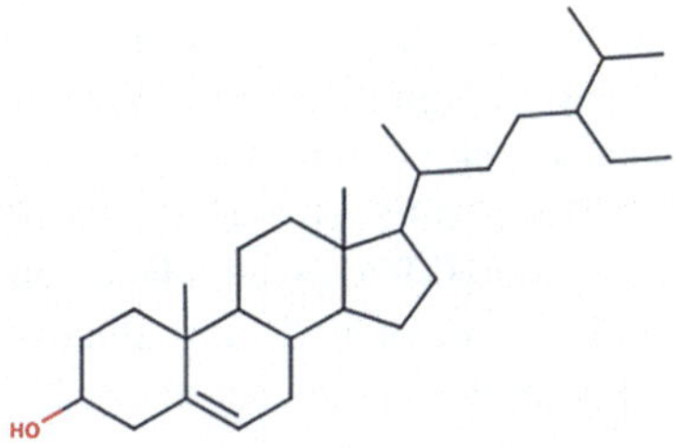

sitosterol

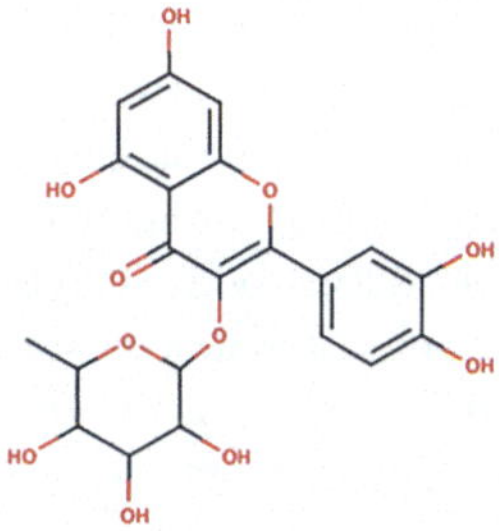

quercitrin

Fig. 8.7 Chemical compounds of *Bauhinia forficata* subsp. pruinosa (Vogel) Fortunato & Wunderlin (Fabaceae)

sitosterol 3-Oα-D-xyluronofuranoside (Fig. 8.7) (Barboza et al. 2009; Castro dos Anjos et al. 2013; Toloza-Zambrano et al. 2015).

Biological activities The extracts from dried leaves have been used as hypoglycemic, antioxidant, hypocholesterolemic, diuretic, antimicrobial, and antinematode (Sertié et al. 1990; Silva et al. 2002, Menezes et al. 2004; Arigony 2005). Also, it is referred a cardiovascular (tachycardia) and calmative effect. The antibacterial activity of stems and fruits containing seeds was reported. The aqueous, ethanolic, and butanolic leaf extracts and kaempferitrin were effective against the oxidative damage resulting from hyperglycemia (Jorge et al. 2004; Sousa et al. 2004) supporting the main use of *B. forficata* tea in folk medicine to treat diabetes *mellitus*. The mechanism of action was hypothesized to be related to the inhibition of insulin catabolism, reabsorption of glucose in the kidney, and/or stimulating glucose uptake by peripheral tissues and not by stimulating insulin secretion, nor by the inhibition of glucose absorption at intestinal level. Kaempferol did not promote glycemia reduction; hence, the authors concluded that the aglycone-bound sugars are essential for the antidiabetic activity (Cazarolli et al. 2008).

More recently, Ferreres et al. (2012) demonstrated that the glycemia reduction promoted by *B. forficata* resulted from the inhibition of α-glycosidase, the enzyme responsible for catalyzing the final process in the digestion of carbohydrates. Such action was ascribed to the presence of the quercetin and kaempferol derivatives of the species (Tadera et al. 2006; Silvestre Marques et al. 2013). The hypoglycemic activity was tested *in vitro* and *in vivo*. The aqueous extract (80 g l^{-1}) was administrated to normotensive (5, 10, 20, 40 mg kg^{-1}; i.v.) and hypertensive-induced (200 and 400 mg kg^{-1}) male Wistar rats (Goldblatt et al. 1934). The oral acute treatment (400 mg kg^{-1}) induced blood pressure reduction (19%) without modifying heart rate (Castro dos Anjos et al. 2013). Studies made in euglycemic mice treated with the aqueous extract (7 mg kg^{-1}) were also successful (Martínez et al. 2009). The effect of the infusion on oxidative stress and liver damage in streptozotocin (STZ)-induced diabetic mice (313 mg kg^{-1}) was tested by the determination of the activities of ũFF;-aminolevulinate dehydratase (ũFF;-ALA-D), superoxide dismutase (SOD), and catalase (CAT) in the liver and NF-E2-related nuclear factor erythroid-2 (Nrf2), NADPH quinone oxidoreductase 1 (NQO-1), and heat shock protein 70 (HSP70) in the liver and in the pancreas. These enzymes related to high levels of oxygen reactive species (ROS) are usually increased in diabetes. The treatment did not change the increased liver/body weight ratio and AST, ALT, and ũFF;-ALA-D enzyme level characteristics of diabetic rats. Also, Nrf2, NQO-1, and HSP70 protein levels in the liver were not altered. On the contrary, the treatment attenuated the reduction of CAT and the increase of pancreatic NQO-1 (Fernandes Salgueiro et al. 2016). Similar results were obtained in alloxan-diabetic rats with the aqueous, ethanol, or hexane extracts (200, 400 mg kg^{-1}) (de Sousa Lino et al. 2004).

Diabetic patients have erythrocytes with increased lipid peroxidation. It was studied the effect of *B. forficata* leaf tea on oxidative damage and antioxidant markers in an *in vitro* experimental model of hyperglycemia in human erythrocytes. A significant DPPH scavenging activity, the reduction of the deoxyribose degradation induced by the Fenton reaction, an increase of thiobarbituric acid reactive substances (that reflect the lipid peroxidation), and a decrease in nonprotein SH levels in human erythrocytes incubated with high-glucose concentrations were found. Those results were attributed to the high level of phenolics and flavonoids acting as DPPH+ radical scavenger and as iron chelators (Salgueiro et al. 2013).

The effects of oral treatment with a decoction of *B. forficata* on characteristic metabolic parameters of streptozotocin (STZ)-diabetic and nondiabetic rats was also reported (Pepato et al. 2002). After 31 days of chronic treatment with *B. forficata* (mean daily dose 35.29/7.8 ml^{-1}/100 g^{-1} body weight), the diabetic group showed a significant reduction in plasma glucose, urinary glucose, and urinary urea levels without any alterations in body weight, water or food intake, urinary volume, serum lipid, or hepatic glycogen either in nondiabetic or STZ-diabetic rats. Those results suggested that the antidiabetic activity is the result of inhibition of gluconeogenesis and a reduction of counterregulatory hormones similar to biguanides. It was also seen a sharp reduction in epididymal and retroperitoneal adipose tissue in the STZ-diabetic rats as compared to the nondiabetic rats (Pepato et al. 2002). The ethanolic extract (40 mg l^{-1}) interfered with hepatic glycolysis, glycogenesis, and lactate production from L-glutamine. *In vivo*, a dose of 300 mg $l^{-1}$of ethanolic extract produced a minimal increase in the liver glycogen content in STZ-diabetic rats, but it did not restore them to control levels. The treatment did not change glycemia (Olivera Gomes et al. 2016). A diuretic action was also observed, and no signs of toxicity were found (Toloza-Zambrano et al. 2015).

The continuous intake of a decoction (50 g l^{-1}) by male Wistar rats (150 ml decoction per 100 g feed) resulted in a modification of the exploratory condition and a decrease in locomotor activity without changes in the glycemic index (Gellert Moranza et al. 2017).

Antioxidant, anticoagulant, and antifibrinogenolytic activities against *Bothrops jararacussu* Lacerda, 1884 (Ophidia, Viperidae) venom and its isolated thrombin-like serine protease enzyme were induced by the aqueous extract from *B. forficata* (Damasceno et al. 2004; Sousa et al. 2004; Oliveira et al. 2005). The aqueous extract neutralized *Bothrops jararacussu* and *Crotalus durissus terrificus* snake venom-induced coagulation, possibly by inhibition of serine protease enzymes involved in blood coagulation disorders. There is also a significant reduction in the *C. durissus terrificus* venom-induced edema, and *Tityus serrulatus* scorpion venom electrolyte level changes; however, it was not able to counteract the glycogenolysis, plasma catecholamines increase, and serum insulin change produced by the venoms (Vasconcelos et al. 2004).

The proteinase activity of the acetone precipitate of *B. forficate* leaves corresponded to the protein baupain (Andrade et al. 2011).

The antiulcer properties of an infusion of fresh leaves (150 g l^{-1}) was demonstrated in three different models of gastric ulcer: 0.3 M HCl: 60% EtOH (Mizui and Doteuchi 1983), indomethacin/bethanechol (30 mg kg^{-1}, 5 mg kg^{-1}) (Rainsford 1978), and absolute ethanol induction (Morimoto et al. 1991). Moreover, the previous administration of the aqueous extract (100 mg kg^{-1}) resulted in an increase of gastric mucosa, possibly associated with the extract flavonol content (Costa Mazzeo et al. 2015).

The ethanolic extract (250 mg ml^{-1}) had antimicrobial activity against *Streptococcus mutants* (ATCC 9927) and *S. sanguis* (ATCC 10557) from oral cavity (Batista de Souza et al. 2014). The hexane, di-chloromethane, and ethyl acetate extracts from leaves, stems, and barks evaluated for their activity against fungi and pathogenic yeasts were not effective against *Candida albicans*, *C. tropicalis*, *Saccharomyces cerevisiae*, *Cryptococcus neoformans*, *Aspergillus flavus*, *A. fumigatus*, *A. niger*. In contrast, the ethyl acetate extract of the stems was moderately active against *Epidermophyton floccosum* (MIC < 750 mg ml^{-1}); the hexane extract from the bark was active against *Microsporum canis*, *Trichophyton mentagrophytes*, *T. rubrum*, and *Epidermophyton floccosum* (MIC < 1000 mg ml^{-1}); and the dichloromethane extract from bark was active against *T. rubrum* and *E. floccosum* (MIC 500 μg ml^{-1}) (Silva & Cechinel-Filho 2002).

HeLa cell cycle progression was inhibited by a compound isolated from *B. forficate* (HY52 = $C_{17}H_{30}O_2N_2$) (Lima et al. 2006).

Seeds contain a lectin (BfL), a glycoprotein with 6.24% carbohydrate content, that has anticoagulant and antiplatelet aggregation properties (Silva et al. 2012).

Toxicity *B. forficata* aqueous extract (4.65 g l^{-1}) was tested against cyclophosphamide, a mutagenic chemotherapeutic agent, using the model of Wistar rat bone marrow cells from *Rattus norvegicus*. It was not detected chromosomal alterations or modification in the mitotic index concluding that the aqueous extract of *B. forficata* is not cytotoxic but showed an effective antimutagenic activity endorsed to its total phenolic and flavonoid content (Düsman et al. 2013). On the other hand, the oral administration of a decoction to rats did not show any modification of the toxicity markers amylase (pancreas toxicity), creatine kinase (muscle toxicity), lactate dehydrogenase (muscle and liver toxicity), bilirubin (liver and biliary toxicity), and angiotensin-converting enzyme (renal microcirculation and kidney toxicity) (Pepato et al. 2004).

Dry extracts obtained using colloidal silicon dioxide at 125 and 250 mg ml^{-1} showed hepatotoxicity signs (AST and alkaline phosphatase ALP increase) in hyperglycemic rats after a 35-day oral treatment. A formation of toxic derivatives resulting from the interaction between colloidal silicon dioxide and compounds from the vegetable extract or modifications of the colloidal silicon dioxide caused by the high temperature of the drying process were presumed responsible of that effect (Pepato et al. 2010). No significant quantities of the toxic metals Pb, Cr, and Ni were detected in *B. forficata* coming from different producers (Franco et al. 2011).

***In vitro* cultures** Axenic plants were obtained from disinfected seeds in MS (Murashige and Skoog 1962) culture medium with 30 g l^{-1} sucrose, 2.3 g l^{-1} agar, and 4.0 mg l^{-1} BAP. Calli were initiated from hypocotyl explants on ½ MS medium, containing 30 g l^{-1} sucrose, 2.3 g l^{-1} agar and supplemented with 3.0 mg l^{-1} BAP concentration. Shoot buds appeared in media with 2.0 and 4.0 mg l^{-1} BAP after 60 days in culture (Mello et al. 2000). Cell suspension cultures were established in MS liquid medium with 30 g l^{-1} sucrose and 17.7 μM BAP (Mello et al. 2001).

8.5 *Caesalpinia gilliesii* var. gilliesii (Wall. ex Hook.) D. Dietr. (Fabaceae)

Synonymy *Erythrostemon gilliesii* Hook Klotzsch var. gilliesii, *Poinciana gilliesii* Wall. ex Hook, *Caesalpinia macrantha.*

Vernacular name Lagaña de perro, algarrobillo, piscala, barbón, barba de viejo, barba de chivo, mal de ojos, flor del indio, chosme, espiga de amor, mal de ojos, yellow bird of paradise, bird of paradise bush (Eng.), desert bird of paradise (Eng.), uccello del paradiso (Italian).

General information *C. gilliesii* is a native species to Argentina that is worldwide cultivated (Osman et al. 2008). It grows sub-spontaneously in the Mesopotamian region, Uruguay, Chile, and Southwest of North America. It is endemic to the Chaco and Monte's phytogeographic formations, and its center of dispersion seems to be the province of COR (Burkart 1936). It grows in dry, stony soils, plains, or hillsides, between 50 and 2500 m.a.s.l. in the provinces of BAI, CAT, CHA, COR, COS, ERI, LPA, LRI, MEN, NEU, RNE, SAL, SDE, SFE, SJU, SLU, and TUC (Cocucci et al. 1992; Barboza et al. 2009). The flowering period lasts 5 months, from September to January, and it is variable among populations and individuals. Pods and seeds are poisonous. Like many members of the legume family, *C. gilliesii* can fix its own nitrogen from the soil because of specialized bacteria living in root nodules.

Botanical characteristics *C. gilliesii* is a shrub with perfect and staminate flowers, arranged in variable numbers in simple clusters. Leaves are bi-pinnately compound with small bluish-green leaflets that give the foliage a fernlike look. Branches are pubescent and glandular, with reddish hairs mixed with single hairs (Macaya 1999). Flowers are cyclic, pentameric, zygomorphic, and hermaphrodite with imbricated pre-flowering. They are grouped into glomeruli; calyx have five hairy glandular sepals; the hairless, yellow corolla has five petals, 2–3 cm long; the androecium commonly has 10 free red stamens with filaments 2–3 times longer than the petals. Anthers have two thecae, and pollen is usually tricolporate; the unicarpelar, superimposed, linear, unilocular gynoecium has filiform ovules in a row in the placenta (Fig. 8.8). The fruit is an oblong and flat dehiscent pod with membranous consistency, 6–9 cm length, with transverse grooves between the seeds; in maturity the

Fig. 8.8 *Caesalpinia gilliesii* var. gilliesii (Wall. ex Hook.) D. Dietr. (Fabaceae). Buenos Aires Botanical Garden. MA Alvarez personal collection

fruit separated into two twisted valves. Ovoid chestnut-olive seeds measure 10 mm in length (Steibel and Troiani 2000). Nectar is found on the inner surface of the hypanto, without a containing structure, always accompanying the vascular bundles of the stamens, being secreted mostly at the beginning of the anthesis (Jausoro and Galetto 2000; Jausoro and Galetto 2001). Pollination is zoophilic (entomophile or ornithophile) (Viglierchio et al. 2000).

Ethnobotanical information The dried flower and leaf infusion are febrifuge, purgative, and against cavities and toothache. Washings made with flowers are used for eye conditions and decoction of the aerial part to treat earaches (Agüero 2017). The dried seed is taken as anthelminthic and insecticidal (Viglierchio et al. 2000).

Chemical data The dried flowers contain phenyl alanine and arabinose. The flowers contain β-sitosterol-3-O-butyl, sterol β-sitosteryl-3-O-glucopyranoside (daucosterol), isorhamnetin and isorhamnetin-3-O-rhamnoside, 12,13,16-trihydroxy-14(Z)-octadecenoic acid, luteolin-7,4'-dimethyl ether, genistein-5-methyl ether, luteolin-7-O-rhamnoside, isovanillic acid, and *p*-methoxy benzoic acid (Fig. 8.9). The seeds contain cesalin, glutamic acid, pipecolic acid, and 3-hydroxyproline (Fig. 8.9). The leaves contain gossypetin-3-O-β-D-robinobioside, quercetin-3-O-β-D-glucoside, quercetin-3-O-β-D-galactoside, rutin, kaempferol-3-O-β-D-rutinoside, and luteolin-7-O- β-D-glucoside (Suarez et al. 1984; Viglierchio et al. 2000; Osman et al. 2008, 2013; Barboza et al. 2009; Emmam et al. 2019)

Biological activities There are reports about antitumor activity, respiratory stimulant effect, and smooth muscle relaxant activity (Barboza et al. 2009).

As for the antioxidant activity, scavenging activity was seen in extracts from leaves (IC50 95.3), flowers (IC50 38.7), and pods (IC50 44.7 μg ml^{-1}). The dichloromethane extract of *C. gilliesii* flowers showed antioxidant activity (IC50 = 45.5 μg ml^{-1}) comparable to standard rutin (24 μg ml^{-1}). Flower ethyl acetate extract had the highest radical scavenging activity (IC50 = 7.43 μg ml^{-1}), even higher than rutin (Osman et al. 2013). Isolated isorhamnetin and isorhamnetin-3-O-rhamnoside possess moderate antioxidant activity with IC50 values of 370 and 372.5 μg ml^{-1}, respectively. On the other hand, β-sitosterol-3-O-butyl and daucosterol possess variable prooxidant activity against DPPH radicals. Such an antioxidant activity was attributed to the presence of phenolic compounds as tannins and flavonoids (Ulubelen et al. 1967; Osman et al. 2008, 2016).

The methanol extract counteracted the liver injury induced by carbon tetrachloride (CCl_4) in mature male albino Wistar rats (reduction of ALT and AST and increase of TAC) protecting the liver architecture at a dose of 300 mg kg^{-1}. The oral administration of the leaf methanol extract significantly diminished carrageenan-induced paw edema in rats (100 and 200 mg kg^{-1}), exhibited substantial gastroprotective properties, and diminished ulcer number and ulcer severity in a dose-dependent way. Those activities were attributed to the presence of antioxidant polyphenolics and phytosterols (Osman et al. 2016; Emmam et al. 2019).

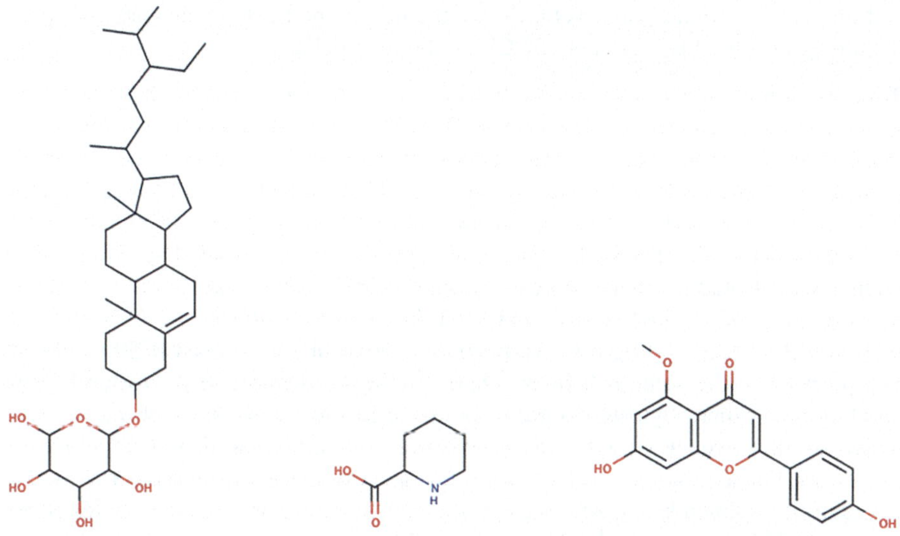

daucosterol pipecolic acid 2-genistein-5-methyl ether

Fig. 8.9 Some chemical compounds from *Caesalpinia gilliesii* var. gilliesii (Wall. ex Hook.) D. Dietr. (Fabaceae)

It was reported the green synthesis of metallic silver nanoparticles (AgNPs) containing *C. gilliesii* leaf extracts. The AgNPs were synthesized by the reduction of 10 ml of $AgNO_3$ solution (1 mM) with different concentrations of *C. gilliesii* hydroalcoholic extract (100 μl to 300 μl, stock solution 0.04 g extract 10 ml^{-1} solvent). The antimicrobial activities of *C. gilliesii* extract and AgNPs were tested against Gram-positive and Gram-negative bacterial pathogens and yeast. Synthesized *C. gilliesii* extract-AgNPs showed highest activity against all tested pathogens than the hydroalcoholic extract (Scott 1989, Emam et al. 2017). The cytotoxicity of $AgNO_3$, *C. gilliesii* hydroalcoholic extract, and AgNPs was evaluated against both normal skin fibroblast (BJ-1) and breast cancer cell line (MCF-7). Cytotoxicity against MCF-7 was 4.2% for the hydroalcoholic extract and 96.5% for the combination of extract with AgNPs. They also produced cytotoxicity against BJ-1 (33.3% and 70.5%) at 100 μg ml^{-1}, respectively. $AgNO_3$ showed high cytotoxicity ≥90% at 12.5 μg ml^{-1} (Emmam et al. 2017).

The aqueous extract from leaves, flowers, and pods (60 g 500 ml^{-1}) was fractionated using solvent-solvent extraction with n-hexane, methylene chloride, ethyl acetate, and n-butanol and tested on Wistar male albino rats (1, 2, 3, 4, and 5 g kg^{-1}) to study the protective effect on CCl_4-induced hepatotoxicity (Osman et al. 2013). Aspartate aminotransferase (AST), alanine aminotransferase (ALT), and alkaline phosphatase (ALP) blood levels showed a significant protection of the hepatic cells. Rats pretreated with leaves or flower extracts of *C. gilliesii* showed normal hepatocytes with minimum fatty changes, portal vain congestion, and mild inflammatory cell infiltration around portal vein. Pretreatment with pod extract showed mild protection against the toxic effects of CCl_4 on the liver cells.

Toxicity Total extract of the leaves, flowers, and pods was tested for their cytotoxic against different cancer cell lines by SRB (sulforhodamine B) cell viability assay (Skehan and Storeng). The flower extract exhibited a potent inhibition of cell growth in different cancer cell lines (MCF7, HepG2, HCT116, HeLa, and PC3 cells) with IC50 values of 10 and 15.6 μg ml^{-1}, respectively. Also the extracts from leaves and pods were cytotoxic. The dichloromethane extract and the isolated steroidal compounds were tested for their cytotoxicity against MCF7 and HepG2 cells. The most active compound against HepG2 and MCF7 cells was β-sitosterol-3-O-butyl, and IC50 was 13.1 and 14.4 μg ml^{-1}, respectively. Doxorubicin showed IC50 values of 3–4 μg ml^{-1} in the same cell lines. These findings indicate that β-sitosterol-3-O-butyl and the other phytosterols are responsible for the cytotoxicity of the extracts which would correlate with the known abortifacient, antimalarial, and anthelmintic properties. Galactomannan isolated from the aqueous extract from seeds revealed a concentration-dependent potency to induce cell death of HepG2 cells with IC50 value of 0.375 μg ml^{-1} (Abdel-Megeed et al. 2018).

Cesalin, an antitumor and antineoplastic protein isolated from *C. gilliesii*, inhibited carcinosarcoma growth by binding to its plasma membrane (Elting and Montgomery 1979).

8.6 *Pilocarpus pennatifolius* Lemaire (Rutaceae)

Synonymy *Pilocarpus selloanus* Engl., *Pilocarpus pennatifolius* Lem. var. selloanus, *Pilocarpus selloanus* Engl. var. gracilis, *Pilocarpus selloanus* Engl. f. brevipedicellata, *Pilocarpus pennatifolius* Lem. var. genuinus, *Pilocarpus pennatifolius* Lem. f. typicus, *Pilocarpus pennatifolius* Lem. f. gracilis, *Pilocarpus pennatifolius* Lem. f. latifoliolatus, *Pilocarpus pennatifolius* Lem. f. brasiliensis, *Pilocarpus pennatifolius* Lem. f. intermedius, *Pilocarpus pennatifolius* Lem. f. paraguariensis.

Vernacular name jaborandí, jaborandí del Paraguay, cutia, yaguarundí, camba, canilla, yvyrá tai.

General information *P. pennatifolius* grows in the provinces of CHA, COS, FOR, and MIS.

Botanical characteristics *P. pennatifolius* is an evergreen shrub about 6 m high with alternate, imparipinnate leaves with ribbed petiole and rachis, 2–3 leaflet pairs, narrowly elliptic or oblong, obtuse and emarginated at apex, a revolute margin. Flowers are 5-merous, 9–11 cm diameter, numerous, disposed in long axillary and terminal racemes, subsessile, pendulous, purple, and with a very small teethed calyx, and the petals are deciduous and reflexed when flowering. Stamens are purple, and ovary is superior, with carpels inserted on gynophores. The fruit is divided into mericarps united at the base, truncate, and mucronate at the apex (Miranda et al. 2000; Spegazzini et al. 2002).

Ethnobotanical information The *Mbyá* native community referred the use of the infusion made with the whole plant to treat diarrhea and parasitosis. The *guaraníes* used it against vomiting and intestinal parasitosis and as diaphoretic. The infusion of leaves, dried leaves, and extracts is used in popular medicine for the treatment of asthma, diabetes, liver diseases, and in the form of cosmetic for hair treatment. The dried leaves possess gastrostimulant, peristaltic, and secretagogue activities, suitable for the treatment of urinary, gastrointestinal, and pulmonary disorders. A not specified part is referred with activity as odontalgic, abortive, and to treat ear disorders. Also it is used as repellent, acaricide, and insecticide (Schultz 1997; Martínez Crovetto 1981; Rodríguez Barbosa 1985; Müller 1997; Pochettino et al. 2003; Barboza et al. 2009; Ferreres et al. 2015)

Chemical data The stem contains pilocarpine and isopilocarpine. The EOs from stems contain 2-tridecanone, 2-pentadecanone, caryophyllene oxide, germacrene D, E-β-ocimene, Z- β-ocimene, δ-cadinene, β-cubebene, γ-muurolene, selinene, β-elemene, bicyclogermacrene, α- humulene, α-copaene, β-bourbonene, β-caryophyllene, and spathulenol, among others (Fig. 8.10) (Lucio et al. 2002; Santos and Hrihorowitsch Moreno 2004; Barboza et al. 2009)

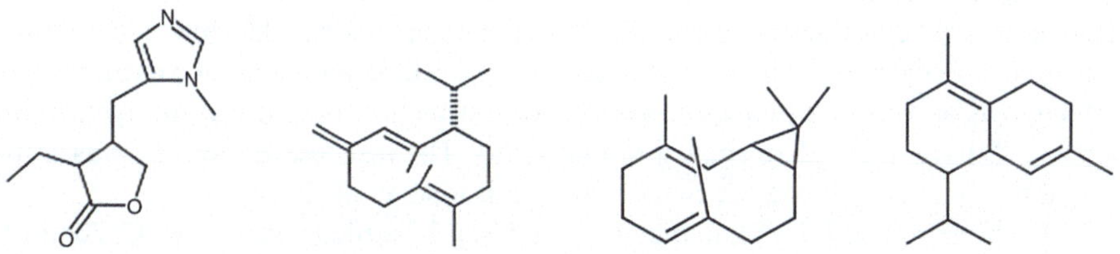

pilocarpine germacrene D bicyclogermacrene δ- cadinene

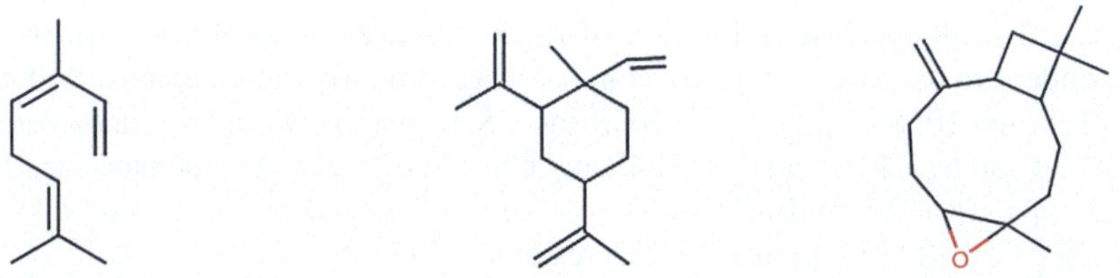

E-β-ocimene β-elemene caryophyllene oxide

Fig. 8.10 Chemical compounds from *Pilocarpus pennatifolius* Lemaire (Rutaceae)

Biological activities It is well known the activity of pilocarpine, an imidazole alkaloid extracted from the leaves, as a direct cholinergic, stimulating the parasympathetic system. It is used for the treatment of glaucoma and also for the stimulation of sweat and lachrymal glands (Goodman and Gilman 2001; Valdez et al. 1993).

As for the study of its antimicrobial activity, leaves (2 kg) and stem bark (98 g) (both dried and crushed) were extracted with methanol. Subsequently, the aqueous residue was thoroughly extracted with ethyl ether rendering acidic and basic ether fractions. The acid stem bark solution was successively extracted with ethyl ether and ethyl acetate. Antibacterial and antifungal assays were performed with the crude leaf methanol extract (CLME), crude bark methanolic extract (CBME), leaf acidic ether fraction (LAEF), and leaf basic acetate fraction (LBAF) and with the isolated compounds.

The fractions showed activity against the bacteria tested, the most sensitive was *Shigella flexneri* (MIC 7.8 μg ml^{-1}), being the CBME and LAEF fractions actives also against *Candida krusei* (MIC 15.6 μg ml^{-1} and 31.2 μg ml^{-1}, respectively). Some of the isolated compounds were active against *Shigella flexneri* (MIC 3.12 μg ml^{-1}), *Salmonella typhimurium* (MIC 12.2 μg ml^{-1}), *Enterococcus* spp. (MIC 25 μg ml^{-1}, 50 μg μg ml^{-1}), *E. faecalis* (MIC 1.56 μg ml^{-1}), and *Salmonella enteritidis* (MIC 1.56 μg ml^{-1}) (do Carmo et al. 2018).

The leaves extract obtained by exhaustive maceration (200 mg ml^{-1}) and its six sequenced serial 1:2 dilutions (starting at 20,000 ppm) were tested for their *in vitro* activity against the ectoparasites *Ctenocephalides felis felis*, *Rhipicephalus sanguineus*, and *R. microplus*. No insecticidal activity was observed after 24 h of treatment. However, insecticidal activity was observed from 1250 ppm after 48 hours. A drop in tick's weight was observed with concentrations between 625 and 10,000 ppm. Hatching and reproductive efficiency of ectoparasites were also lower. The most effective concentration was 10,000 ppm (59.06% efficiency) being the extract more effective as an anti-tick than as an insecticide. As an anti-tick, it was more effective against *R. sanguineus* than against *R. microplus*. The insecticidal activity was associated to 2-pentadecanone (de Sousa Oliveira Batista et al. 2013).

A tea prepared with 3 g plant material per 500 ml distilled water (100 °C, 20 min), a commercial sample (CS) (6-methoxy-quercetin-3-O-robinobioside), and an authenticated sample (BGS) (hesperetin-7-O-rutinoside) were tested for their *in vitro* ability to inhibit acetyl-cholinesterase (AChE), butyl-cholinesterase (BuChE), and α-glucosidase. Inhibition of α-glucosidase and cholinesterase was dose-dependent in all cases. The two samples were similarly active against AChE (IC25 872.02 and 1056.03 μg ml^{-1} for BGS and CS, respectively) and α-glucosidase (IC50 604.34 and 573.48 μg ml^{-1} for BGS and CS, respectively). On the other hand, a stronger capacity to inhibit BuChE was observed for BGS (IC50 1112.14 μg ml^{-1}) than for CS (IC50 1743.96 μg ml^{-1}) (Ferreres et al. 2015).

In vitro culture Calli and cell suspension cultures were established from leaves on MS medium supplemented with 5 mg l^{-1} 2,4-D and 1.0 mg l^{-1} kin. Both callus and cell suspension cultures produce the alkaloid pilocarpine at a concentration of 1.01 μg g^{-1} and 0.045 μg g^{-1} at day 30 and 19 of culture, respectively (Andrade Salles et al. 2004).

8.7 *Jodina rhombifolia* (Hook. & Arn.) Reissek (Cervantesiaceae)

Synonymy *Celastrus rhombifolius* Hook. & Arn., *Ilex cuneifolia* L. var. bonariensis, *Celastrus jodina* Steud., *Jodina ruscifolia* Hook. & Arn., *Jodina cuneifolia* Miers, *Jodina bonariensis* Kuntze.

Vernacular name Quebracho flojo, peje, sombra de toro, quebrachillo.

General information *J. rhombifolia* is a native species from Bolivia, Brazil, Paraguay, Uruguay, north and center of Argentina. It is one of the eight species that conform the forest of *tala* (*C. spinosa*) in the Talares district. Its soft wood is used for construction, yokes construction, and axles of carts; its bark is used to clean rawhide harnesses (Pérez Meroni et al. 2010). It is also used as an ornamental in gardens and to form defensive fences (Cabrera and Zardini 1978; De la Peña and Pensiero 2004; Dimitri 2004). It is also one of the adulterants of yerba mate (Lahitte et al. 1999). Blooming is from January to July. Pollination is entomophile.

Botanical characteristics *J. rhombifolia* is a tree of medium to small texture, 3–5 meters high, with a globose, open, and irregular crown. The bark is thick and reddish-brown. Leaves are alternate with a rhomboidal shape, hard, with spines on the apices and lateral vertices, coriaceous, bright, dark green (Fig. 8.11). The flowers

Fig. 8.11 *Jodina rhombifolia* (Hook. & Arn.) Reissek (Cervantesiaceae). Buenos Aires Botanical Garden. MA Alvarez personal collection

are bisexual, formed by 5 floral pieces arranged in axillary groups of about 7–8 yellowish green little fragrant flowers. The fruit is a pseudodrupe, rough, brownish-green with a modified nectary disc (white fleshy layer). During the development of the fruit, the persistent petals became redish. Seeds are subspherical, flattened, top depressed, and chestnut clear of approximately 5 mm in diameter (Luna et al. 2017).

Ethnobotanical information Leaf and bark preparations are used in folk medicine as antidiarrheal, antidysenteric, anti-inflammatory, antitussive, abortive, and hypotensive, for sore throats, kidney stones or pain, stomach indisposition, and articular and muscular pain. The leaves are used against alcoholism, as digestive, and for carcinoma. Tisanes made with leaves and branches are taken to treat stomach ailments and colds. The decoction is used as astringent and antidysenteric; the leaves are used to heal wounds and to cure nasal polyps. The infusion of leaves is drunk as digestive, hepatic, antihidrotic, pectoral, antiasthmatic, and anti-alcoholic. The decoction of the ingested bark is taken against dysentery (Ratera and Ratera 1980; Toursarkissian 1980; Martínez Crovetto 1981; Adzet et al. 1991; Filipov 1994; Demaio et al. 2002; Davicino et al. 2007). Hieronymus (1882) attributed to the oil extracted from fruits the property of healing buboes and venereal sores. In the Buenos Aires-La Plata conurbation, *J. rhombifolia* leaves are marketed as having a cholesterol-lowering effect (Hurrel et al. 2015).

Chemical data The seed oil contains 9-octadec-ynoic acid (stearolic acid), *trans*-10-heptadecen-8-ynoic hydroxytransl (-heptadecen-8-)-ynoic acid, *trans*-10,16-heptadecadien-8-ynoic acid, 7-hydroxytrans-10,16-heptadecadien-ynoic acid, *trans*-ll-actadecen-yn oic acid (ximenynic acid), 8-hydroxytranslloctadecen-ynoic acid, *trans*-ll, 17-octadecadien-9-ynoic acid, 8-hydroxytransll, 17-octadecadien-9-ynoic acid, 9-hydroxystearic acid, and 9,10-epoxystearic acid. The leaves contain phenolic compounds, organic acids, tannins, flavonoids, steroids, gum, and mucilage. The aerial parts contain steroids, triterpenoids, alkaloids, coumarins, saponins, vicenin-2, vitexin, orientin, and swertisin (Fig. 8.12) (Spitzer et al. 1994; Barboza et al. 2009).

Biological activities The activity of the aqueous (250–750 mg kg^{-1}) and hydroalcoholic extracts (250–750 mg kg^{-1}) from the aerial parts were tested against HCl/ethanol-induced gastric lesions modified model (Yesilada et al. 1997). Female Wistar rats, pretreated with the aqueous extract (10 mg kg^{-1}), had gastric lesions in all the cases. On the contrary, animals treated with 250 and 750 mg kg^{-1} of the hydroalcoholic extract had gastric ulcers in 75% and 62.5% of cases, respectively. It was postulated that the antiulcer activity was related to the presence of the C-glycosyl flavonoids vicenin-2, vitexin, isovitexin, orientin, and swertisin (Montanha et al. 2009).

As for the diuretic activity of aerial parts, when methanolic extracts of leaves, bark, phloem, and branches were administered to Wistar rats (125, 250, and 500 mg $kg^{-1}$of weight) in normal saline solution, a diuretic activity was observed in all cases. The more potent was the 250 mg kg^{-1} dose of leaf methanolic extract

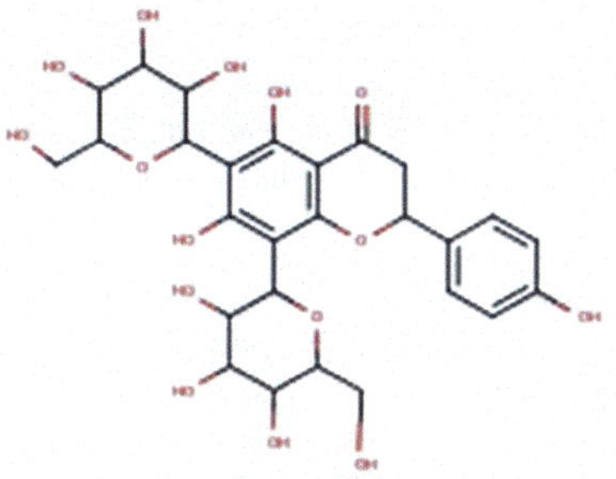

vicenin-2

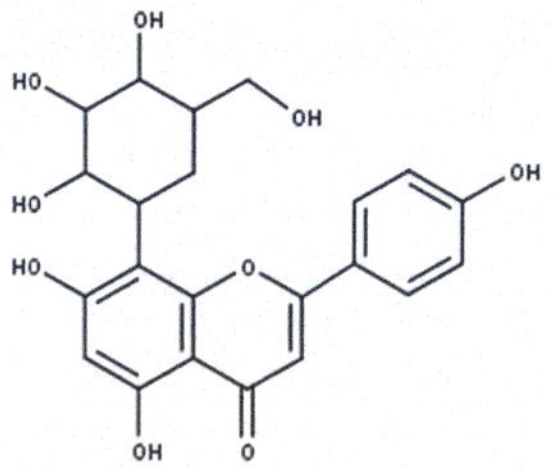

vitexin

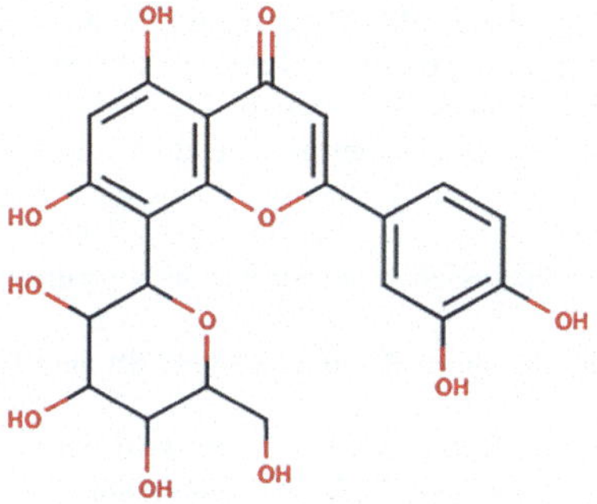

orientin

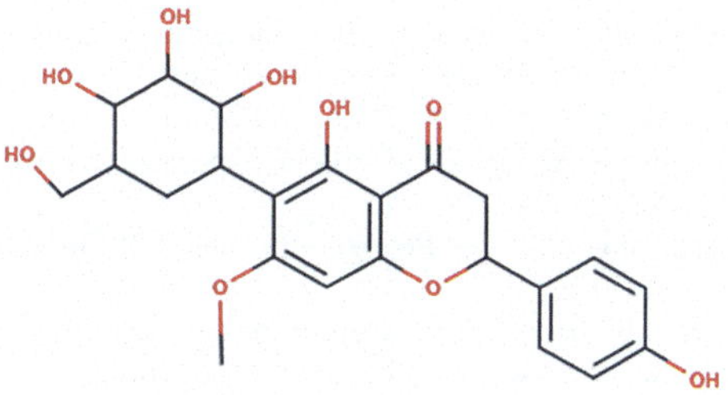

swertisin

Fig. 8.12 Chemical compounds found in *Jodina rhombifolia* (Hook. & Arn.) Reissek, (Cervantesiaceae) aerial parts

(diuretic index = 2.01), followed by 500 mg kg^{-1} dose of leaf methanolic extract (diuretic index = 1.98). The minimal diuretic activity was observed with the bark methanolic extract at all concentrations tested (Teves et al. 2018).

The anti-alcoholic activity of *J. rhombifolia* lyophilized aqueous extract was tested in male Wistar rats. Rats received the extract twice daily (1 ml 200 g^{-1}), one of the doses of extract (125, 250, and 500 mg kg^{-1} weight) for 10 consecutive days. The daily ethanol intake fall was evidenced from the first day of treatment and persisted throughout the entire treatment period. The daily water intake neither the body weight gain was affected; however, daily food intake was higher in the group of treated rats (Teves et al. 2015a). When the treatment (62.5, 125, 250 mg kg^{-1}) was applied to adolescent Wistar rats (postnatal day 29; 83–105 g of weight), it markedly reduced ethanol voluntary intake on a dose-dependent base (Teves et al. 2015b).

Repeating administration of *Jodina rhombifolia* lyophilized extract markedly reduced alcohol voluntary intake in rats in a self-administration model treated during 10 consecutive days, without signs of tolerance or apparent side effects (Teves et al. 2015a, b, Wang et al. 2016).

Toxicity There are no trials that demonstrate toxicity (Alonso and Desmarchelier 2015).

References

Abdel-Megeed RM, Hamed AR, Matloub AA (2018) Regulation of apoptotic and inflammatory signaling pathways in hepatocellular carcinoma via *Caesalpinia gilliesii* galactomannan. Mol Cell Biochem. https://doi.org/10.1007/s11010-018-3404-y

Adzet T, Marin E, Gene RM (1991) Estudio de la actividad antiinflamatoria de especies vegetales de origen centro y sudamericano. Dominguezia 9(1):17–23

Agüero E (2017) Plantas medicinales silvestres del centro de Argentina. Guía para su reconocimiento y uso terapéutico. Primera edición, Córdoba ECOVAL

Alipieva K, Korkina L, Orhan IE, Georgiev MI (2014) Verbascoside–a review of its occurrence, (bio)synthesis and pharmacological significance. Biotechnol Adv 32:1065–1076

Alonso J, Desmarchelier CJ (2015) Plantas medicinales autóctonas de la Argentina. Bases científicas para su aplicación en atención primaria de la salud. 1a ed. – CiudadAutónoma de Buenos Aires: Corpus Libros Médicos y Científicos

Andrade SS, Silva-Lucca RA, Santana LA, Gouvea IE, Juliano MA, Carmona AK, Araújo MS, Sampaio MU, Oliva MLV (2011) Biochemical characterization of a cysteine proteinase from *Bauhinia forficata* leaves and its kininogenase activity. Process Biochem 46:572–578

Arana MD, Bianco CA (2011) Helechos y licófitas del centro de la Argentina, 1st edn. Universidad Nacional de Río Cuarto, Río Cuarto

Arana MD, Ponce M, Morrone JJ, Oggero AJ (2013) Patrones biogeográficos de los helechos de las Sierras de Córdoba (Argentina) y sus implicancias en la conservación. Gayana Bot 70(2):357–376

Argentina National Pharmacopeia (ANP) Codex Medicamentarius Argentino VIth edition (1978) 111–112

Arigony ALV (2005) Determinação química e biológica de *Bauhinia forficata* Link subespécie pruinosa (pata-de-vaca – Leguminosae). [Dissertação]. Porto Alegre: Programa de Pós-Graduação em Ciências Farmacêuticas, Universidade Federal do Rio Grande do Sul

Bagher Hashem SM, Khaneghah AM, Koubaa M, Barba FJ, Abdi E, Niakpusari M, Tavakoli J (2018) Extraction of essential oil from *Aloysia citriodora* Palau leaves using continuous and pulsed ultrasound: kinetics, antioxidant activity and antimicrobial properties. Process Biochem 65:197–204

Bahramsoltani R, Rostamiasrabadi P, Shahpiri Z, Marques AM, Rahimi R, Far-zaei MH (2018) *Aloysia citriodora* Paláu (Lemon verbena): a review of phytochemistry and pharmacology. J Ethnopharmacol 222:34–51

Bandoni A (2000) Los recursos vegetales aromáticos en Latinoamérica, su aprovechamiento industrial para la producción de aromas y sabores. Editorial Universidad Nacional de La Plata

Bandoni AL, Nigueral C, Associació C (2003) Los Recursos vegetales aromáticos en Latinoamérica su aprovechamiento industrial para la producción de aromas y sabores, 2nd edn, Buenos Aires, p 418

Barboza GE, Bonzani, N, Filipa, E, Lujan M, Morero R, Bugatti M, Decolatti N, Ariza-Espinar L (2001) Atlas Histo-Morfológico de plantas de interés medicinal de uso corriente en Argentina. Museo de Botánica. Fac de Cs Ex, Fcas y Nat y Fac de Cs Qcas de la Universidad Nacional de Córdoba

Barboza GE, Cantero JJ, Núñez C, Pacciaroni A, Ariza EL (2009) Medicinal plants: a general review and a phytochemical and ethnopharmacological screening of the native Argentine Flora. Kurtziana 34(12):7365

Barrios I, Miltos V, Piris A, Piris G, Centurión C (2016) Antidepressant-like effect of Hidroalcoholic extract of *Aloysia citriodora* (Paraguayan Kedron) using tail suspension test in Swiss Albino Mice. Rev Sci 14(1):9–12

Batista de Souza F, Frederico de Souza MM, Duarte Damasceno PV, Gomes de Barros MC, Oliveira Lima Loiola C, de Souza Monteiro A, Freitas Ferreira Corrêa G (2014) Avaliação *in vitro* das atividades antimicrobianas de extratos etanólicos de *Arnica montana* L., *Bauhinia forficata* e *Plantago major* sobre *Streptococcus mutans* e *Streptococcus sanguis*. J Appl Pharmaceut Sci (2):18–28

Bayoub K, Baibai T, Mountassif D, Retmane A, Soukri A (2010) Antibacterial activities of the crude ethanol extracts of medicinal plants against *Listeria monocytogenes* and some other pathogenic strains. Afr J Biotechnol 9(27):4251–4258

Benelli G, Pavela R, Canale A, Cianfaglione K, Ciaschetti G, Conti F, Nicoletti M, Senthil-Nathan S, Mehlhorn H, Maggi F (2017) Acute larvicidal toxicity of five essential oils (*Pinus nigra*, *Hyssopus officinalis*, *Satureja montana*, *Aloysia citriodora* and *Pelargonium graveolens*) against the filariasis vector *Culex quinquefasciatus*: synergistic and antagonistic effects. Parasitol Int 66:166–171

Bresciani LFV, Priebe JP, Yunes RA, Magro JD, Monache FD, Campos FD, Souza MM, Cechinel-Filho V (2003) Pharmacological and phytochemical evaluation of *Adiantum cuneatum* growing in Brazil. Z Naturforsch 58c:191–194

Burkart A (1936) Las especies argentinas y uruguayas del género Caesalpinia. Rev Argent Agron 3:67–112

Cabrera AL (1971) Fitogeografía de la República Argentina. Boletín de la Sociedad Argentina de Botánica XIV (1–2): 1–58

Cabrera AL, Zardini EM (1978) Manual de la Flora de los alrededores de Buenos Aires, 2nd edn. Editorial ACME SACI, Buenos Aires

Calzada-Sánchez EV, Aguilar-Rodríguez S, López-Villafranco ME, Aguilar-Contreras A (2014) Anatomía de hoja de tallo de Verbenaceae medicinales empleadas en México. Bot Sci 92(4):469–480

Carnat A, Carnat AP, Fraisse D, Lamaison JL (1999) The aromatic and polyphenolic composition of lemon verbena tea. Fitoterapia 70:44–49

Castro dos Anjos PJ, Pereira PR, Alves Moreira IJ, Russo Serafini M, Antunes Souza Araújo A, Amaral da Silva F, Santos Ramos C, Antoniolli AR, Viana dos Santos MR (2013) Antihypertensive effect of *Bauhinia forficata* aqueous extract in rats. J Pharmacol Toxicol. https://doi.org/10.3923/jpt.2013

Cazarolli LH, Zanatta L, Jorge AP, Sousa E, Horst H, Woehl VM, Pizzolatti MG, Szpoganicz B, Silva FRMB (2008) Follow-up studies on glycosylated flavonoids and their complexes with vanadium: their anti-hyperglycemic potential role in diabetes. Chem Biol Interact 163:177–191

Cocucci AA, Galetto L, Sersic A (1992) El sindrome floral de *Caesalpinia gilliesii* (Fabaceae-Cesalpinioideae) Darwiniana 31(1–4): 11–135

Costa Mazzeo GC, de Oliveira SP, Lopes Guimarães L, Monteiro Souza Brito A, Toma W (2015) Evaluation of anti-ulcerogenic activity in an Aqueous Extract obtained from *Bauhinia forficata* leaves. Rev Ciênc Farm Básica Apl 36(1):21–26

Damasceno DC, Volpato GT, Calderon IMP, Aguilar R, Rudge MVC (2004) Effect of *Bauhinia forficata* extract in diabetic pregnant rats: maternal repercussions. Phytomedicine 11:196–201

Davicino R, Mattar MA, Casali YA, Correa SG, Pettenati EM, Micalizzi B (2007) Antifungal activity of plant extracts used in folk medicine in Argentina. Rev Peru Biol 14(2):247–251

de Andrade Salles L, de Oliveira Lopes S, Zuanazzi JAS, Rech SB, Henriques AT (2004) Pilocarpine accumulation on *Pilocarpus pennatifolius* tissue culture. Braz J Pharm Sci 40(3):437–439

De Feo V (1992) Medicinal and magical plants in the northern Peruvian Andes. Fitoterapia 63:417–440

De la Peña MR, Pensiero JF (2004) Plantas Argentinas. Catálogo de nombres comunes. Buenos Aires, Argentina: L.O.L.A., Buenos Aires

de la Sota ER, Luna ML, Giudice GE, Ramos Giacosa JP (2009) Sinopsis de las Pteridofitas de la Provincia de San Luis (Argentina). Bol Soc Argent Bot 44(3–4):367–385 de Sousa LC, Diógenes JPL, Azevedo Pereira B, Pelegrine Gomes Faria RA, Andrade Neto M, Sousa Alves R, Rodriguez de Queiroz MG, Florenço de Sousa FC, Barros Viaja GS (2004) Antidiabetic activity of *Bauhinia forficata* extracts in Alloxan-Diabetic Rats. Biol Pharm Bull 27(1):125–127

de Souza Oliveira Batista LC, do Nascimento CF, Peluso Cid Y, de Sousa Ma-galhães V, Siqueira de Almeida Chaves D, Coumendouros K (2013) Bioprospecão de extractos de jaborandi contra *Ctenocephalides felis felis*, *Rhipicephalus sanguineus* e *Rhipicephalus microplus*. Rev Bras Med Vet 35(Supl.2):113–118

de Souza MM, Pereira MA, Ardenghi JV, Mora TC, Bresciani LF, Yunes RA, Delle Monache F, Cechinel-Filho V (2009) Filicene obtained from *Adiantum cuneatum* interacts with the cholinergic, dopaminergic, glutamatergic, GABAergic, and tachykinergic systems to exert antinociceptive effect in mice. Pharmacol Biochem Behav 93:40–46

Demaio P, Karlin UO, Medina M (2002) Árboles nativos del Centro de Argentina. L.O.L.A, Buenos Aires

Dev A, Shyam K, Prashon PKJ, Thomas A (2018) Antimicrobial potential of bio-synthesized silver nanoparticle against human pathogen *Pseudomonas aeruginosa* and *Klebsiella pneumoniae*. Innov J Life Sci 6(1):1–4

Di Leo LP, van Baren CM, Retta D, Bandoni AL, Gil A, Gattuso M, Gattuso S (2007) Characterization of Lemon Verbena (*Aloysia citriodora* Palau) from Argentina by the essential oil. J Essent Oil Res 20(4):350–353

Dimitri MJ (2004) Enciclopedia Argentina de Agricultura y Jardinería, 3rd edn. ACME AGENCY S.A.C.I. y F, Buenos Aires, pp 657–1161

Do Carmo G, Fernandes TS, Pedroso M, Ferraz A, Neto AT, Silva UF, Mo-stardeiro MA, Back DF, Dalcol II, Morel AF (2018) Phytochemical and antimicrobial study of *Pilocarpus pennatifolius* Lemaure. Fitoterapia 131:1–8

Duarte MC, Leme EE, Delarmelina C, Soares AA, Figueira GM, Sartoratto A (2007) Activity of essential oils from Brazilian medicinal plants on *Escherichia coli*. J Ethnopharmacol 111:197–201

Düsman E, de Almeida IV, Coelho AC, BAlbi TJ, Düsman Tonin LT, Pimenta Vicentini VE (2013) Antimutagenic effect of medicinal plants *Achillea millefolium* and *Bauhinia forficata In Vivo*. Evidence-based complementary and alternative medicine Volume 2013, Article ID 893050, 6 pages. https://doi.org/10.1155/2013/893050

Elechosa MA, Di Leo LP, Juárez MA, Viturro CI, Heit CI, Molina AC, Martínez AJ, López S, Molina AM, van Baren CM, Bandoni AL (2017) Essential oil chemotypes of *Aloysia citriodora* (Verbenaceae) in Northwestern Argentina. Biochem Syst Ecol 74:19–29

Elting J, Montgomery R (1979) Biological effects of cesalin, an anti-tumor protein, on cultured mammalian cells. J Biol Chem 254(2):453–458

Emam M, El Raey MA, Eusa WH, El-Haddad AE, Osman SM, El-Ansari MA, Rabie AGM (2017) Green synthesis of silver nanoparticles from *Caesalpinia gilliesii* (Hook) leaves: antimicrobial activity and *in vitro* cytotoxic effect against BJ-1 and MCF-7 cells. J Appl Pharmaceut Sci 7(08):226–233

Emmam M, El Raey M, El Haddad AE, El Awdan SA, Rabie A-GM, l-Ansari MA, Sobeh M, Osman S, Wink M (2019) A new Polyoxygenated Flavonol Gossypetin-3-O-β-D-Robinobioside from *Caesalpinia gilliesii* (Hook.) D. Dietr. and *in vivo* Hepatoprotective, anti-inflammatory, and anti-ulcer activities of the leaf methanol extract. Molecules 24:138. https://doi.org/10.3390/molecules24010138

Felgines C, Fraisse D, Besson C, Vasson M-P, Texier O (2014) Bioavailability of lemon verbena (*Aloysia triphylla*) polyphenols in rats: impact of colonic inflammation. Br J Nutr 111:1773–1781

Fernandes Salgueiro AC, Folmer V, Pies da Silva M, Loureiro Mendez AS, Pegoraro Zemolin AP, Posser T, Franco JL, Puntel RL, Orione Puntel G (2016) Effects of *Bauhinia forficata* tea on oxidative stress and liver damage in diabetic mice. Oxidative Medicine and Cellular Longevity Volume 2016, Article ID 8902954, 9 pages. https://doi.org/10.1155/2016/8902954

Ferreres F, Gil-Izquierdo A, Vinholes J, Silva ST, Valentão P, Andrade PB (2012) *Bauhinia forficata* link authenticity using flavonoids profile: relation with their biological properties. Food Chem 134:894–904

Ferreres F, Grosso C, Gil-Izquierdo A, Fernandes A, Valentão P, Andrade PB (2015) Comparing the phenolic profile of *Pilocarpus pennatifolius* Lem. byHPLC–DAD–ESI/MSn with respect to authentication and enzyme inhibition potential. Ind Crop Prod 77:391–401

Filipov A (1994) Medicinal plants of the Pilagá of Central Chaco. J Ethnopharmacol 44:181–193

Franco MJ, Caetano ICS, Caetano J, Dragunski DC (2011) Determinação de metais em plantas medicinais comercializadas na região de Umuarama-PR. Arq Ciênc Saúde UNIPAR 15(2):121–127

Fujino DW, Reid MS, Vandermolen GE (1983) Identification of vascular blockages in rachides of cut maidenhair (*Adiantum raddianum*) fronds. Sci Hortic 21:381–388

Garibotti Lusa M, Bona C (2009) Análise morfoanatômica comparativa da folha de *Bauhinia forficata* Link e B. variegata Linn. (Leguminosae, Caesalpinioideae). Acta Bot Bras 23(1):196–211

Gattuso S, van Baren CM, Gil A, Bandoni A, Ferraro G, Gattuso M (2008) Morpho-histological and quantitative parameters in the characterization of lemon verbena (*Aloysia citriodora* Palau) from Argentina. Boletín Latinoamericano y del Caribe de Plantas Medicinales y Aromáticas 7(4):190–198

Gellert Moranza H, Scanavaca-Júnior L, Mataqueiro MI, Queiroz-Neto A, de Ca-margo Ferraz G (2017) Chronic Administration of *Bauhinia forficata* link reduced exploratory activity without affecting blood glucose in healthy rats. Plant 5(5–1):27–32

Gillij YG, Gleiser RM, Zygadlo JA (2008) Mosquito repellent activity of essential oils of aromatic plants growing in Argentina. Bioresour Technol 99:2507–2515

Goldblatt H, Lynch J, Hanzel RF, Summerville WW (1934) Studies on experimental hypertension. The production of persistent elevation of systolic blood pressure by means renal ischemia. J Exp Med 59:347–379

Gonzalez-Audino P, Picollo MI, Gallardo A, Toloza A, Vassena C, Mougabure-Cueto G (2011) Comparative toxicity of oxygenated monoterpenoids in experimental hydroalcoholic lotions to permethrin-resistant adult head lice. Arch Dermatol Res 303:361–366

Goodman LS, Gilman A (2001) The pharmacological basis of therapeutics, 10th edn. McGraw Hill co Inc., New York

Guerra Santos M, Kelecom A, Ribeiro de Paiva S, Gomez de Moraes M, Rocha L, Garret R (2010) Phytochemical studies in pteridophytes growing in Brazil: a review. Am J Plants Sci Biotechnol 4(special issue 1):113–125

Gupta M (1995) 270 Plantas Medicinales Iberoamericanas. CYTED, Santa Fé de Bogotá D.C. Colombia. p. 617

Hausmann M, Obermeier F, Paper DH, Balan K, Dunger N, Menzel K, Falk W, Shoelmerich J, Hearfarth H, Rogler G (2007) *In vivo* treatment with the herbal phenylethanoid acteoside ameliorates intestinal inflammation in dextran sulphate sodium-induced colitis. Clin Exp Immunol 148:373–381

Herranz-López M, Barrajón-Catalán E, Segura-Carretero A, Menéndez JA, Joven J, Micol V (2015) Lemon verbena (*Lippia citriodora*) polyphenols alleviate obesity-related disturbances in hypertrophic adipocytes through AMPK-dependent mechanisms. Phytomedicine 22:605–614

Hieronymus G (1882) Plantae Diaphoricae Florae Argentinae. Boletín de Ciencias en Córdoba 337

Hurrel JA, Puentes JP, Arenas PM (2015) Medicinal plants with cholesterol-lowering effect marketed in the Buenos Aires-La Plata conurbation, Argentina: an urban Ethnobotany study. Ethnobiol Conserv 4(6):1–19

Jausoro M, Galetto L (2000) Estudio comparativo de nectario y néctar en flores perfectas y estaminads de *Caesalpinia gilliesii* (Fabaceae, Caesalpinioideae). Bol Soc Argent Bot 35(1–2):107–114

Jausoro M, Galetto L (2001) Producción de flores y frutos en una especie andromonoica: *Caesalpinia gilliesii* (Fabaceae). Kurtziana 29(1):15–25

Jorge AP, Horst H, De Souza E, Pizzolatti MG, Silva FRMB (2004) Insulinomimetic effects of kaempferitrin on glycaemia and on 14C-glucose uptake in rat soleus muscle. Chem Biol Interact 149:9

Juárez de Varela F (2001) Flora del valle de Lerma. Fabaceae Lidl. Subfam. Caesalpinioideae (R.Br.) A.DC. Aportes Botánicos de Salta - Ser. Flora 6(10):1–5

Kumar NK, Kumar KS, Raman B, Reddy I, Ramarao M, Rajagopal S (2008) Antibacterial activity of *Lippia citriodora* a folklore plant. JPAM 2(1):249–252

Lahitte HB, Hurrell JA, Haloua MP, Jankowski LS, Belgrano MJ (1999) Árboles Rioplatenses. Ed L.O.L.A, Buenos Aires

Lai HY, Lim YY (2011) Evaluation of antioxidant activities of the methanolic extracts of selected ferns in Malaysia. Int J Environ Sci Dev 2(6):442–447

Lamaison JL, Petitjean FC (1993) Verbascoside, major phenolic compound of the leaves of ash (*Fraxinus excelsior*) and vervain (*Aloysia triphylla*). Plantes Medicinales et Phytotherapie 26(3):225–233

Lenoir L, Joubert-Zakeyh J, Texier O, Lamaison JL, Vasson MP, Felgines C (2012) *Aloysia triphylla* infusion protects rats against dextran sulfate sodium induced colonic damage. J Sci Food Agric 92:1570–1572

Lima H, Kim M, Lim Y, Cho YH, Lee CH (2006) Inhibition of cell-cycle progression in HeLa cells by HY52, a novel cyclin-dependent kinase inhibitor isolated from *Bauhinia forficate*. Cancer Lett 233:89–97

Lucio EMR de A, Sharapin N, Franá HS (2002) Estudo de alcalóides de *Pilocarpus pennatifolius* Lemaire. Rev Bras Farmacogn 12(supl):130–131

Luna ML, Giudice GE, Grossi MA, Gutiérrez DG (2017) Development and morphology of the fruit and seed of the hemiparasite genus *Jodina* (Cervantesiaceae). An Jardín Bot Madrid 74(1):e051

Macaya J (1999) Leguminosas arbóreas y arbustivas cultivadas en Chile. Chloris Chilensis. Año 2. N° 1

Martínez Crovetto RN (1981) Las plantas utilizadas en medicina popular en el Noroeste de Corrientes (Argentina). Miscelánea 69: 1–140. Inst. M. Lillo, Tucumán

Martínez N, Cayama E, Goncalvez L, Labrador S, Espino C, Pérez L (2009) Hipoglicemic effect of *Bauhinia purpurea* L., in euglicemic mice. Comunidad y Salud 7(2):43–49

Mazzon E, Esposito E, Di Paola R, Riccardi L, Caminiti R, Dal Toso R, Pressi G, Cuzzocrea S (2009) Effects of verbascoside biotechnologically produced by *Syringa vulgaris* plant cell cultures in a rodent model of colitis. Naunyn Schmiedebergs. Arch Pharmacol 380:79–94

Mello MO, Melo M, Appezzato-da-Gloria B (2000) *Bauhinia forficata* link shoot regeneration: histological analysis of organogenesis pathway. Braz Arch Biol Technol 43(4). https://doi.org/10.1590/S1516-89132000000400013

Mello MO, Santos Dias CT, Campos Amaral AF, Melo M (2001) Growth of *Bauhinia forficata* link, *Curcuma zedoaria* Roscoe and *Phaseolus vulgaris* L. cell suspension cultures with carbon sources. Sci Agric 58(3):1–12

Menezes PR, Schwarz EA, Santos CAM (2004) *In vitro* antioxidant activity of species collected in Paraná. Fitoterapia 75(3–4):398–400

Miranda DE, Bohren AV, Keller H, Grance LA, Gartland HM (2000) A key to recognize woody species of the Rutaceae family in the Paranaense Forest, Argentina, through dendrological features. Quebracho 8:47–55

Mizui T, Doteuchi M (1983) Effect of polyamines on acidified ethanol induced gastric lesion in rats. Jpn J Pharmacol 33:939–945

Molina AC, Viturro CI, Heit CI (2003) Estudio químico de aceites esenciales de *Aloysia triphylla*. Resúmenes del XIV Congreso de la Sociedad Argentina de Investigación en Química Orgánica (SAIQO), Santa Fe, Argentina

Montanha JA, Schenkel EP, Cardoso, Taketa AT, Dresch AP, Langeloh A, Dallegrave E (2009) Chemical and antiulcer evaluation of *Jodina rhombifolia* (Hook. & Arn.) Reissek extracts. Braz J Pharmacogn 19(1A):29–32

Morimoto Y, Shimohara K, Oshima S, Sukamoto T (1991) Effects of the new antiulcer agent KB-5492 on experimental gastric mucosal lesions and gastric mucosal defensive factors, as compared to those of teprenone and cimetidine. Jpn J Pharmacol 57:495–505

Múlgura de Romero ME, Rotman AD, Atkins S (2012) Verbenaceae Adans. In: Flora Argentina. Disponible en http://www.floraargentina.edu.ar/publicaciones/VERBENACEAE.pdf

Müller F (1997) Drogas y medicamentos de los indios Guaraní (Mbyá, Pai y Chiripá) en las regiones orientales de la selva del Paraguay. Parodiana 10(1–2):197–209

Muñoz López de Bustamante F (1996) Plantas medicinales y aromáticas. Estudio, cultivo y procesado. Ed. Mundi Prensa-Madrid, p 365

Murashige T, Skoof F (1962) A revised medium for rapid growth and bioassay with tobacco tissue cultures. Physiol Plant 15:473–497

Nishitani Y, Yamamoto K, Yoshida M, Azuma T, Kanazawa K, Hashimoto T, Mizuno M (2013) Intestinal anti-inflammatory activity of luteolin: role of the aglycone in NF-kB inactivation in macrophages co-cultured with intestinal epithelial cells. Biofactors 39:522–533

Ocazionez RE, Meneses R, Torres FA, Stashenko E (2010) Virucidal activity of Colombian Lippia essential oils on dengue virus replication *in vitro*. Mem Inst Oswaldo Cruz 105(3):304–309

Oliva M, Beltramino E, Gallucci M, Casero C, Zygadlo J, Demo M (2010) Antimicrobial activity of essential oils of *Aloysia triphylla* (L'Her.) Britton from different regions of Argentina. BLACPMA 9(1):29–37

Oliva M, Carezzano E, Callucci N, Freytes S, Zygaldo JA, Demo MS (2015) Growth inhibition and morphological alterations of *Staphylococcus aureus* caused by the essential oil of *Aloysia triphylla*. Boletín Latinoamericano y del Caribe de Plantas Medicinales y Aromáticas 14(2):83–91

Oliveira CZ, Maiorano VA, Marcussi S, Sant'Ana CD, Januário AH, Lourenço MV, Sampaio SV, França SC, Pereira PS, Soares AM (2005) Anticoagulant and antifibrinogenolytic properties of the aqueous extract from *Bauhinia forficata* against snake venoms. J Ethnopharmacol 98:213–216

Olivera Gomes L, Reis Lívero FA, Stolf AM, Alves de Souza E, Pontes Werneck M, Vizioli de Castro Ghizoni C, Verdan MH, Querioz Telles JE, Locatelli Dittrich R, Alves Stefanello ME, Comar JF, Acco A (2016) Ethanolic extract of *Bauhinia forficata* metabolic effects in diabetic and normoglycemic rats. Int J Diabetes Clin Res 3(2):1–8

Osman SM, El-Khalik SMA, El Haddad AE, Wink M (2008) A new steroidal compound (β-sitosterol-3-O-butyl) isolated from *Caesalpinia gilliesii* flowers. Int J Appl Res Nat Prod 8(2):14–19

Osman SM, Alazzouni AS, Khakek SMA, Koheil MA, El-Haddad AE (2013) Phytoconstituents and biological activities of the aerial parts of *Caesalpinia gilliesii* (Hook) family Caesalpiniaceae growing in Egypt. Life Sci J 3:2418–2430

Osman SM, El-Haddad AE, El-Raey MA, El-Khalik SMA, Koheil MA, Wink M (2016) A new octadecenoic acid derivative from *Caesalpinia gilliesii* flowers with potent hepatoprotective activity. Pharmacog Mag 12:S332

Pan C, Chen YGM, Ma XY, Jiang JH, He F, Zhang Y (2011) Phytochemical constituents and pharmacological activities of plants from the genus *Adiantum*: a review. Trop J Pharm Res 10(5):681–692

Parodi TV, Cunha MA, Becker AG, Zeppenfeld CC, Martins DI, Koakoski G, Barcellos LG, Heinzmann BM, Baldisserotto B (2013) Anesthetic activity of the essential oil of *Aloysia triphylla* and effectiveness in reducing stress during transport of albino and gray strains of silver catfish, Rhamdia quelen. Fish Physiol Biochem 40:323–334

Pepato MT, Keller EH, Baviera AM, Kettelhut IC, Vendramini RC, Brunetti IL (2002) Anti-diabetic activity of *Bauhinia forficata* decoction in streptozotocin-diabetic rats. J Ethnopharmacol 81:191–197

Pepato MT, Baviera AM, Vendramini RC, Brunetti IL (2004) Evaluation of toxicity after one-months treatment with *Bauhinia forficata* decoction in streptozotocin-induced diabetic rats. BMC Complement Altern Med 4:article 7

Pepato MT, Conceição CQ, Gutierres VO, Vendramini RC, Souza CRF, Oliveira WP, Brunetti IL (2010) Evaluation of the spouted bed dried leaf extract of *Bauhinia forficata* for the treatment of experimental diabetes in rats. Afr J Biotechnol 9(42):7165–7173

Pérez Meroni M, Paleo MC, Lema VS (2010) Procesamiento y consumo de vegetales por grupos cazadores-recolectores del Holoceno tardío, en los partidos de Magdalena y Punta indio, Provincia de Buenos Aires. In: Berón M, Luna L, Bonomo M, Montalvo C, Aranda C, Aizpitarte MC (eds) Pasado y presente desde la arqueología pampeana. Mamul Mapü, pp 215–226

Pérez Zamora C, Torres CA, Núñez MB (2018) Antimicrobial activity and chemical composition of essential oils from Verbenaceae species growing in South America. Molecules 23:544. https://doi.org/10.3390/molecules23030544

Pochettino ML, Martínez MR, Arenas P, Crivos M, Navone G, Digiani MC, Teves L, Remorini C, Sy A, Illkow C, Delorenzini N (2003) Fitoterapia Mbya: análisis y evaluación del tratamiento de las parasitosis. Delpinoa J Ethnobot 46:301–310

Ponce-Montaner H, Fernández-Martínez E, Ortiz MI, Ramírez-Montiel ML, Cruz-Elizalde D, Pérez-Hernández N, Cariño-Cortés R (2010) Spasmolytic and anti-inflammatory effects of *Aloysia triphylla* and citral *in vitro* and *in vivo* studies. J Smooth Muscle Res 46(6):309–319

Ragone MI, Sella M, Conforti P, Volonté MG, Consolini AE (2007) The spasmolytic effect of *Aloysia citriodora*, Palau (South American cedrón) is partially due to its vitexin but not isovitexin on rat duodenums. J Ethnopharmacol 113:258–266

Rainsford KD (1978) The effect of 5-lipoxygenase inhibitors and the leukotriene antagonists on the development of gastric lesions induced by non-steroidal anti-inflammatory drugs in mice. Agents Actions 21:316–319

Rastogi S, Pandey MM, Rawat AKS (2018) Ethnopharmacological uses, phytochemistry and pharmacology of genus *Adiantum*: a comprehensive review. J Ethnopharmacol 215(6):101–119

Ratera EL, Ratera MO (1980) Plantas de la flora argentina empleadas en medicina popular. Ed. Hemisferio sur S.A.

Ricco RA, Wagner ML, Gurni AA (2011) Dinámica de polifenoles de "Cedrón" (*Aloysia citriodora* Palau -Verbenaceae-) en relación al desarrollo foliar. BLACPMA 10(1):67–74

Rodríguez Barbosa N (1985) Pojhá Ñaná. Recetario de plantas medicinales usadas en el Paraguay. 2da. Ed. Asunción, Paraguay (sin datos tipográficos)

Rodríguez Vaquero MJ, Tomassini Serravalle LR, Manca de Nadra MC, Strasser de Saad AM (2010) Antioxidant capacity and antibacterial activity of phenolic compounds from Argentinean herbs infusions. Food Control 21:779–785

Rojas Armas J, Palacios Agüero O, Ortiz Sánchez JM, López de la Peña L (2015) *In vivo* evaluation of *Aloysia triphylla* britton (lemon verbena) essential oil toxicity and citral anti-*Trypanosoma cruzi* activity. An Fac Med 76(2):129–134

Rojas J, Solís H, Palacios O (2010) *In vitro* anti-*Trypanosoma cruzi* activity of essential oils of ten medicinal plants. An Fac Med 71(3):161–165

Rojas J, Palacio O, Ronceros S (2012) The effect of the essential oil from *Aloysia triphylla* Britton (lemon verbena) on *Trypanosoma cruzi* in mice. Rev Peru Med Exp Salud Publica 29(1):61–68

Rzedowski GC, Calderón de Rzedowski G (2002) Verbenaceae. Flora del Bajío y de regiones adyacentes 100:1–145

Salgueiro ACF, Leal CQ, Bianchini MC, Prado IO, Mendez ASL, Puntel RL, Folmer V, Soares FA, Ávila DS, Puntel GO (2013) The influence of *Bauhinia forficata* Link subsp. pruinosa tea on lipid peroxidation and non-protein SH groups in human erythrocytes. J Ethnopharmacol. https://doi.org/10.1016/j.jep.2013.03.070i

Santos AP, Hrihorowitsch Moreno PR (2004) *Pilocarpus* spp.: a survey of its chemical constituents and biological activities. Braz J Pharm Sci 40(2):115–134

Santos-Gomes PC, Fernandes-Ferreira M, Vicente MAS (2005) Composition of the essential oils from flowers and leaves of vervain [*Aloysia triphylla* (L'Herit.) britton] grown in Portugal. J Essent Oil Res 17:73–78

Santos-Silva F, Mastroberti AA, de Araujo ME (2011) Development of the epidermal cells of the Pinnules of *Adiantum raddianum* C. Presl (Pteridaceae): environmental adaptive plastidial characteristics. Am Fern J 101(3):172–181

Sartoratto A, Machado A, Delarmelina C, Figueira G, Duarte M, Rehder L (2004) Composition and antimicrobial activity of essential oils from aromatic plants used in Brazil. Braz J Microbiol 35:275–280

Schultz AG (1997) Algunas plantas usuales del Nordeste Argentino. Parodiana 10(1–2):211–241

Scott AC (1989) Laboratory control of antimicrobial therapy. In: Collee JG, Duguid JP, Fraser AG, Marmion BP (eds) Mackie and MacCartney practical medical microbiology, vol 2, 13th edn. Churchill Livingstone, pp 161–181

Sertié JAA, Basile AC, Panizza S, Matida AK, Zelnik R (1990) Anti-inflammatory activity and sub-acute toxicity of artemetin. Planta Med 56:36–40

Severin C, Bruzzese D, Di Sapio O, Giubuleo MG, Gattuso S (2005) Regeneración *in vitro* de plantas de *Aloysia citriodora* Palau (Verbenaceae). Revista de Investigaciones de la Facultad de Ciencias Agrarias 8:1–7

Shiojima K, Arai Y, Nakane T, Ageta H, Cai S-Q (1997) Fern constituents: *Adiantum cuneatum*. III. Four new triterpenoids, 4,23- Bisnor-3,4-secofilic 5(24)-en-3-al, 4,23-bisnor-3,3-dimethoxy-3,4-secofilic- 5(24)-ene, 7,25-epoxyfern-9(11)-en-8α-ol and 7α,8α epoxyfernan-25-ol. Chem Pharm Bull 45:1608–1610

Shrivastava AK, Pandey B, Chauhan AD (2014) Phytochemical analysis of *Adiantum* and *Pteris* ferns & its role as antioxidant. Indian J Sci Res 4(1):31–38

Silva KL, Cechinel-Filho V (2002) Plantas do gênero *Bauhinia*: composição química e potencial farmacológico. Quím Nova 25(3):449–454

Silva FRMB, Szpoganicz B, Pizzolatti MG, Willrich MAV, Sousa E (2002) Acute effect of *Bauhinia forficata* on serum glucose levels in normal and alloxan induced diabetic rats. J Ethnopharmacol 83:33–37

Silva FSA, Mastroberti AA, Mariath JEA (2007) Aspectos anatômico-funcionais das células epidérmicas de pínulas de *Adiantum raddianum* Presl. (Pteridaceae). Rev Bras Bio 5(supl.1):831–833

Silva MCC, Santana LA, Mentele R, Ferreira RS, de Miranda A, Silva-Lucca RA, Sampaio MU, Correia MTS, Oliva MLV (2012) Purification, primary structure and potential functions of a novel lectin from *Bauhinia forficata* seeds. Process Biochem 47(7):1049–1059

Silvestre Marques G, Araújo Rolim L, Santos Alves LD, de Andrade Rodríguez Silva CC, Lira Soares LA, Rolim-Neto PJ (2013) Estado da arte de *Bahuinia forficata* link (Fabaceae) como alternativa terapêutica para o tratamento do Diabetes mellitus. Ver Ciênc Farm Básica Apl 34(3):313–320

Singh NP, McCoy MT, Tice RR, Schneider EL (1988) A simple technique for quantification of low levels of DNA damage in individual cells. Exp Cell Res 17:184–191

Sousa E, Zanatta L, Seifriz I, Creczynskipasa TB, Pizzolatti MG, Szpoganicz B, Silva FRMB (2004) Hypoglycemic effect and antioxidant potential of kaempferol-3,7-O-(a)-dirhamnoside from *Bauhinia forficata* leaves. J Nat Prod 67:829–832

Spegazzini ED, Castro MT, Rossi J, Carpano SM, Nájera MT (2002) Micrografía foliar cuali-cuantitativa de las especies de *Pilocarpus* (Rutaceae): *P. jaborandi* Holmes, *P. microphyllus* Stapf ex Wardleworth y *P. pennatifolius* Lem. Dominguezia 18(1):5–16

Spitzer V, Bordignon SA de L, Schenkel EP, Marx F (1994) Identification of nine acetylenic fatty acids, 9-hydroxystearic acid and 9,10-epoxistearic acid in the seed oil of *Jodina rhombifolia* Hook et Arn. (Santalaceae). JOACS 71(12):1343–1348

Steibel P, Troiani H (2000) Curso de Actualización: Identificación de Malezas y Forrajeras. 1–353

Suarez SS, Cabrera lL, Juliani HR (1984) Flavonoides, Aminoácidos e Hidratos de Carbono en *Caesalpinia gilliesii*. An Asoc Quím Argent 72(3):261–263

Tadera K, Minami Y, Takamatsu K, Matsuoka T (2006) Inhibition of α-Glucosidase and α-Amylase by Flavonoids. J Nutr Sci Vitaminol 52:149–153

Teves MR, Wendel GH, Pelzer LE (2015a) *Jodina rhombifolia* leaves lyophilized aqueous extract decreases alcohol intake and preference in adolescent male Wistar rats. J Ethnopharmacol 174:11–16

Teves MR, Wendel GH, Plzer LE (2015b) Reduction in voluntary ethanol intake following repeated oral administration of *Jodina rhombifolia* lyophilized aqueous extract in male Wistar rats. J Ethnopharmacol 161:170–174

Teves MR, Pacheco PH, Bazán C, Wendel GH (2018) Diuretic activity of methanolic extracts from *Jodina rhombifolia* aerial parts of Wistar rats. J Pharm Pharmacogn Res 6(5):402–411

Thomas T (2014) A study on antibacterial and phytochemical evaluation of fronds of *Adiantum raddianum* C. Presl. Int J Pharmacol Screen Methods 4(2):85–88

Toloza-Zambrano P, Avello M, Fernández P (2015) Determination of rutin and trigonelline in extracts of *Bauhinia forficata* subsp. pruinosa and hypoglycemic effect on diabetic and prediabetic patients humans. BLACPMA 14(1):21–32

Toursarkissian M (1980) Plantas Medicinales de la Argentina. Ed. Hemisferio Sur, Buenos Aires

Ulubelen A, McCaughey W, Cole J (1967) Proteinaceous antitumor substances from plants III. *Caesalpinia gilliesii* (Leguminosae). J Pharm Sci 56:914–916

Valdez IH, Wolff A, Atkinson JC, Macynski A, Fox PC (1993) Use of pilocarpine during head and neck radiation therapy to reduce xerostomia and salivary dysfunction. Cancer 71(5):1848–1851

Valentao P, Fernándes E, Carvahlo F, Andrade PB, Seabra RM, de Lourdes BM (2002) Studies on the antioxidant activity of *Lippia citriodora* infusión: scavenging effect on superoxide radical, hydroxyl radical and hypochlorous acid. Biol Pharm Bull 25(10):1324–1327

Vasconcelos F, Sampaio SV, Garófalo MAR, Guimarães LFL, Giglio JR, Arantes EC (2004) Insulin-like effects of *Bauhinia forficata* aqueous extract upon Tityus serrulatus scorpion envenoming. J Ethnopharmacol 95:385–392

Viglierchio M, Toribio M, Alonso G, Bruni M, Fernández M, Ferreyra M, Toso R (2000) Relevamiento fitoquímico de *Caesalpinia gilliesii* (Leguminosae) y de otras especoes de Caesalpinias. Anuario 2000 Facultad de Ciencias Veterinarias, Universidad Nacional de la Pampa 34–46

Wang F, Li Y, Zhang Y-J, Zhou Y, Li S, Li H-B (2016) Natural products for the prevention and treatment of hangover and alcohol use disorder. Molecules 21:64. https://doi.org/10.3390/molecules21010064

Winter SL de Sousa, da Silva Sylvestre L, Prado J (2011) O gênero *Adiantum* (Pteridaceae) no estado do Rio de Janeiro, Brasil. Rodriguesia 62(3):663–681

Yesilada E, Gurbuz I, Ergun E (1997) Effects of *Cistus laurifolius* L. flowers on gastric and duodenal lesions. J Ethnopharmacol 55:201–211

Zamorano-Ponce E, Fernández J, Vargas G, Rivera P, Carballo MA (2004) Protective activity of cedron (*Aloysia triphylla*) infusion over genetic damage induced by cisplatin evaluated by the comet assay technique. Toxicol Lett 152:85–90

Zamorano-Ponce E, Morales C, Ramos D, Sepúlveda C, Cares S, Rivera P, Fer-nández J, Carballo MA (2006) Antigenotoxic effect of *Aloysia triphylla* infusion against acrylamide-induced DNA damage as shown by the comet assay technique. Mutat Res 603:145–150

Chapter 9
Patagonia

9.1 Introduction

9.1.1 The Steppe and the Andine Forest

Patagonia Argentina is an extensive region (Fig. 9.1) located to the South of the American continent that includes the provinces of the LPA, NEU, RNE, CHU, SCR, TDF, Argentina, Antarctica, and South Atlantic Islands. The borders of the region are to the North in East-West direction, the line that integrates the mouth of the Colorado River into the Mar Argentino, the channels of *Rio Grande* and *Rio Barrancas* that are the border between the provinces of NEU and MEN, until the limit with the Chilean Patagonia and the mountainous formation of the *Cordillera de los Andes*. Towards the South end, it is the Cape Horn in the Drake Passage that is the boundary between the Atlantic and Pacific Oceans, and to the East is the Mar Argentino in the South Atlantic Ocean.

Ecologically, Patagonia could be divided into the cordillera de los Andes sector, which is more humid and has forests and a great variety of lakes of glacial origin, and the Patagonia desert or Patagonian steppe with semiarid characteristics and predominance of shrubs.

The highest peak in Patagonia is the Domuyo Volcano (4709 m.a.s.l), and the lowest point is the *Laguna del Carbón* (108 m.b.s.l.). This area has numerous volcanoes such as *Catedral*, *Piltriquitrón*, *Chapelco*, *Caviahue*, *Bayo*, and *La Hoya*. Binational volcanoes (Argentina-Chile) are *Lanin*, *Copahue*, *Pantojo*, *Tronador*, Fitz Roy, *Cerro Torre*, *Murallón*, and *San Lorenzo*. Some of the lakes of the region are *Nahuel Huapi*, *Puelo*, *Argentino*, Strobel, *Futalaufquen*, *Traful*, and Mascardi, and the important rivers are *Colorado*, *Negro*, *Chubut*, *Deseado*, *Santa Cruz*, *Gallegos*, and *Limay*. There are also glaciers such as the *Perito Moreno* and *Vespignani*.

M. A. Alvarez, *Pharmacological Properties of Native Plants from Argentina*,
https://doi.org/10.1007/978-3-030-20198-2_9

Fig. 9.1 Patagonia Argentina. (**a**) Ushuaia, Vespignani glacier viewpoint, (**b**) Patagonian steppe, (**c**) Perito Moreno Glacier, (**d**) MA Alvarez personal collection

There are wide temperature amplitude with a difference between the North and the South of the region. The *cordillera de los Andes* retains humidity from the Pacific Ocean, beneficial for the Patagonian forest; on the other hand, the arid Patagonian steppe is crossed by dry winds.

As for the economic activities, Patagonia has natural forest resources (*Nothofagus pumilio*, cypresses, and *Lomatia hirsuta*) that have to be preserved. The agricultural produce includes apples, pears, cherries, fodder, potatoes, peaches, and vegetables.

Livestock includes breeding of bovine, sheep, and goat cattle. Mining includes deposits of gold, silver, gas, oil, coal, lead, and uranium, among others. Sea products include silverside, anchovy, haddock, shrimp, hake, king crab, crustaceans, rainbow trout, and Pacific salmon. Also, Patagonia produces hydroelectric power. As for the fauna, the marine fauna includes southern right whale, orca, dolphins, sea lions, and sea elephants. Among the birds, in the oceanic coast, there are the Magallanes penguin, cormorants, seagulls, and Antarctic dove. In the Andean Patagonia, the most witnessed birds are raptors such as the *condor*. The South area has foxes, wild cats, Patagonian hares, armadillos, Pampean deer, huemul, puma, guanaco, and *ñandú*, among others.

As for the flora, in the sea coast of the provinces of RNE and CHU, there are grasslands, shrubs, and trees. The central region of Patagonia has steppe vegetation with more sparse grassland, mounts of *Larrea* spp., *Mulinum spinosum*, and *Festuca gracillima*. However, this region is going through a desertification process as a consequence of livestock grazing (sheep and goats) and the effect of volcanic ashes. Small oases, with *Salix humboldtiana*, and other species are located in the riverbanks from the central plateau; in the mountain ranges and the Andean Patagonia forest, there are coniferous, *Nothofagus pumilio* (lenga), *N. antarctica* (ñires), and *N. dombeyi* (coihue), shrubs such as *Berberis darwinii* (michay) and *B. microphylla* (calafate), ferns and mosses in muddy grounds, and fungus such as *Cyttaria hariotii* (llao llao) growing in the coihue bark. Among the medicinal species from Patagonia are *Acantholippia seriphioides* (A. Gray) Moldenke (Verbenaceae), *Adesmia boronioides* (Fabaceae), *Adiantum chilense* (Pteridaceae), *Apium panul* (Bertero ex DC.) (Apiaceae), *Araucaria araucana* Molina K. Koch (Araucariaceae), *Arjona tuberosa* (Santalaceae), *Atriplex lampa* (Moq.) D. Dietr. (Chenopodiaceae), *Baccharis crispa* Spreng. (Asteraceae), *Baccharis sagittalis* (Asteraceae), *Berberis darwinii* Hook. (Berberidaceae), *Berberis heterophylla* (Berberidaceae), *Buddleja globosa* (Buddlejaceae), *Colliguaja integerrima* Gillies & Hook. (Euphorbiaceae), *Coriaria ruscifolia* L. Deu (Coriariaceae), *Chenopodium ambrosioides* (Chenopodiaceae), *Cyttaria darwinii* (Cyttariaceae), *Equisetum bogotense* (Echicetaceae), *Empetrum rubrum* (Empetraceae), *Ephedra ochreata* (Ephedraceae), *Euphorbia serpens* (Euphorbiaceae), *Fabiana imbricata* (Solanaceae), *Festuca argentina* (Poaceae), *Fragaria chiloensis* (Rosaceae), *Fuchsia magellanica* Lam. (Onagraceae), *Grindelia chiloensis* (Cornel.) Cabrera (Asteraceae), *Gunnera tinctoria* (Gunneraceae), *Larrea divaricata* Cav. (Zygophyllaceae), *Lycium chilensis* (Solanaceae), *Maytenus boaria* (Celastraceae), *Mulinum spinosum* (Cav.) Pers. (Apiaceae), *Nothofagus antarctica* (G. Forst.) Oerst (Nothofagaceae), *Nothofagus betuloides* (Nothofagaceae), *Oxalis enneaphylla* Cav. (Oxalidaceae), *Plantago lanceolata* L. (Plantaginaceae), *Prosopis denudans* (Fabaceae), *Ribes magellanicum* (Grossulariaceae), *Rubus geoides* Sm. (Rosaceae), *Rumex crispus* (Polygonaceae), *Salix humboldtiana* (Salicaceae), *Clinopodium darwinii* (Benth.) Kuntze (Lamiaceae), *Schinus marchandii* (Anacardiaceae), *Senecio filaginoides* (Asteraceae), *Urtica dioica* (Urticaceae), *Usnea barbata* (Usnaceae), *Valeriana carnosa* (Valerianaceae), and *Viola magellanica* (Violeaceae) (Díaz 2010; Toledo and Kutschker 2012; Molares and Ladio 2015).

9.2 *Adesmia boronioides* Hook. f. (Fabaceae)

Synonymy *Patagonium boronioides* (Hook. f.) Kuntze.

Vernacular name Paramela, yagneu, lonckotrevo, té pampa, té silvestre, yerba carmelita, and éter.

General information *A. boronioides* is an aromatic and medicinal plant native to Argentina and Chile. In Argentina, it is found in the provinces of CHU, NEU, RNE, SCR, and TDF; it inhabits sites with little irrigation. It is widely valued by the original Patagonian peoples that have used it both as medicinal and in rituals. Some *mapuche-tehuelche* communities cultivated and protected the species. The aerial parts have a large amount of resin and a sweet-woody licorice-spicy odor. The EO has become an important economic resource for the fragrance industry. The leaves are used as cattle sheep and horse fodder (Burkart 1967; Ulibarri and Burkart 2000; González et al. 2016; Ladio and Molares 2017).

Botanical characteristics *A. boronioides* is a medium-sized shrub reaching 1 m in height, very branched, with glandular ascending branches, fragrant and very resinous. Leaves are petiolate, with numerous elliptical leaflets (10–20). Leaflets have an irregular margin (smooth-wavy). Scented flowers are arranged in terminal clusters. The calyx is glandulous and pubescent, with tooth equal in length to the calyx tube. The corolla is yellow, with a heart-shaped banner. The fruit is a legume with 3–5 strangulations (Peltz 1962; Dimitri 1974; Correa 1984; Nájera et al. 2000; Vidal 2007).

Ethnobotanical information The aerial organs, fresh or dried, are used as masticatory and to prepare infusions against rheumatic pain, cold, alopecia, and gastrointestinal and renal problems, to treat burns and bruises, and to relieve menstrual discomforts (Montes and Peltz 1963; Martínez Crovetto 1980; Montes and Wilkomirsky 1985; Silva et al. 2004; Backhouse et al. 2008). Also, it is considered having bronchodilator properties and aphrodisiacs (Montes and Wilkormirsky 1985; Nájera et al. 2000).

Chemical data The EO have a high percentage of bisnorsesquiterpenes as esquel-6-en-9-one and esquel-7-en-9-one. Also, it has α-copaen-11-ol, δ-cadinene, 10-epi-γ-eudesmol, 4α-hydroxydihydroagarofuran, 1-epicubenol, guaia-6,9-diene, and furopelargones A and B (Fig. 9.2). The relative proportion of the main compounds was directly related to the season in which the plant material was collected. Autumn appeared as the optimal harvest time for industrial purposes of this species (González et al. 2002, 2004; Barboza et al. 2009; Gastaldi et al. 2016).

Biological activities The antioxidant activity (DPPH and BCB assays) was evaluated in a 5% infusion (Backhouse et al. 2008; Gastaldi et al. 2016, 2018). The polar

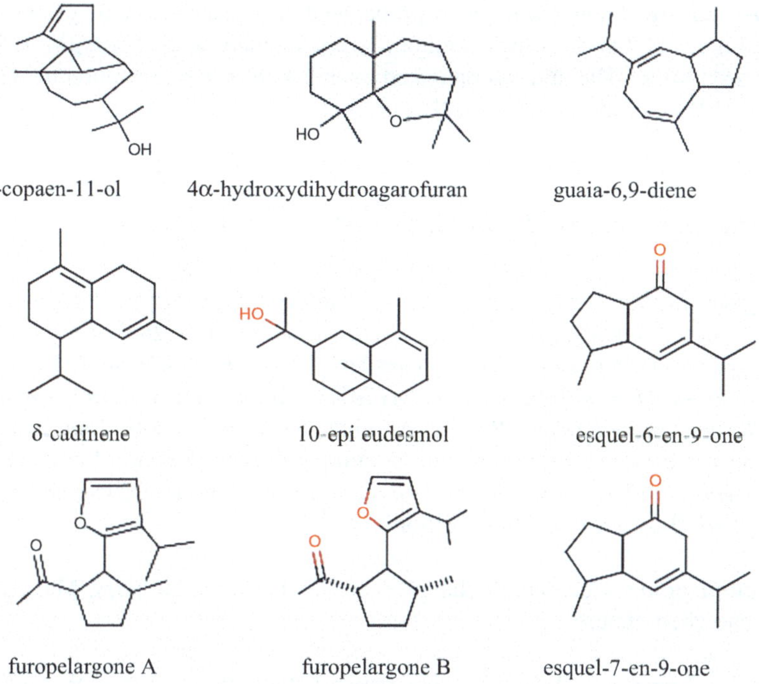

Fig. 9.2 Chemical compounds from *Adesmia boronioides* Hook. f.

fractions (methanol and aqueous extract, 0.5 ml 100 g^{-1}) and EOs obtained by hydrodistillation of *A. boronioides* aerial parts were *in vitro* tested at concentrations of 15 and 50 μg ml^{-1} for their effects on cyclooxygenases (COX) and 5-lypoxygenases (LOX) pathways of eicosanoid generation such as thromboxane-B2 (TXB2), prostaglandin E2 (PGE2), and leukotriene B4 (LTB4) in stimulated rat peritoneal leukocytes. The methanol extract caused a potent inhibition of LTB4 and TXB2 generation. The EO displayed strong inhibition of LTB4 but was less active on TXB2 generation. The aqueous extract was inactive on LTB4 and with very low activity on TXB2 generation. The effect on PGE2 production was poor, suggesting that the site of action is on thromboxane synthetase rather than on COX (González et al. 2003).

The lyophilized infusions (5 mg ml^{-1}) showed an antiproliferative activity (MTT assay) on the T84 (tumoral) and HTR-8/SVneo (non-tumoral) cell lines. The percentage of viable cells decreased with the increase in the concentration; however, the highest antiproliferative effect was on T84 cells (Kogiannou et al. 2013; Gastaldi et al. 2018).

Toxicity *A. salina* bioassay showed toxic effects (LC50 5–17 mg ml^{-1}); additional studies have to be performed in order to define the safety and long-term effects of the extracts (Gastaldi et al. 2018).

In vitro culture Plants were micropropagated from sterile seeds, grown in MS (Murashige and Skoog 1962) medium supplemented with 2219 μM BAP and 0.053 μM NAA. The micropropagated plants had 1 mg g^{-1} polyphenolic and 2 mg g^{-1} FW anthocyanin content (Estomba et al. 2010).

9.3 *Berberis microphylla* G. Forest (Berberidaceae)

Synonymy *Berberis buxifolia* Lam. var. antucoana, *Berberis buxifolia* Lam., *Berberis heterophylla* Juss. ex Poir., *Berberis michay* Job, *Berberis parodii* Job, *Berberis inermis* Pers., *Berberis cuneata* DC., *Berberis antucoana* C.K. Schneid., *Berberis barilochensis* Job, *Berberis buxifolia* Lam. var. nuda, *Berberis marginata* Gay, *Berberis spinosissima* (Reiche) Ahrendt, *Berberis buxifolia* Lam. var. spinosissima, *Berberis buxifolia* Lam. var. gracilior, *Berberis buxifolia* Lam. var. papillosa, *Berberis dulcis* Sweet, *Berberis buxifolia* Lam. var. antarctica, and *Berberis heterophylla* Juss. ex Poir. var. pluriflora.

Vernacular name Calafate, michay, box-leaved barberry, barberry, Magellan barberry, and cheu-cheu.

General information *B. buxifolia* is a native Patagonian species with a wide geographical distribution. In Argentina, it grows from the Patagonian woods to Tierra del Fuego, in the provinces of CHU, NEU, RNE, SCR, and TDF (Bottini et al. 2000; Bottini and Arena 2000). It also could be found in the Patagonian steppe and sub Antarctic forests growing in the *cordillera de los Andes*, from MEN and NEU to TDF (Rapoport et al. 1999; Roig 2001). The fruits were consumed by the *onas* from TDF that also had used *calafate* wood to make arrows. The *tehuelches* prepared a not fermented drink with the crushed and macerated fruits (Martínez Crovetto 1982). Recently, the fruits are having a high demand both to be consumed fresh and for the elaboration of foodstuff (Arena and Cabana 2008). It is also used as ornamental, in parks and gardens (Bottini 2000).

Botanical characteristics *B. buxifolia* is a perennial rhizomatous bush, highly branched, with an axillary main central thorn and two low developed thorns at each side. The leaves are simple, fasciculate, elliptic, coriaceous, and mucronate, with a smooth margin. The adaxial side of the leaves is bright and the abaxial one is opaque. The orange-yellowish flowers are solitary or grouped in inflorescences, umbels with 2–3 flowers. The perianth has six sepals and six petals in alternating cycles of three (Arena et al. 2011). The calyx has leathery and yellow velvety petals with two basal nectaries. Androecium has six retractable stamens, opposite to the petals, with two dithecae-anthers with valvar dehiscence. Fruits are subglobose berries purplish blue when mature with 6–10 dark seeds (Fig. 9.3) (Orsi 1984; Landrum 1999; Arena and Curvetto 2008; Arena et al. 2011, 2013; Arena and Radice 2014). Blooming is in November; the fruiting stage is between January and March (Tell et al. 1997; Rapoport et al. 1999).

Fig. 9.3 *Berberis microphylla* G. Forest (Berberidaceae) branches with leaves and fruits. MA Alvarez personal collection

Ethnobotanical information This plant was used by native communities to treat fever, inflammation, stomachache, diarrhea, urinary tract infection, throat infection, gingivitis, and liver problems. Also, as an antiseptic and as a source of natural dyes (Zin and Weiss 1998). The fruit was taken by the *Kawésqar* people as food; the *Aonikenk* used the yellow scraping of the bark as tobacco for its hallucinogenic effect; and the *Mapuche* community uses the fruit to treat colds and intermittent fever (Domínguez et al. 2012). Hieronymus (1882) attributed to *B. buxifolia* the astringent and substitutive properties of quinine. The extracts are employed in folk medicine mainly for their antibacterial and antidiarrheal properties. The decoction made with leaves and stems is used for stomachache and other intestinal irritations. The fruits are considered as astringent, digestive, and antidiarrheal, while the roots as antidiarrheal, astringent, febrifuge, digestive, hepatic, anti-inflammatory, and tonic. A non-specified part was referred as antipyretic, and for cold. The decoction of dried bark is used for urinary and respiratory tract infections (Ratera and Ratera 1980; Conticello et al. 1997; Villinski et al. 2003; Caraballo and Caraballo 2004; Alvarez et al. 2009; Barboza et al. 2009; Alonso and Desmarchelier 2015).

Chemical data The dried entire plant has the alkaloids chillanamine, chiloenamine, chiloenamine A, B, chiloenine, calafatamine, calafatine, calafatine 2α-N-oxide, calafatine 2-β-N-oxide, curacutine, and talcamine. It also has resins, oil, starch, and malic acid (Pomilio 1973; Podestá et al. 1987). The fruits have the anthocyanins peonidin, malvidin, petunidin and delphinidin, peonidin-3-glucoside, malvidin-3-

rutinoside, malvidin-3-glucoside, petunidin-3-rutinoside, petunidin-3-glucoside, peonidin-3-rutinoside-5-glucoside, delphinidin-3-rutinoside, delphinidin-3-glucoside, petunidin-3-rutinoside-5-glucoside, and petunidin-3-gentiobioside (Pomilio 1973). The dried root has argemonine and berberine. The roots have berberine, allocryptopine, calafatine, jatrorrhizine, palmatine, protopine, reticuline, and thalifendine. The stems have berberine, allocryptopine, isocorydine, jatrorrhizine, protopine, scoulerine, and thalifendine. Leaves have berberine and tetrahydroberberine (Fig. 9.4) (Fajardo et al. 1979; Podestá et al. 1987; Manosalva et al. 2014).

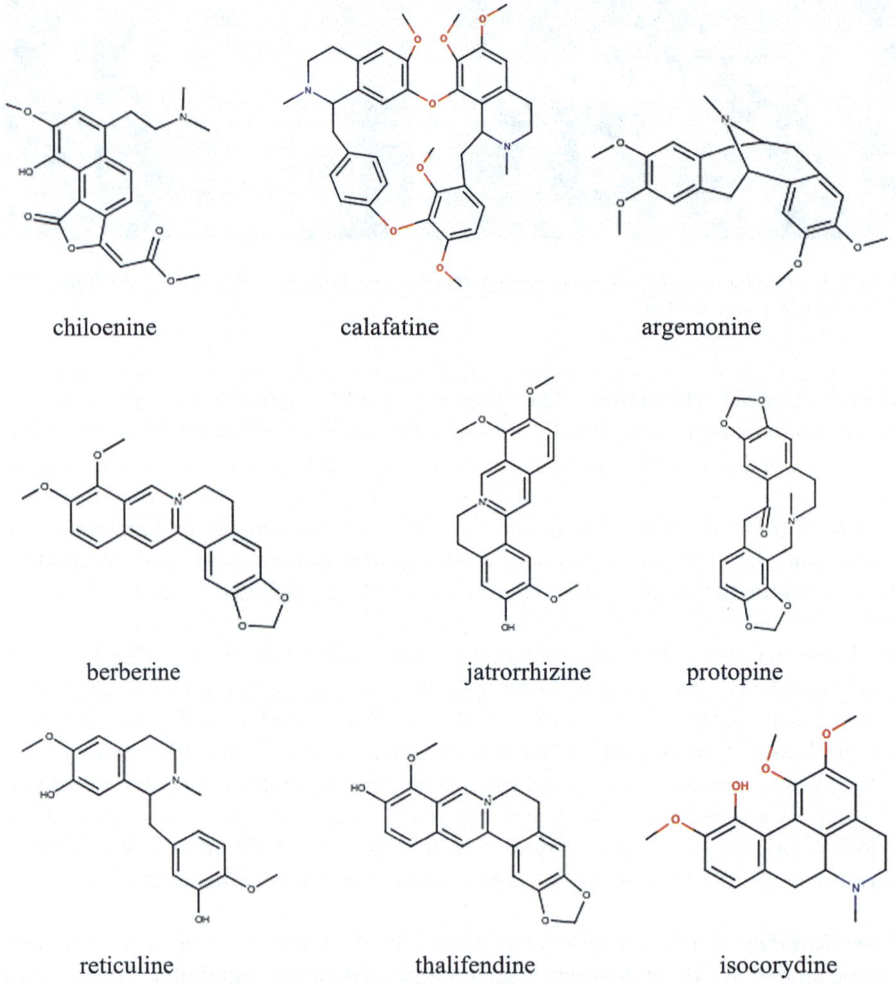

Fig. 9.4 Chemical compounds from *Berberis microphylla* G. Forest (Berberidaceae)

Biological activities Berberine has been defined as a potential drug with relevant pharmacological properties (Singh et al. 2010; Furrianca et al. 2015). The antimicrobial activity was attributed to the alkaloids berberine, jatrorrhizine, and tetrahydroberberine (Iwasa et al. 1996

Manosalva et al. 2014). *B. buxifolia* fruit appeared to possess excellent antioxidant activity as it ripens, which may contribute to its functional value, acquiring relevance and applicability for nutraceutical purposes (Arena et al. 2013). The co-administration of *B. buxifolia* aqueous extract in combination with chloramphenicol was proposed to contribute to the detoxification of oxidized metabolites originated by antibiotic treatment (Albrecht et al. 2009).

When the methanolic extracts of leaves, stems, and roots were tested against *Escherichia coli* (ATCC 25922), *Salmonella typhimurium* (ATCC 13311), *Listeria monocytogenes* (ATCC 13932), *Enterobacter aerogenes* (ATCC 13084), *Staphylococcus aureus* (ATCC 25923), *Bacillus cereus* (ATCC 11778), *Staphylococcus epidermidis* (ATCC 12228), and *Bacillus subtilis* (ATCC 6633), the antibacterial activity was only observed against Gram-positive bacteria (Manosalva et al. 2016). The antibacterial activity of the extract differs according to the plant tissue and the bacteria, which would be related to their different alkaloid compositions, which seems to act synergically in stems and roots. On the contrary, the lower activity found in leaf extracts could be related to the presence of the inactive antimicrobial tetrahydroberberine.

A synergistic effect was found against *B. cereus*, *B. subtilis*, and *S. epidermidis* with the association of the alkaloid extract with ampicillin (AMP). A similar effect was found with the associations of the stem alkaloid extract and AMP against *S. aureus* and *S. epidermidis* and root alkaloid extract and AMP against *B. subtilis*. Also, a synergistic effect was shown between the leaf alkaloid extract and cephalothin (CFL) against *S. aureus*, *B. cereus*, and *S. epidermidis*, the stem alkaloid extract and CFL against *S. aureus* and *B. subtilis*, and the root alkaloid extract with CFL against *B. cereus* and *S. epidermidis* (Manosalva et al. 2016). Berberine, which is the principal component of the alkaloid extracts, has been shown to *in vitro* enhance the antibacterial activity of selected antibiotics against coagulase-negative *Staphylococcus* strains (Wojtyczka et al. 2014). On the other hand, berberine demonstrated *in vitro* activity against *Staphylococcus epidermidis*, *Streptococcus* spp., *Corynebacterium diphtheriae*, *Trichomonas vaginalis*, *Neisseria gonorrhoeae*, *N. meningitidis*, *Treponema pallidum*, *Entamoeba histolytica*, and *Leishmania donovani* (Kowalewski et al. 1972).

Berberine is highly effective to treat human diarrhea, with a potent activity against *Escherichia coli* enterotoxins, as well as *Salmonella typhi*, *Shigella dysenteriae*, *Giardia lamblia*, *Salmonella paratyphi* B, *Klebsiella pneumoniae*, and *Vibrio cholerae* (Rabbani et al. 1987). However, the aqueous extract from leaves and cortex did not present antimicrobial action against *Staphylococcus aureus*, *E. coli*, *Aspergillus niger* (Anesini and Perez 1993), and *Salmonella typhi* (Pérez and Anesini 1994).

It was reported the use of berberine against *Candida albicans*, to fight trachoma, an ocular infection produced by *Chlamydia trachomatis* (Pizzorno and Murray 1985), the *in vitro* inhibition of influenza A and B virus replication (Lesnau 1990), and the inhibition of the α-amylase enzyme, essential for the production of aflatoxins by the human pathogenic fungus *Aspergillus flavus* (Tintu et al. 2012).

Berberine has hypoglycemic effect; early studies arise from 1986; since then, berberine was introduced as an alternative medicine for the treatment of diabetes in Asian countries (Yin et al. 2012). The alkaloid jatrorrhizine had also hypoglycemic activity, causing a pronounced glycemia decrease both in normal and hyperglycemic mice mediated by an improvement of aerobic glycolysis (Yan et al. 2005). However, the response was lower than that obtained with berberine at the same dose (Fu et al. 2005). Palmatine also lowered glycemia in normal rats (Patel and Mishra 2011; Sangeetha et al. 2013).

In vitro culture Rooted shoots (89%) were obtained on 1/2 MS medium and 1.25 pM IBA for 7 days, followed by IBA-free medium until day 28 (Arena et al. 2000). Callus was induced in1/2 MS medium with 2-inositol phosphate N6-(2-isopentenyl) adenine (2iP) as plant growth regulator. Cell suspension cultures and *in vitro* shoot cultures capable of producing berberine were obtained by a two-stage culture, using in the first step picloram (PIC) and thidiazuron (TDZ) as plant growth regulators for producing biomass and in the second step by transferring the cultures to MS medium containing BAP (4.4 mM) as plant growth regulator. Berberine yields were significantly lower in cell suspension cultures (102 mg g^{-1} DW) and shoots (200 mg g^{-1} DW) than in plants growing in the field (416 mg g^{-1} DW). Furthermore, berberine was secreted into the medium which is advantageous for conducting biotechnological processes in bioreactors. Extracts from *in vitro* cultures have demonstrated antimicrobial activity against *Escherichia coli* and *Staphylococcus aureus* RN 4222 and KLE820 (Alvarez et al. 2009).

Legal Status

The Argentine Food Code has incorporated berberis as a native fruit of the Andean zone. It is authorized the use of berberis in foodstuffs such as sweets, jams, liqueurs, ice cream, and confectionery (ANMAT Resolutions 22/2006 and 409/2006, Res SPReI 93/2008, and SAGPyA 356/2008). The United States Department of Agriculture (USDA) has approved its entry into the USA.

9.4 *Buddleja globosa* Hope (Scrophulariaceae)

Synonymy *Buddleja capitata* Jacq., *Buddleja connata* Ruiz & Pav., and *Buddleja globifera* Duhamel.

Vernacular name Matico, pañil, panguil, matica, palquín, matico, pallín, and ballín.

General information *B. globosa* grows in dry and moist forests, from sea level up to 2000 m.a.s.l. Blooming is from November to February and fructification from December to February. It grows in the provinces of CHU, NEU, and RNE. The flowers are used to prepare fermented cold drinks and the fruits to make foodstuff. Traditionally, it is grown in home gardens for domestic use (Norman and Ariza Espinar 1995; Vogel et al. 2010).

Botanical characteristics *B. globosa* is a dioecious bush, 1–5 m high, with tomentose branches. Leaves are subsessile, with conspicuous stipulation, lanceolate or elliptic, subcoriaceous to membranoceous, glabrescent, and decurrent to the base. Flowers are clustered in pairs in pedunculated heads. The tubular tomentose calix is tubular, with acute lobes. The intense yellow to orange corolla is glabrous on the outside (Fig. 9.5). Stamens are sessile; ovary is tomentose in the upper half with a claviform style. The fruit is an ellipsoidal capsule, tomentose and glandular, with septicidal dehiscence and rounded valves (Norman and Ariza Espinar 1995).

Ethnobotanical information The infusion made with leaves has a pleasant taste and is consumed by the *Mapuche* to treat ulcers and wounds. The infusion made with dried aerial parts is taken as digestive and hepatic, the decoction to treat urinary and respiratory tract infections. The infusion made with dried leaves is used as urogenital antiseptic; to treat chronic dysentery; for stomach ulcers; as hemostatic, antirheumatic, spasmolytic, antidiarrheal, antibacterial, anti-inflammatory, and antihemorroidal; and for treating external wounds, ulcers, and acne (Houghton 1984; Backhouse et al. 2008; Barboza et al. 2009; Gastaldi et al. 2016, 2018; Gallegos Zurita and Gallegos Zurita 2015).

Chemical data It has coumaric acid, p-methyl ester, echinacoside, buddledins A and B, dehydrobuddledin A, zerumbone, buddledones A and B, acubin, catalpol, and their related compounds, α- amyrin, linarin, verbascoside, lupeol, and maytenone (Fig. 9.6) (Houghton 2003; Barboza et al. 2009; Vogel et al. 2010).

Fig. 9.5 *Buddleja globosa* Hope (Buddlejaceae). MA Alvarez personal collection

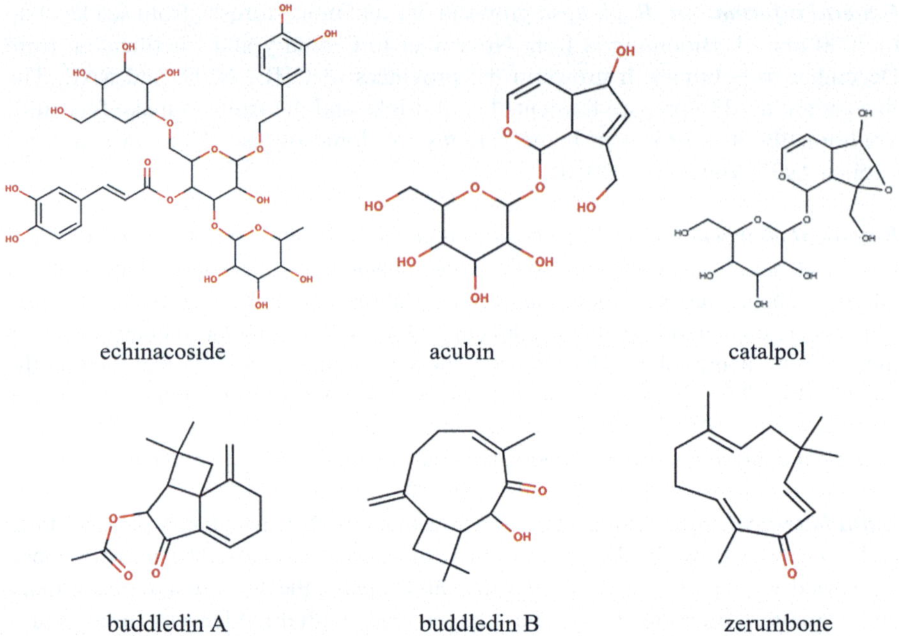

Fig. 9.6 Some chemical compound found in *Buddleja globosa* Hope (Scrophulariaceae)

Biological activities The antioxidant activity (DPHH and BCB assays) of a 5% infusion was reported to be comparable to that of green tea (Mensah et al. 2001; Sünter et al. 2012; Bustamante et al. 2015; Gastaldi et al. 2016, 2018).

Anti-inflammatory, cicatrizing, and analgesic activities as well as the regeneration of the mucous membrane in case of ulcers were reported (Liao et al. 1999; Mensah et al. 1998, 2001; Vargas 2001; Rodríguez et al. 2001; Yañez 2001).

A recent *in vitro* study tested the aqueous extract of leaves on growth and protection from oxygen-free radical attack induced by hydrogen peroxide of cultured fibroblasts. A weak and not statistically significant increase in growth was observed at concentrations below 1 g ml^{-1}; concentrations above 10 g ml^{-1} inhibited growth and appeared to be cytotoxic. Antioxidant activity was significant at 50 g ml^{-1}. The activity was attributed to flavonoids and phenylethanoids, being verbascoside, echinacoside, and linarin, twice as active as flavonoid aglycones (Mensah et al. 2001).

As for wound healing, the *in vivo* effect of this extract was tested in adult male Sprague-Dawley rats showing no alteration of hemogram and clinical chemistry parameters after 12 days of oral treatment. In addition, the topical treatment with the same extract accelerated wound healing presumably by shortening the inflammatory stage, as assessed by COX-2 levels, and by accelerating cellular proliferation rate, as shown by the date of the proliferation marker Ki-67. This wound healing effect was also attributed to the extract antioxidant activity (Letelier et al. 2012). *B. globose* extract was a potent inhibitor of collagen-mediated platelet aggregation, probably for its high content of antiplatelet compounds such as luteolin and verbascoside. The cellular mechanism of that effect might be mediated by the inhibition of PLC-g2 and PKC-b2 phosphorylation (Fuentes et al. 2017). The authors postulate a potential use of *B. globosa* preparations to prevent thrombosis-related diseases (Bustamante et al. 2015; Fuentes et al. 2017).

The leaf aqueous extract (1 mg ml^{-1}) protected hepatocytes challenged with carbon tetrachloride, galactosamine, and complement-mediated cytotoxic medium (CMC). From the isolated compounds, linarin was the most protective against carbon tetrachloride (40% reduction) and galactosamine (75% reduction). Echinacoside and aucubin were also hepatoprotective (Houghton and Hikino 1989; Chang 1998; Houghton 2003).

Polyphenol content was related to the inhibition of lipid peroxidation. It was found that leaves from cultivated plants had the same or even higher polyphenol content than that from wild plants. Extracts from young leaves showed the best protection on thiol groups. Less watered plants showed higher polyphenol contents and protection on thiol groups, whereas the inhibition of lipid peroxidation was not affected by irrigation (Vogel et al. 2010).

The antiproliferative activity of the infusion (5 mg ml^{-1}) was tested by the MTT assay on T84 (tumoral) and HTR-8/SVneo (non-tumoral) cell lines. *B. globosa* had a higher antiproliferative activity on HTR-8/SV neo cells than on T84 cells (Gentile et al. 2016; Kogiannou et al. 2013; Gastaldi et al. 2018).

The sesquiterpenes buddledins A–C 15–17 (50 mg ml^{-1}) inhibited both cyclooxygenase (COX) and 5-lipoxygenase (5-LOX) enzymes being the activity stronger against 5-LOX. Buddledin A15 showed the greatest inhibitory activity (COX 89%, 5-LOX 98% inhibition), while zerumbone 19 and dihydrobuddledin A 18 showed no activity against COX and 5-LOX at the concentrations tested (Liao et al. 1999).

Lipophilic root extracts were tested against pathogenic, dermatophytes, and soil fungi; the sesquiterpene components were selectively antifungal against dermatophytes (*Epidermophyton floccosum*, *Trichophyton interdigitale*, and *E. rubrum*) and soil fungi with little effect on *Penicillium notatum*, *Aspergillus niger*, and *Candida albicans* (Mensah et al. 2000). Aqueous and methanol extracts of *B. globosa* aerial parts showed antiplasmodial activity, but the compounds responsible for this activity have not yet been isolated (Debenedetti et al. 2002). Also, a nematicidal effect was reported (Bohm et al. 2009).

The antinociceptive activity of sequential ethanol extracts, verbascoside, and the global methanol extract was tested using complementary chemical and thermal mice models of acute pain. The analgesic effect was probably produced through a mechanism partially linked to either lipoxygenase or cyclooxygenase enzymes via the arachidonic acid cascade and/or opioid receptors. When verbascoside and luteolin glycoside were tested, they showed similar topical analgesic effects, but verbascoside exhibited a much stronger oral analgesic effect than the flavonoid (Merfort et al. 1994; Backhouse et al. 2008; Küpeli et al. 2007).

Leaf extracts were also used to green synthesis of silver nanoparticles with potential biotechnological applications in medicine, agriculture, and bioengineering among others (Carmona et al. 2017).

Finally, the hydroalcoholic extract was effective to prevent the oxidation of lipids and thiol groups in *Oncorhynchus mykiss* (rainbow trout) stakes mediated by polyphenols and terpenes (Gormaz Araya 2005).

9.4.1 Toxicity

Mutagenic and antimutagenic activity of the aqueous extract (3.83–15.32 mg ml^{-1}) was tested by the wing-spot test in *Drosophila melanogaster*. At 15.32 mg ml^{-1}, toxicity was high, but at lower concentrations (1.91, 3.83, and 7.66 mg ml^{-1}), it did not induce a significant increase in the frequency of mutant spot formation. When a combined treatment, leaf extract/ethyl methanesulfonate (EMS), was tested, it was reported a relevant reduction of single, large, and total mutant spots in somatic wing cells in comparison to those induced by the EMS alone (Carmona et al. 2016).

9.5 *Verbena litoralis* Kunth var. litoralis (Verbenaceae)

Synonymy *Verbena approximata* Briq., *Verbena caracasana* Kunth, *Verbena cordobensis* Briq., *Verbena litoralis* Kunth var. caracasana, *Verbena litoralis* Kunth var. leptostyachya, *Verbena litoralis* Kunth var. pycnostachya, *Verbena affinis* M. Martens & Galeotti, *Verbena litoralis* Kunth var. glabrior, *Verbena nudiflora* Nutt. ex Turcz., *Verbena litoralis* Kunth f. angustifolia, *Verbena litoralis* Kunth var. albiflora, *Verbena integrifolia* Sessé & Moç. f. albiflora, *Verbena bonariensis* L. var. litoralis, *Verbena integrifolia* Sessé & Moç., and *Verbena litoralis* Kunth f. albiflora.

Vernacular name Verbena del litoral, yerba del acero, verbena encañutada, verbena, gervãozinho-do-campo, and erva-de-pai-caetano (Portuguese).

General information *V. litoralis* is widespread in the country, growing in the provinces of BAI, CAT, CHA, COR, COS, ERI, FOR, JUJ, LPA, LRI, MEN, MIS, NEU, RNE, SAL, SDE, SFE, SJU, SLU, and TUC. Flowering is from mid-spring until mid-autumn.

Botanical characteristics *V. litoralis* is a perennial herb with a quadrangular glabrous stem, opposite, narrow, ovate, ovate-lanceolate, or lanceolate leaves with irregular serrate margins. The blue, purple, or lavender small flowers are arranged in groups of three in terminal cylindrical thin spikes, being the central larger than the two laterals. Bracts are lanceolate; the tubular calix has short, triangular, uneven teeth. Corolla has tubular lobules. Mericarps are cylindrical, brownish, with rounded apex and base (Correa 1999; Tasso de Souza et al. 2005, Alonso and Desmarchelier 2015).

Ethnobotanical information The decoction is used in cases of skin allergy and the aerial parts as cicatrizing and anti-inflammatory and for neonatal and pediatrics care, menorrhagia, and menopause. The decoction of dried branches and leaves is used to treat diarrhea, and the decoction from leaves as hepatic, digestive, emmenagogue, anti-cephalea, febrifuge, and contraceptive to treat angina and pharyngitis and to expel intestinal tapeworms. The root is considered as antitussive (Di Lullo 1929; Marzocca 1997; Toursarkissian 1980; Tasso de Souza et al. 2005; Barboza et al. 2009).

Chemical data The chemical compounds found in *V. litoralis* are verbenalin, verbenaloside, brasosid, aurantiamide acetate, jionoside, and verbenachalcone: dimeric dihydrochalcon-60 (Fig. 9.7) (Barboza et al. 2009; Rodríguez Morcelle 2015).

Biological activity Antioxidant activity was tested in *V. litoralis* hexanol extracts by DPPH test, *and* (IC50 = 169.3 μg ml^{-1}) (Braga et al. 2012). The hydroalcoholic

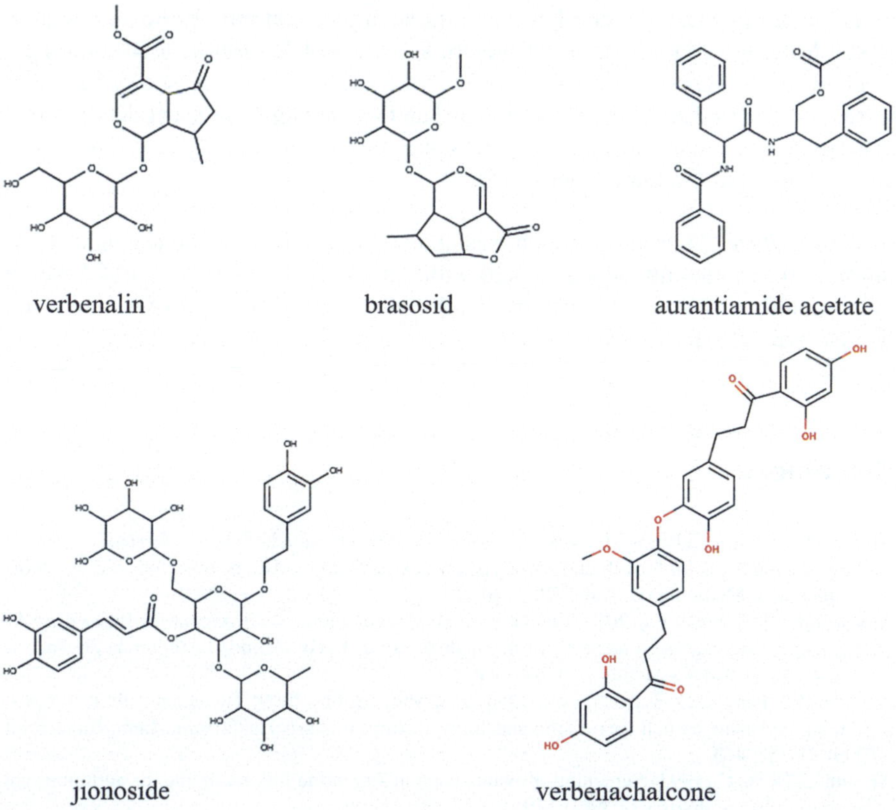

Fig. 9.7 Chemical compounds from *Verbena litoralis* Kunth var. litoralis

(1: 9 v/v) leaf extract showed activity against *Staphylococcus aureus* (ATCC 25923) with an inhibition hallo of 11 mm diameter (Toribio et al. 2009).

Antinociceptive activity was tested on male Swiss mice using *V. litoralis* hexane (100, 300 and 500 mg kg^{-1}) extracts by writhing and hot plate tests. A good antinociceptive potential was observed, but it was not possible to establish a dose-dependent relation. However, the potential supraspinal analgesic activity was only proved to the highest dose (Braga et al. 2012). The anticonvulsant activity of the ethanolic extract was tested for anticonvulsant activity in pentylenetetrazol (PTZ) and maximal electroshock (MES) induced seizures in mice. The extract significantly increased the period taken before the onset and decreased the duration of the seizures induced by PTZ without signs of toxicity on animals (Rashidin et al. 2017).

The treatment of corn crops with different concentrations of *V. litoralis* (0.4, 0.8, and 1.6 g per every 10 g of corn) showed an inhibitory activity on *Sitophilus granarius* (maize beetle) with a 50% mortality rate at 1.6 g (Silva et al. 2001; Arango Gutiérrez and Vásquez Villegas 2008).

Toxicity Mice treated with an infusion (>25 g kg^{-1}) for 6 months did not have micro or macroscopic histopathological lesions. A dose of 500 mg kg^{-1} of β-myrcene was reported to cause skeletal fetus alterations in pregnant rats (Paumgartten et al. 1988). Subacute administration of the crude extract of *V. litoralis* at 400 mg kg^{-1} resulted in an increase in AST in males, whereas ALT enzyme showed a small increase in males that received 200 mg kg^{-1} and 400 mg kg^{-1} of the extract (de Lima et al. in press). Small farm animals that consumed *V. litoralis* suffered diarrhea or died (Alonso and Desmarchelier 2015).

In vitro culture Micropropagation was attained transferring nodal segments of *V. litoralis* to MS medium supplemented with 7.5 μM BAP and 0.0005 μM NAA to induce multiple shoots and transferring to MS medium with 0.2 μM IAA for rooting (Braga et al. 2012).

References

Albrecht C, Pellarin G, Rojas MJ, Albesa I, Eraso AJ (2009) Beneficial effect of *Berberis buxifolia* Lam, *Ziziphus mistol* Griseb and *Prosopis alba* extracts on oxidative stress induced by chloramphenicol. Medicina (B. Aires) 70(1):65–70

Alonso J, Desmarchelier CJ (2015) Plantas medicinales autóctonas de la Argentina. Bases científicas para su aplicación en atención primaria de la salud. 1a ed. – Ciudad Autónoma de Buenos Aires: Corpus Libros Médicos y Científicos

Alvarez MA, Fernández Eraso N, Pitta-Alvarez S, Marconi PL (2009) Two-stage culture for producing berberine by cell suspension and shoot cultures of *Berberis buxifolia* Lam. Biotechnol Lett 31:457–463

Anesini C, Perez C (1993) Screening of plants used in Argentine folk medicine for antimicrobial activity. J Ethnopharmacol 39(2):119–128

Arango Gutiérrez GP, Vásquez Villegas MC (2008) Efecto tóxico de*Verbena officinalis* (familia Verbenaceae) en *Sitophilus granarius* (coleoptera: curculionidae). Revista Lasallista de Investigación 5(2):74–82

Arena ME, Cabana JA (2008) Frutas finas nativas de la Patagonia austral: los Berberis. In: de Sesar D et al (eds) Avances en cultivos frutales no tradicionales: Arándanos-Cerezas-Frutillas-Granadas. FAUBA, Buenos Aires, pp 237–244

Arena ME, Martínez Pastur G, Vater G (2000) *In vitro* propagation of *Berberis buxifolia* Lam. Biocell 24(1):73–80

Arena ME, Giordani E, Radice S (2011, New York) Flowering, fruiting and leaf and seed variability in *Berberis buxifolia*, a native Patagonian fruit species. In: Marin L, Kovac D (eds) Native species: identification, conservation and restoration, Nova Sciences Pub-lishers, pp 117–136

Arena ME, Postemsky P, Curvetto NR (2013) Accumulation patterns of phenolic compounds during fruit growth and ripening of *Berberis buxifolia*, a native Patagonian species. N Z J Bot 50(1):15–28

Arena ME, Radice S (2014) Shoot growth and development of *Berberis buxifolia* Lam. in Tierra del Fuego (Patagonia). SciHortic 165:5–12

Arena ME, Curvetto N (2008) *Berberis buxifolia* fruiting: kinetic growth behavior and evolution of chemical properties during the fruiting period and different growing seasons. Sci Hortic 118:120–127

Backhouse N, Rosales L, Apablaza L, Goïty L, Erazo S, Negrete R, Theodoluz C, Rodríguez J, Delporte C (2008) Analgesic, anti-inflammatory and antioxidant properties of *Buddleja globosa*, Buddlejaceae. J Ethnopharmacol 2(5):263–269

Barboza GE, Cantero JJ, Núñez C, Pacciaroni A, Ariza Espinar L (2009) Medicinal plants: a general review and a phytochemical and ethnopharmacological screening of the native Argentine Flora. Kurtziana 34(12):7365

Bohm L, Arismendi N, Ciampi L (2009) Nematicidal activity of leaves of common shrub and tree species from Southern Chile against *Meloidogyne hapla*. Ciencia e Investigacion Agraria 36:249–258

Bottini M, Greizerstein EJ, Aulicino MB, Poggio L (2000) Relationships among genome size, environmental conditions and geographical distribution in natural populations of NW Patagonian species of *Berberis L.* (Berberidaceae). Ann Bot 86:565–573

Bottini MCJ, Arena M (2000) Aportes al conocimiento de *Berberis buxifolia* (Berberidaceae). Bol Soc Arg Bot 35(1.2):161–162

Bottini MCJ (2000) Estudios multidisciplinarios en las especies patagónicas argentinas del género *Berberis* L. (Berberidaceae). Tesis doctoral. Universidad de Buenos Aires, Facultad de Ciencias Exactas y Naturales. pp. 223

Braga VF, Mendes GC, Oliveira RTR, Soares CQG, Resende CF, Pinto LC, de Santana R, Viccini LF, Raposo NRB, Peixoto PHP (2012) Micropropagation, antinociceptive and antioxidant activities of extracts of *Verbena litoralis* Kunth (Verbenaceae). An Acad Bras Cienc 84(1):139–147

Burkart A (1967) Sinopsis del género sudamericano de Leguminosas Adesmia D.C. Darwin 14:463–572

Bustamante S, Álvarez N, Mendiburen R, Vergara F, Zárate I, Collado C, Morales M (2015) Fundación pré-clínica del uso etnomédico de matico (*Buddleja globosa* Hope). Revista de Fitoterapia 15(1):37–51

Caraballo A, Caraballo B (2004) Preliminary assessment of medicinal plants used as antimalarials in the southeastern Venezuelan Amazon. Rev Soc Bras Med Trop 37:186–188

Carmona ER, Escobar B, Obando V (2016) Antimutagenic activity of *Buddleja globosa* extracts in the Drosophila wing-spot test. Fresen Environ Bull 25(12a):5758–5764

Carmona ER, Benito N, Plaza T, Recio-Sánchez G (2017) Green synthesis of silver nanoparticles by using leaf extracts from the endemic *Buddleja globose* Hope (matico) extract in Sprague-Dawley rats. Green Chem Lett Rev 10(4):250–256

Chang IM (1998) Liver-protective activities of aucubin derived from traditional oriental medicine. Res Commun Mol Pathol Pharmacol 102:189–204

Conticello L, Gandullo R, Bustamante A, Tartaglia C (1997) El uso de plantas medicinales por la comunidad Mapuche de San Martín de los Andes, Provincia de Neuquén. Parodiana 10(1–2):165–180

Correa MN (1984). Flora Patagónica. Parte IV-b. Dicotiledóneas dialipétalas. Colección científica del INTA. Ed. INTA. Buenos Aires

Correa MN (1999) Flora Patagónica. Dicotyledones Gamopétalas (Ericaceae a Calyceraceae). Colleción científica del INTA, Buenos Aires Tomo VIII parte VI: 193–195, 202

Debendetti S, Muschetti L, van Baren C, Clavin M, Broussalis A, Martino V, Houghton PJ, Warhurst D, Steele J (2002) *In vitro* antiplasmodial activity of extracts of Argentinian plants. J Ethnopharmacol 80:163–166

Di Lullo O (1929) La Medicina popular en Santiago del Estero. Talleres Gráficos El Liberal, Buenos Aires

Dimitri M (1974) Pequeña Flora Ilustrada de los Parques Nacionales Andino-Patagónicos. Anales de Parques Nacionales, Buenos Aires. Tomo XIII: 1–122

Domínguez E, Aguilera O, Villa-Martínez R, Aravena JC, Henríquez JM (2012) Ethnobotanical survey of Kalau Island, ancestral Kawesqar territory, Magallanes Región, Chile. Inst Patagon Ann 40:19–35

Domínguez Díaz E (2010) Flora de interés etnobotánico usada por los pueblos originarios: Aónikenk, Selk'nam, Kawésqar, Yagan y Haush en la Patagonia Austral. Dominguezia 26(2):19.29

Estomba D, Matters Fernandez H, Stella AM (2010) Antioxidants and porphyrin from in vitro cultures of *Adesmia boronioides*, *Larrea divaricata* and *Atriplex lampa*. Rev FCA UNCuyo 42(2):135–142

Fajardo V, Urzúa A, Torres R, Cassels BK (1979) Metabolitos secundarios de *Berberis buxifolia*. Rev Latinoamer Quím 10:131–134

Fu Y, Hu B, Tang Q, Fu Q, Zhang Q, Xiang JZ (2005) Effect of jatrorrhizine, berberine, Huanglian decoction and compound-mimic prescription on blood glucose in mice. Chin Tradit Herb Drug 36:548–551

Fuentes M, Sepúlveda C, Alarcón M, Palomo I, Fuentes E (2017) *Buddleja globosa* (matico) prevents collagen-induced platelet activation by decreasing phospholipase C-gamma 2 and protein kinase C phosphorylation signaling. J Tradit Complement Med 8(1):66–71

Furrianca MC, Alvear M, Zambrano T, Barrientos L, Fajardo V, Salazar LA (2015) Medicinal value of the *Berberis genus* as hypoglycemic agent. Boletín Latinoamericano y del Caribe de Plantas Medicinales y Aromáticas 14(5):423–441

Gallegos-Zurita M, Gallegos-Z D (2015) Medicinal plants used for treatment of skin diseases in rural communities in Los Ríos province, Ecuador. An Fac Med 78(3):315–321

Gastaldi B, Assef Y, van Baren C, Di LeoLira P, Retta D, Bandoni AL, González SB (2016) Antioxidant activity in teas, tinctures and essential oils of native species from Patagonia Argentina. Rev Cubana Pl Med 21(1):51–62

Gastaldi B, Marino G, Assef Y, Silva Sofrás FM, Catalán CAN, González SB (2018) Nutraceutical properties of herbal infusions from six native plants of argentine Patagonia. Plant Foods Hum Nutr 73:180. https://doi.org/10.1007/s11130-018-0680-3

Gentile C, Reig C, Corona O, Farina V et al (2016) Pomological traits, sensory profile and nutraceutical properties of nine cultivars of loquat (Eriobotrya japonica Lindl.) fruits grown in Mediterranean area. Plant Foods Hum Nutr 71(3):330–338

González SB, Bandoni AL, van Baren C, Di Leo LP, García Rojas CMC, Natham PJ (2002) Structure, conformation and absolute configuration of novel bisnorsesquiterpenes from the *Adesmia boronioides* essential oil. Tetrahedron 58:3065–3071

González SB, Houghton PJ, Hoult JRS (2003) The activity against leukocyte eicosanoid generation of essential oil and polar fractions of *Adesmia boronioides* Hook.F. Phytother Res 17:290–293

González SB, Contardi LT, Pasquini NM (2016) Reproduction of Adesmia boronoides, a medicinal and aromatic plant currently harvested from natural populations in Patagonia Argentina. Naturalia patagónica 9:50–60

González SB, Bandoni A, van Baren C, Di Leo LP, García C, Joseph-Nathan P et al (2004) The essential oil of the aerial parts of *Adesmia boronioides* Hook f. J Essent Oil Res 16:513–516

Gormaz Araya JG (2005) Extracto hidroalcohólico de *Buddleja globosa* y extracto seco de *Rosmarinus officinalis* como preservantes de filetes de Onchorhynchus mykiss. Tesis de pregradro. Facultad de Ciencias Químicas y Farmacéuticas, Universidad de Chile. http://repositorio.uchile.cl/handle/2250/105446

Hieronymus J (1882) Plantas diafóricas. Flora Argentina, Editorial Atlántida

Houghton PJ (1984) Ethnopharmacology of some *Buddleja* species. J Ethnopharmacol 11(3):293–308

Houghton PJ (2003) *Buddleja globosa*: a medicinal plant of Chile, their chemistry, biological activity and traditional uses. Boletín Latinoamericano y del Caribe de Plantas Medicinales y Aromáticas 2(3):36–41

Houghton PJ, Hikino H (1989) Antihepatotoxic activity of extracts and constituents of *Buddleja* species. Planta Med 55:123–126

Iwasa K, Kamigauchi M, Ueki M, Taniguchi M (1996) Antibacterial activity and structure-activity relationships of berberine analogs. Eur J Med Chem 31:469–478

Küpeli E, Sahin FP, Yesilada E, Calis I, Eze N (2007) *In vivo* anti-inflammatory and antinociceptive activity of phenolic compounds from *Sideritis stricta*. Zeitschrift für Naturforschung C 62:519–525

Kogiannou D, Kalogeropoulus N, Kefalas P, Polissiou MG, Kaliora A (2013) Herbal infusions; their phenolic profile, antioxidant and anti-inflammatory effects in HT29 and PC3 cell. Food Chem Toxicol 61:152–159

Kowalewski Z, Kedzia W, Mirska I (1972) Effect of berberine sulphate on Staphylococci. Arch Immunol Ther Exp 20:353–360

Ladio AH, Molares S (2017) Etnoconservacionismo y prácticas locales en Patagonia: avances y perspectivas. In: Casas A, Torres Guevara J, Parra F (eds) Domesticación en el Continente Americano. Historia y perspectivas del manejo de recursos genéticos en el Nuevo Mundo, Chapter 29. Universidad Agraria La Molina. IIES, Lima, pp 649–672

Landrum LR (1999) Revision of *Berberis* (Berberidaceae) in Chile and adjacent Southern Argentina. Ann Missouri Bot Gard 86(4):793–834

Letelier ME, Jones R, López C, Palma K, Aracena P, Razmilic I, Polanco X, Vogel H (2012) Safety profile and wound healing properties of a standardized *Buddleja globose*. Rev Farmacol Chile 5(1):13

de Lima R, Gaube Guex C, Haas da Silva AR, Lima Lhamas C, de Santos Moreira KL, Casoti R, Castro Dornelles R, da MIUM R, da Veiga ML, de Freitas Bauermann L, Palermo Manfron M (in press) Acute and subacute toxicity and chemical constituents of the hydroethanolic extract of *Verbena litoralis* Kunth. J Ethnopharmacol 224:76. https://doi.org/10.1016/j.jep.2018.05.012

Liao YH, Houghton PJ, Hoult JR (1999) Novel and known constituents from *Buddleja* species and their activity against leukocyte eicosanoid generation. J Nat Prod 62(9):1241–1245

Manosalva L, Mutis A, Díaz J, Urzúa A, Fajardo V, Quiroz A (2014) Identification of isoquinoline alkaloids from *Berberis microphylla* by HPLC ESI-MS/MS. Bol Latinoam Caribe Plant Med Aromat 13:323–334

Manosalva L, Mutis A, Urzúa A, Fajardo V, Quiroz A (2016) Antibacterial activity of alkaloid fractions from *Berberis microphylla* G. Forst and study of synergism with ampicillin and Cephalothin. Molecules 21:76. https://doi.org/10.3390/molecules21010076

Marzocca A (1997b) Vademecum de malezas de la Argentina. Indígenas y exóticas. Orientación Gráfica, Buenos Aires

Mensah AY, Houghton PJ, Hughes M, Cherry G (1998) *In vitro* investigation of the wound-healing properties of Buddleja globosa. J Pharm Pharmacol 50:83

Martínez Crovetto R (1982) Breve panorama de las plantas utilizadas por los indios de la Patagonia y Tierra del Fuego. Supl Antropol 17:61–97

Martínez Crovetto R (1980) Apuntes Sobre la Vegetación del Lago Cholila, Publ. Técnica n°1 Universidad Nac Del Nordeste. UNNE: Corrientes. Argentina 1:1–22

Mensah AY, Houghton PJ, Bloomfield S, Vlietinck A, Vanden Berghe D (2000) Known and novel terpenes from *Buddleja globosa* displaying selective antifungal activity against dermatophytes. J Nat Prod 63:1210–1213

Mensah AY, Sampson J, Houghton PJ, Hylands PJ, Westbrook J, Dunn M, Hughes MA, Cherry GW (2001) Effects of *Buddleja globosa* Hope and its constituents relevant to wound healing. J Ethnopharmacol 77(2–3):219–226

Merfort I, Heilmann J, Hagedorn-Leweke U, Lippold BC (1994) *In vivo* skin penetration studies of camomileflavon estructures. Pharmazie 49:509–511

Molares S, Ladio AH (2015) Edible and medicinal plant complexes in Patagonia Argentina: components and possible associated processes. Blacpma 14(3):237–250

Montes A, Peltz L. (1963) Esencias de plantas aromáticas del parque Nacional Nahuel Huapi y sus aledaños. 2. *Adesmia boronioides* Hooker, o paramela. Anales Asoc Científica Argentina 91–101

Montes M, Wilkomirsky T (1985) Medicina Tradicional Chilena. Ed. de la Univ. de Concepción, Concepcion, pp 104–105

Murashige T, Skoog F (1962) A revised medium for rapid growth and bio assays with tobacco tissue cultures. Physiol Plant 15(3):473–497

Nájera M, Spegazzini E, Castro M, Guerra P, González S (2000) Anatomía Foliar y Caulinar de *Adesmia boronioides* Hook. f. (Leguminosae -Papilionoideae). Acta Farm Botlaeretzse 19(4):245–250

Norman EM, Ariza Espinar L (1995) Buddlejaceae Wilh. Flora Fanerogámica Argentina pp:1–14

Orsi MC (1984) Berberidaceae. In: Correa MN (ed) Flora Patagónica Sección 4ª. INTA, Buenos Aires, pp 325–348

Paumgartten F, De Carvalho R, Souza C, Madi K, Chahoud I (1988) Study of the effects of beta-myrcene on rat fertility and general reproductive perfomance. Braz J Med Biol Res 3:955–965

Pizzorno J, Murray M (1985) *Hydrastis canadiensis, Berberis vulgaris, Berberis aquitolium* and other berberine containing plants. In: Textbook of natural medicine. J Bastyr Coll Publ, Seattle (EEUU)

Podestá F, Fajardo V, Urzua A (1987) Chemical constituents of *Berberis buxifolia.* Lam Ans Inst Oat Cs Ns Punta Arenas (Chile) 17:99–102

Patel MB, Mishra S (2011) Hypoglycemic activity of alkaloidal fraction of *Tinospora cordi*folia. Phytomedicine 18:1045–1052

Peltz L (1962) Estudio de la composición del aceite esencial y resinoide de la *Adesmia boronioides* Hooker (Paramela). Thesis. Biblioteca digital FCEN-UBA, p.67

Pomilio AB (1973) Berberidaceae. Anthocyanins in fruits of *Berberis buxifolia.* Phytochem Rep 12:218–220

Rabbani GH, Butler T, Knight J, Sanyal SC, Alam K (1987) Randomized controlled trial of Berberine sulfate therapy for diarrhea due to Enterotoxigenic *Escherichia coli* and *Vibrio cholerae.* J Infect Dis 155(5):979–984

Rapoport E, Ladio A, Sanz E (1999) Plantas nativas comestibles de la Patagonia andina argentino-chilena. Dpto de Ecología, Centro Regional Universitario Bariloche. Ediciones de Imaginaria. Bariloche

Rashidin A, Kazemi F, Mehrzadi S, Reza Dehpour A, Ejtemai Mehr S, Mahdi Rezayat S (2017) Anticonvulsivant effects of aerial parts of *Verbena officinalis* extract in mice: involvement of benzodiazepine and opioir receptors. J Evid Based Compl Alt Med 22(4):632–636

Ratera EL, Ratera MO (1980) Plantas de la flora argentina empleadas en medicina popular. Ed. Hemisferio sur S.A

Rodríguez J, Theoduloz C, Yáñez T, Astudillo L, Schmeda-Hirschmann G (2001) Actividad biológica de *Buddleja globosa* Hope. IV Congreso Internacional de Plantas Medicinales. Talca, Chile

Rodríguez Morcelle MI, Rossi AL, Gatusso M, Apóstolo NM (2015) Composición de los aceites esenciales de especies de Verbena nativas de la provincia de Buenos Aires bajo cultivo. Dominguezia 31(2):25–32

Roig F (2001) Flora medicinal mendocina. Universidad Nacional de Cuyo, EDUNC, Mendoza

Sangeetha MK, Priya CDM, Vasanthi HR (2013) Antidiabetic property of *Tinospora cordifolia* and its active compound is mediated through the expression of Glut-4 in L6 myotubes. Phytomedicine 20:246–248

Silva F, Ullrich T, Hartman P, Medina H, Moraga L, Saini G (2004) Plantas medicinales de la región de Aysen, Chile. BLACPMA 3(2):36–45

Silva GA (2001) Evaluación de polvos vegetales solos y en mezcla con inertes minerales para el ombate de *Sitophilus zeamais* Motschulsky en maíz almacenado. Montecillo, Texcoco. pp. 75 Tesis (Magister en Ciencias). Instituto de Fitosanidad. Colegio de Postgraduados en Ciencias Agrícolas

Singh A, Duggal S, Kaur N, Singh J (2010) Berberine: alkaloid with wide spectrum of pharmacological activities. J Nat Prod 3:64–75

Sünter I, Akkol EK, Nahar L, Sarker SD (2012) Wound healing and antioxidant properties: do they coexist in plants? Free Radic Antioxid 2(2):1–7

Tasso de Souza TJ, Manfron MP, Zanetti GD, Hoelzel SCM, Pagliarin VP (2005) Análise Morfo-Histológica e Fitoquímica de Verbena litoralis Kunth. Acta Farm Bonaer 24(2):209–214

Tell G, Izaguirre, Quintana RD (1997) Flora y fauna patagónicas. Primera Edición, Ediciones Caleuche 175 pp; San Carlos de Bariloche, Río Negro, Argentina

Tintu I, Dileep KV, Augustine A, Sadasivan AC (2012) An Isoquinoline alkaloid, Berberine, can inhibit fungal alpha amylase: enzyme kinetic and molecular modeling studies. Chem Biol Drug Des 80(4):554–560

Toledo C, Kutschker A (2012) Plantas medicinales en el Parque Nacional Los Alerces, Chubut, Patagonia Argentina. Bol Soc Argent Bot 47(3–4):461–470

Toribio MS, Oriani SD, Toso RE, Tortone CA, Fernández JG (2009) *Staphylococcus aureus* sensible a extractos metanólicos obtenidos de plantas nativas de la provincial de. La Pampa, Argentina (2009) Ciencia veterinaria 11(1):14–17

Toursarkissian M (1980) Plantas Medicinales de la Argentina. Ed. Hemisferio Sur. Buenos Aires

Ulibarri EA, Burkart A (2000) Sinopsis de las especies de Adesmia (Leguminosae, Adesmiae) de la Argentina. Darwin 38(1–2):59–126

Vargas L (2001) Evaluación de las actividades analgésica y antiinflamatoria de *Buddleja globosa*, Buddlejaceae. Formulación de un preparado dermatológico. Tesis Facultad Ciencias Químicas y Farmacéuticas, Universidad de Chile, Santiago, Chile

Vidal O (2007) Flora Torres del Paine, Guía de Campo. Fantástico Sur Ltda., Chile

Villinski JR, Dumas ER, Chai HE, Pezzuto J, Angerhofer C, Gafner S (2003) Antibacterial activity and alkaloid content of *Berberis thunbergii*, *Berberis vulgaris* and *Hydrastis canadensis*. J Pharm Biol 41:551–557

Vogel H, Razmilic Bonilla I, Polanco González X, Letelier Muñoz ME (2010) Effect of different provenances and production conditions on antioxidant properties in *Buddleja globosa* leaves. Boletín Latinoamericano y del Caribe de Plantas Medicinales y Aromáticas 9(5):333–342

Wojtyczka R, Dziedzic A, Kępa M, Kubina R, Kabała-Dzik A, Mularz T, Idzik D (2014) Berberine enhances the antibacterial activity of selected antibiotics against coagulase-negative *Staphylococcus* strains *in vitro*. Molecules 19:6583–6596

Yan F, Benrong H, Qiang T, Qin F, Jizhou X (2005) Hypoglycemic activity of Jatrorrhizine. J Huazhong Univ Sci Technolog Med Sci 25:491–493

Yánez T (2001) Matico, otro arbusto chileno con potencial exportador: Efecto de la infusión de hojas de matico en la úlcera gástrica. Memoria Instituto de Biología Vegetal y Biotecnología, Universidad de Talca

Yin J, Ye J, Jia W (2012) Effects and mechanisms of berberine in diabetes treatment. Acta Pharm Sin B 2:327–334

Zin J, Weiss C (1998) La Salud por Medio de Plantas Medicinales, 8th edn. Don Bosco, Santiago, p 148

Index

M. A. Alvarez, *Pharmacological Properties of Native Plants from Argentina*,
https://doi.org/10.1007/978-3-030-20198-2

E

F

G

H

I

J

K

L

The manufacturer's authorised representative in the EU is Springer Nature Customer Service Centre GmbH, Europaplatz 3, 69115 Heidelberg, Germany. If you have any concerns regarding our products, please contact ProductSafety@springernature.com

Printed and bound by CPI Group (UK) Ltd, Croydon, CR0 4YY
05/10/2026
02228145-0008